Discovering Cell Mechanisms

The Creation of Modern Cell Biology

Between 1940 and 1970, pioneers in the new field of cell biology discovered the operative parts of cells and their contributions to cell life. They offered mechanistic accounts that explained cellular phenomena by identifying the relevant parts of cells, the biochemical operations they performed, and the way in which these parts and operations were organized to accomplish important functions. Cell biology was a revolutionary science in its own right, but in this book, it also provides fuel for yet another revolution, one that focuses on the very conception of science itself. Laws have traditionally been regarded as the primary vehicle of explanation, but in the emerging philosophy of science it is mechanisms that do the explanatory work. William Bechtel emphasizes how mechanisms were discovered by cell biologists, focusing especially on the way in which new instruments – the ultracentrifuge and the electron microscope – made these inquiries possible. He also describes how scientists organized new journals and professional societies to provide an institutional structure to the new enterprise.

William Bechtel is professor of philosophy and science studies at the University of California, San Diego. He is the author and editor of many books, including *Discovering Complexity* (with Robert C. Richardson, 1993) and *Connectionism and the Mind* (with Adele Abrahamsen, 2002), and is editor of the journal *Philosophical Psychology*. He is past president of the Society for Philosophy and Psychology and the Southern Society for Philosophy and Psychology.

CAMBRIDGE STUDIES IN PHILOSOPHY AND BIOLOGY

General Editor
Michael Ruse   *Florida State University*

Advisory Board
Michael Donoghue   *Yale University*
Jean Gayon   *University of Paris*
Jonathan Hodge   *University of Leeds*
Jane Maienschein   *Arizona State University*
Jesús Mosterín   *Instituto de Filosofía (Spanish Research Council)*
Elliott Sober   *University of Wisconsin*

Alfred I. Tauber   *The Immune Self: Theory or Metaphor?*
Elliott Sober   *From a Biological Point of View*
Robert Brandon   *Concepts and Methods in Evolutionary Biology*
Peter Godfrey-Smith   *Complexity and the Function of Mind in Nature*
William A. Rottschaefer   *The Biology and Psychology of Moral Agency*
Sahotra Sarkar   *Genetics and Reductionism*
Jean Gayon   *Darwinism's Struggle for Survival*
Jane Maienschein and Michael Ruse (eds.)   *Biology and the Foundation of Ethics*
Jack Wilson   *Biological Individuality*
Richard Creath and Jane Maienschein (eds.)   *Biology and Epistemology*
Alexander Rosenberg   *Darwinism in Philosophy, Social Science, and Policy*
Peter Beurton, Raphael Falk, and Hans-Jörg Rheinberger (eds.)
The Concept of the Gene in Development and Evolution
David Hull   *Science and Selection*
James G. Lennox   *Aristotle's Philosophy of Biology*
Marc Ereshefsky   *The Poverty of the Linnaean Hierarchy*
Kim Sterelny   *The Evolution of Agency and Other Essays*
William S. Cooper   *The Evolution of Reason*
Peter McLaughlin   *What Functions Explain*
Steven Hecht Orzack and Elliott Sober (eds.)   *Adaptationism and Optimality*
Bryan G. Norton   *Searching for Sustainability*
Sandra D. Mitchell   *Biological Complexity and Integrative Pluralism*
Joseph LaPorte   *Natural Kinds and Conceptual Change*
Greg Cooper   *The Science of the Struggle for Existence*
Jason Scott Robert   *Embryology, Epigenesis, and Evolution*
William F. Harms   *Information and Meaning in Evolutionary Processes*

Discovering Cell Mechanisms

The Creation of Modern Cell Biology

WILLIAM BECHTEL

University of California, San Diego

 CAMBRIDGE
UNIVERSITY PRESS

CAMBRIDGE UNIVERSITY PRESS
Cambridge, New York, Melbourne, Madrid, Cape Town, Singapore, São Paulo

Cambridge University Press
40 West 20th Street, New York, NY 10011-4211, USA

www.cambridge.org
Information on this title: www.cambridge.org/9780521812474

First published 2006

Printed in the United States of America

A catalog record for this publication is available from the British Library.

Library of Congress Cataloging in Publication Data

Bechtel, William.
Discovering cell mechanisms : the creation of modern cell biology / William Bechtel.
    p.   cm. – (Cambridge studies in philosophy and biology)
Includes bibliographical references and index.
ISBN-13: 978-0-521-81247-4 (hardback)
ISBN-10: 0-521-81247-X (hardback)
1. Cytology.   I. Title.   II. Series.
[DNLM: 1. Cytology–history.   2. Cell Physiology.
3. Cytological Techniques.   QU11.1B391d   2006]
QH581.2.B42   2006
571.6 – dc22   2005017960

ISBN-13   978-0-521-81247-4 hardback
ISBN-10   0-521-81247-X hardback

Contents

Contents

Preface

This book is the product of research spanning two decades. In the 1980s I had been investigating the history of cytology in the nineteenth century and biochemistry in the early twentieth century when I responded to an announcement from the American Society for Cell Biology of a fellowship for support of research on the history of cell biology. With their financial support in 1986 and 1987 I began to examine the creation of modern cell biology in the decades after World War II. I am enormously grateful not only for the fellowship funding but also for the invaluable assistance of individuals associated with ASCB. In particular, I thank Robert Trelstad, then Secretary of ASCB, who invited me to society and executive council meetings, gave me encouragement, and provided entrée to senior members of the society. A number of the founders of modern cell biology were still active in the society and I had the opportunity to meet and interview them regarding their own contributions to cell biology and their recollections of the early days of this field. I also had access to the archives of the society, which were then housed at the Society offices. (They have since been transferred to the University of Maryland, Baltimore County.) I have relied heavily on this material in analyzing in Chapter 7 the history of the American Society for Cell Biology.

In the early 1990s I received additional support from the National Endowment for the Humanities. I am most appreciative of the support from NEH as well as from the National Science Foundation, which funded my earlier work on the history of biochemistry. Among other activities, this support enabled me to interview many additional pioneer cell biologists.

I particularly wish to thank the following scientists for taking time to meet with me and provide their reflections and insights on the history of cell biology: Max Alfert, Vincent Allfrey, Helmut Beinert, Britton Chance, Christian de Duve, Morgan Harris, Daniel Mazia, Montrose Moses, George Palade, George Pappas, Keith Porter, Van Potter, Hans Ris, Birgit Satir,

Peter Satir, Philip Siekevitz, Paul Stumpf, and Hewson Swift. I am also deeply appreciative for the guidance Pamela Henson provided me on the techniques for doing aural history with scientists.

The NEH grant also enabled me to carry out research at the Rockefeller Archive Center (RAC). This center holds records of laboratories at the Rockefeller Institute for Medical Research, as well as the archives of the Rockefeller Foundation, which provided funding for many of the early cell biologists. Especially useful for Chapters 5 and 6 were the Annual Reports submitted each April by every laboratory at the Rockefeller Institute. These were incorporated into the *Scientific Reports of the Laboratories to the Board of Scientific Directors* (RG 439, Rockefeller University Archives, RAC). I will refer to these reports throughout the text as simply the *Annual Report* for a laboratory.

While working on this project I have been associated with the philosophy departments of three universities: Georgia State University, Washington University in St. Louis, and the University of California, San Diego. Each has provided invaluable support for which I am most grateful. I particularly appreciate the contributions of several of my graduate students. In particular, at Washington University Jennifer Mundale transcribed many of the oral interviews and engaged in numerous fruitful discussions with me about this and related projects. More recently, at UCSD I have benefited from the ideas and insights of Andrew Hamilton, who provided detailed comments on Chapter 2, and of Cory Wright. I have also benefited from interactions with graduate students who participated in my seminars on mechanism at both Washington University (presented jointly with Carl Craver) and UCSD.

My spouse and frequent collaborator, Adele Abrahamsen, has made an enormous contribution to this project. I thank her for extremely valuable comments on the entire manuscript. I have benefited from our many productive discussions of mechanism and the history of research on cells. As well, various parts of the text draw upon papers we coauthored.

A key component of my analysis is that the knowledge developed in cell biology consisted of the discovery of various cell mechanisms. My understanding of what a mechanism is and how scientists investigate them has benefited from many discussions with Robert Richardson, my coauthor on *Discovering Complexity*, as well as Carl Craver, Lindley Darden, Stuart Glennan, and William Wimsatt.

Finally, I thank Michael Ruse, who initially invited me to submit a proposal for this book to Cambridge University Press and then showed admirable patience while I produced it.

1

Introduction

Cell Mechanisms and Cell Biology

> In order to succeed in solving these various problems, one must so to speak progressively dismantle the organism, as one takes to pieces a machine in order to recognize and study all its works.
>
> (Bernard, 1865, Part II, Chapter 1)

1. A DIFFERENT KIND OF SCIENCE

To many people, cell biology is an unlikely domain to impart impetus for a major shift in the way philosophy of science is practiced. Cells appear to be the object of straightforward empirical observation, not of bold theories that challenge the status quo. In high school or at the museum you look into the microscope and struggle to see what you are told you should see – structures that are never as sharp and well delineated as in the drawings in textbooks. What could cell biology be but a tedious descriptive science? This popular conception, however, is quite erroneous. Cells, as Theodor Schwann first concluded in the 1830s, are the basic units of life. They perform all essential vital functions: extracting energy and building materials from their environment, constructing and repairing themselves and synthesizing products for export, regulating their own internal operations, reproducing themselves periodically by dividing, cleaning up their own waste, and so forth. Beginning in the 1940s an initially small cadre of investigators who were pioneers in the modern discipline of cell biology began to figure out the biochemical mechanisms that enable cells to perform these functions. Although miniaturized, the mechanisms they found to be operative in each cell are staggeringly complex. Their work and that of their successors, primarily in 1940–70, revolutionized our understanding of the basic processes of life. Now, half a century later,

1

I am drawing on that lively period of scientific enterprise to find fuel for yet another revolution, one that focuses on the very conception of science itself.

The science of cell biology is very different from the textbook image of science, including that advanced in traditional philosophy of science. That picture, grounded on some of the great successes of the scientific revolution and subsequent developments in some areas of physics, emphasizes bold unifying generalizations – the laws of nature. Newton's laws of motion promised to explain all motion, both terrestrial and celestial. The laws of thermodynamics and electromagnetism are similarly broad in their sweep. In biology, Darwin's insight that evolution by natural selection occurs when there is heritable variation in fitness (Lewontin, 1970) has provided a similarly powerful unifying generalization. However, most areas of biology – including cell biology – do not fit into this picture. Instead of unifying generalizations, cell biology offers detailed accounts of complex mechanisms in which different component parts perform specific operations, which are organized and orchestrated so that a given type of cell can accomplish the functions essential for its life. Not elegant generalizations, but exquisitely detailed accounts of mechanisms, are the products. This difference in product has broad implications for our overall understanding of science, including the challenges of generating evidence, advancing new hypotheses and theories, and evaluating and revising them.

In proposing an alternative characterization of science as the search for mechanisms, I am not seeking to eradicate the old picture of science as the quest for bold generalizations but to complement it. There are domains in which the Newtonian vision is appropriate – ones in which the aim of inquiry is best served by far-reaching generalizations that can be economically stated, often in a single equation. In many domains, though, the aim of inquiry leads to meticulous accounts of complex mechanisms. This is particularly true in the functional domains of biology – cell biology, molecular biology, physiology, pathology, developmental biology, neurobiology – and also in related areas of physics (biophysics) and chemistry (biochemistry). It does not advance our understanding of these sciences to impose an ill-fitting model. Rather, we need to develop a conception of science that is appropriate for them. Only then can we adequately address some of the traditional questions about science – what it is to explain a phenomenon, how explanations are discovered, and how they are evaluated.

The idea that much of science is a quest to articulate mechanisms is not news to biologists. Frances Crick (1988, p. 138) put it succinctly:

"What is found in biology is *mechanisms*, mechanisms built with chemical components . . ." Biologists do not always reach as far down as chemistry in characterizing biological mechanisms, but they do use the term *mechanism* naturally and often. A search I undertook of titles of articles in *Science* from 1880 to 1998 revealed 656 articles that included *mechanism, mechanisms,* or *mechanistic* in their titles. Only one appeared before 1900, and that concerned a psychological mechanism. Titles referring to biological mechanisms began in 1904 and are far more frequent than articles about non-biological mechanisms. They also outnumber articles that include *theory, theories,* or *theoretical* (584) or *law* or *laws* (165) in their title (the count for law discounted 25 titles clearly referring to political laws). A few of the early papers referring to mechanisms in their titles involved the vitalism–mechanism controversy that was still very active at the turn of the twentieth century. Most, however, focused on specific biological mechanisms. The following are some illustrative examples:

Edwin G. Conklin (1908). The mechanism of heredity.
Frank R. Lillie (1913). The mechanism of fertilization.
E. Newton Harvey (1916). The mechanism of light production in animals.
Jacques Loeb (1917). A quantitative method of ascertaining the mechanism of growth and of inhibition of growth of dormant buds.
W. J. V. Osterhout (1921). The mechanism of injury and recovery of the cell.
John H. Northrop (1921). The mechanism of an enzyme reaction as exemplified by pepsin digestion.
F. H. Pike and Helen C. Coombs (1922). The organization of the nervous mechanism of respiration.
Caswell Grave and Francis O. Schmitt (1924). A mechanism for the coordination and regulation of the movement of cilia of epithelia.

These titles reveal an interesting variation in generality, with Conklin discussing *the* mechanism of heredity while Grave and Schmitt discuss *a* mechanism within a particular cell type. The latter reflects the sort of engagement of individual scientists and research teams that became the norm in the twentieth century. Individual scientists and research teams honed in on much more delimited phenomena at a scale that can be fruitfully investigated in a single laboratory across a period of perhaps a few years. At the general level a broad research community might devote itself to a general phenomenon such as protein synthesis and seek to identify the general nature of the mechanism of protein synthesis. Specific researchers, though, focus on particular components of the mechanism or on the mechanism that is operative in particular cells or particular organisms. This is reflected by considering some typical

3

titles of papers referring to mechanisms and protein synthesis in their titles[1] in the period 1950 to 1970:

Winnick, T. (1950). Studies on the mechanism of protein synthesis in embryonic and tumor tissues. I. Evidence relating to the incorporation of labeled amino acids into protein structure in homogenates. *Archives of Biochemistry*, 27, 65–74.

Novelli, G. D. and Demoss, J. A. (1957). The activation of amino acids and concepts of the mechanism of protein synthesis. *Journal of Cell Physiology,* 50 (Supplement 1), 173–97.

Yoshida, A. (1958). Studies on the mechanism of protein synthesis: bacterial alpha-amylase containing ethionine. *Biochimica et Biophysica Acta*, 29, 213–4.

Goodman, H. M. and Rich, A. (1963). Mechanism of polyribosome action during protein synthesis. *Nature,* 199, 318–22.

Griffin, B. E. and Reese, C. B. (1964). Some observations on the mechanism of the acylation process in protein synthesis. *Proceedings of the National Academy of Sciences, USA*, 51, 440–4.

Carey, N. H. (1964). The mechanism of protein synthesis in the developing chick embryo. The incorporation of free amino acids. *Biochemical Journal,* 91, 335–40.

Mano, Y. and Nagano, H. (1966). Release of maternal RNA from some particles as a mechanism of activation of protein synthesis fertilization in sea urchin eggs. *Biochemical and Biophysical Research Communications,* 25, 210–15.

Although the conception of a mechanism is widely invoked in the biological sciences, it has only recently become the target of philosophical inquiry. Chapter 2 will articulate the conceptions of mechanism and mechanistic explanation that figure in biology, especially cell biology. The quest to understand nature mechanistically has its roots in the scientific revolution and, although challenged by vitalist critics, figured prominently in the attempts to understand physiological systems throughout the eighteenth and nineteenth centuries. The key to the mechanistic approach was not the analogy of physiological systems to human made machines, but the quest to explain the functioning of whole systems in terms of the operations performed by their component parts. Beginning with Bernard, biologists also recognized the importance of the way in which the parts and their operations were organized. Increasingly, biology became a science in which phenomena were explained by discovering the organized parts and operations by which a mechanism performed its function.

---

[1] Another interesting class of papers uses the term mechanism not for the general phenomenon, protein synthesis, but for the way in which a particular substance alters that phenomenon. A characteristic example is the following: de Kloet, S., van Dam, G., and Koningsberger, V. V. (1962). Studies on protein synthesis by protoplasts of Saccharomyces carlsbergensis. III. Studies on the specificity and the mechanism of the action of ribonuclease on protein synthesis. *Biochimica et Biophysuca Acta*, 55, 683–9.

As I explore in Chapter 2, recognizing that the goal of many scientific inquiries is to describe the mechanism responsible for the phenomenon of interest provides a different perspective on many aspects of scientific inquiry. Diagrams often provide the most fruitful way of representing a mechanism, in which case scientists may relate the mechanism to the phenomenon of interest by mentally simulating its operation. In part this involves a reductionistic strategy of decomposing the mechanism into its parts and operations, but equally significant is figuring out how these are organized to work together and how various environmental conditions affect the mechanism's functioning. Finally, although traditional philosophy of science has had little to say about the process of discovery, when the focus is on mechanisms we can set out what the challenge of discovery is and analyze typical experimental strategies – strategies that figured prominently in discovering cell mechanisms.

2. THE ORGANIZATION OF SCIENCE INTO DISCIPLINES[2]

Although a central feature of my discussion will be the discovery of cell mechanisms, my broader focus is on the establishment of cell biology as a discipline. In 1940 no one would have listed cell biology when identifying scientific disciplines. By 1970 it was a well-established discipline. My goal is to trace and account for this change. First, though, a preliminary issue must be addressed. The word *discipline* is familiar enough, but what exactly is a scientific discipline? In the disciplines that analyze science (philosophy of science, history of science, and sociology of science, which are collectively referred to as *science studies*), a variety of criteria have been offered.

Perhaps the most common way in which people identify disciplines is in terms of the objects they investigate. Thus, astronomy is described as the study of suns, planets, and the like, whereas psychology investigates mental activities or behaviors. Dudley Shapere captured this feature of our ordinary conception of disciplines when he introduced the term *domain* for "the set of things studied in an investigation" (Shapere, 1984, p. 320; for his classic treatment of domains, see Shapere, 1974). Shapere's conception of a domain is more sophisticated than the lay conception, however, for he argued that domains are not simply presented to scientists but result from their decision as to what items (his term for the constituents of domains) to group together to constitute a domain. Thus, he showed how during the nineteenth century chemists made facts about basic elements a domain for study, because they

---

[2] The discussion in this section draws in part upon Bechtel (1986a).

construed them as the constituents of ordinary substances. Crucially, Shapere drew attention to the fact that scientists' reasons for grouping items together into a common domain may change over time. Moreover, as Toulmin noted, an item may be grouped with different items into different domains depending on the questions asked, and different investigators may ask different questions:

> If we mark sciences off from one another (using Shapere's term) by their respective 'domains', even these 'domains' have to be identified, not by the types of objects with which they deal, but rather by the questions which arise about them. Any particular type of object will fall in the domain of (say) 'biochemistry' only in so far as it is a topic for corresponding 'biochemical' questions; and the same type of object will fall within the domains of several different sciences, depending on what questions are raised about it. The behavior of a muscle fibre, for instance, can fall within the domains of biochemistry, electrophysiology, pathology, and thermodynamics, since questions can be asked about it from all four points of view . . . (1972, pp. 149)

While the objects of study are an important part of what characterizes a discipline, both Shapere and Toulmin made it clear they are insufficient. To identify the set of objects comprising the domain of a discipline, we need to consider why scientists group them together. Scholars who study science generally split over two approaches to addressing this issue, roughly differentiated by their respective disciplines. Philosophers and historians of ideas adopt what is often characterized as an *internalist* approach to understanding science, emphasizing cognitive factors such as theories and evidence, while sociologists and social historians adopt an *externalist* approach, focusing on social and institutional factors. In the 1970s and 1980s these two approaches were often portrayed as competing and mutually exclusive; more recently, many in science studies have recognized a role for both.

Through most of the twentieth century, philosophers of science focused on the theories advanced by scientists and the relation theories bore to evidence. To the extent that disciplines were considered at all, they were characterized in terms of theories. For example, in discussions of the unity of science – the question of how different sciences related to one another – the logical empiricists identified disciplines with their theories and asked whether they could be related to one another logically. Thus, the question of the relation of biology to physics and chemistry became the question of whether the theories (laws) of biology could, with the aid of bridge principles and boundary conditions, be derived from those of chemistry and physics (Oppenheim & Putnam, 1958; Nagel, 1961). If so, biology was said to be *reduced* to, and thereby unified with, physics.

Finding this singular focus on theory misguided, a number of philosophers have proposed alternative accounts that are more multifaceted and naturalistic. Best known is Kuhn's (1962/1970) notion of a *paradigm* and distinction between *normal science* and times of *paradigm shift*. In its more restricted sense, a paradigm for Kuhn was an exemplar – a solution to a particular problem that became a model for solving other problems. In its more extended sense, a paradigm was a general theory or theory schema that characterized a domain, identified problems to be solved, and specified strategies for solving them and criteria for evaluating proposed solutions. Kuhn's notion of a paradigm (both the restricted sense of an exemplar but especially the extended sense of a general theory) provided a way to characterize a group of scientists engaged in a similar enterprise and to tell a historical narrative of how the enterprise developed.[3] The extended notion of a paradigm was sufficiently vague, however, that it also became the focus of severe criticism and has largely ceased to figure in philosophical accounts of science.

Adopting the term *field* rather than discipline, Lindley Darden and Nancy Maull advanced a multifaceted conception in which no single element dominated (though it did not extend so far as to include externalist elements). Incorporating Shapere's notion of a domain, they defined a *field* as consisting of the following elements:

> a central problem, a domain consisting of items taken to be facts related to that problem, general explanatory facts and goals providing expectations as to how the problem is to be solved, techniques and methods, and sometimes, but not always, concepts, laws and theories which are related to the problem and which attempt to realize the explanatory goals. (Darden and Maull, 1977, p. 144)[4]

Especially relevant here are the explanatory goals, types of accounts offered (e.g., laws or theories), and conceptualization of the central problem, a cluster that I will call, for convenience, the field's *mission*. As I noted in Section 1, in many areas of biology explanation takes the form of an account of the mechanism responsible for a phenomenon. The central problem is then the discovery and refinement of this mechanism.

A second important component of fields, to which Darden and Maull drew attention, is its array of techniques and methods for solving problems. These

---

[3] Kuhn inspired several other attempts to characterize larger-scale units that served to unite the practitioners of a discipline. Two examples were Lakatos' (1970) notion of a research program and Laudan's (1977) notion of a research tradition.

[4] Shapere (1984) largely endorsed Darden and Maull's conception of a field but cautioned that one must be sensitive to the fluidity of fields and to the fact that often different practitioners within a field will not share exactly the same methods.

are not just cognitive or reasoning strategies but include instruments, and techniques for using instruments, that enable the scientist to observe and manipulate objects in the domain. Ian Hacking (1983) pioneered discussion within philosophy of the importance of techniques for intervening in nature. Historians such as Kathryn Olesko have also emphasized the importance of techniques of investigation in delimiting a discipline. She nicely noted that a discipline *disciplines* its practitioners by requiring them to master a particular body of knowledge and techniques of investigation (Olesko, 1991, p. 14). Steven Shapin (1982), focusing on the differences between biometricians and Mendelians, illustrated the importance of differences in techniques and methods in demarcating these two groups of investigators.

In contrast, sociologists and social historians of science tend to focus on the social networks and institutional structures within which scientists work. One role such social units play can be related to the cognitive elements emphasized by philosophers and historians of ideas – they insure compliance with a discipline's mission and accepted methods. Thus, Michael Polanyi introduced "the principle of mutual control," which

> consists . . . of the simple fact that scientists keep watch over each other. Each scientist is both subject to criticism by all others and encouraged by their appreciation of him. This is how *scientific opinion* is formed, which enforces scientific standards and regulates the distribution of professional opportunities. (1966, p. 72)

This aspect of the social structure of disciplines was much emphasized by Robert Merton (1973) and the tradition in sociology of science which he inspired.

Subsequently, though, sociologists of science pushed beyond the Mertonian tradition to address how the institutional structure of disciplines influences the content of scientific research by, for example, focusing a particular scientist's endeavors. Rosenberg commented,

> It is the discipline that ultimately shapes the scholar's vocational identity. The confraternity of his acknowledged peers defines the scholar's aspirations, sets appropriate problems, and provides the intellectual tools with which to address them; finally, it is the discipline that rewards intellectual achievement. (1979, p. 444)

Contemporary sociologists emphasize that the factors that shape a scientist's identity are not limited to ideas internal to the science itself but can include those of the broader society. Thus, Robert Kohler characterized such institutions as "mediat[ing] between science and the political, cultural, and economic institutions on which science depends for material and support" (1982, p. 2). Accordingly, Kohler characterized disciplines such as biochemistry, on which he focused, as "political institutions that demarcate areas of academic territory" (p. 1).

Sometimes the emphasis on the roles played by the broader social, cultural, and economic institutions is presented as repudiating the significance of cognitive factors in shaping science (Barnes, 1977; Bloor, 1991; Collins, 1981; Latour & Woolgar, 1979). Such a stance has provoked equally ardent responses from philosophers who have construed any acknowledgment of social factors as undermining the epistemic warrant of science (Laudan, 1981; Kitcher, 2001; and several of the papers in Koertge, 1998). Other philosophers (Longino, 1990; Longino, 2002; Solomon, 2001) have developed a more moderate response, articulating how social factors figure in the intellectual development of science without sacrificing its epistemic warrant. While my sympathies lie with the last position, I will not advance arguments for it here; instead I will briefly discuss how scientists create institutions and how their decisions help shape research in a discipline.

The candidate political institutions of a discipline are academic departments, professional societies, and journals. Of these, departments are the most problematic for tracking scientific disciplines. There are a plethora of ways in which universities divide faculty into departments, often having to do with very local politics and ease of administration. Especially in the biological sciences, the differentiation into departments depends upon the size of the institution and whether the biological sciences are situated among the arts and sciences or in a medical school (or both). Small undergraduate colleges will typically group all the biological disciplines within one department, although they may have separate tracks for majors that correspond to divisions within biology. Research universities tend to have separate departments for different biological disciplines, although they may be grouped in pairs (e.g., cell and molecular biology, ecology and evolution). Although departments may not correspond exactly to disciplinary units as differentiated by such criteria as professional societies and journals, they ensure the historical continuity of disciplines by training subsequent generations of researchers and thereby securing the cognitive and social allegiances of members of a discipline. Richard Whitley commented, "Educational institutions form the basic commitments of scientists in nearly all fields, and constitute the fundamental unit of social and cognitive identity in the sciences, which is one reason why the term 'discipline' is usually understood to refer to units of organization in universities" (1980, p. 310).

Outside of the context of the local university, professional societies and journals are the major institutions that provide disciplinary identity. Given the importance of publication both in establishing a scientist's career and in disseminating results of research, the availability of journals influences the direction of a field. They determine not only what topics of research can most readily be published, but also what methods investigators can employ in investigating those topics. An important step in developing a new area of research and a research community that will carry out the research is the creation of new journals that will publish the results. In many cases, professional societies manage journals. Societies typically also hold regular meetings that provide a context in which scientists meet formally and informally to share results and formulate directions for future research. Although talks at professional society meetings often receive less credit in terms of professional advancement, they are favored vehicles for rapid communication and provide important opportunities for personal interaction.

Beyond formal institutions, sociologists have also focused on informal networks of scientists. Derek de Solla Price (1961; see also Crane, 1972; Chubin, 1982) coined the concept *invisible college* for groups of researchers who are in regular communication and share a common conceptual framework, problem focus, and set of techniques for dealing with a problem, although they may disagree on empirical claims or proposed theories. Sociologists identify such networks using such techniques as tracking citations and identifying clusters (Garfield, 1979). A variety of quantitative techniques have been developed for identifying and graphing social networks (Wasserman & Faust, 1994). A recent approach focuses on collaboration networks characterized in terms of coauthorship of papers (if two scientists have coauthored a paper, they are directly linked; if two scientists have not coauthored a paper but have each coauthored a paper with another scientist they are linked through that scientist, etc.). These networks have been shown to constitute structures known as *small worlds* in which randomly chosen pairs of scientists are typically separated by only a short path of intermediate collaborators (Newman, 2001). Such networks can also be studied more qualitatively and in detail to reveal the interactions that shape the direction of science. Jean-Paul Gaudillière (1996), for example, studied collaborative networks of scientists in France in the 1960s as specimens that could be revealing of the relationship between biochemistry and the emerging molecular biology.

As it turns out, despite the frequent conflict between theorists pursuing cognitive and social accounts of science, the various cognitive and social criteria for delineating units in science tend to converge on the same units. That is, institutional structures, methods of inquiry, domains of inquiry, and missions align closely enough that they can be treated as different aspects of the same unit. The contributors to a particular journal or members of a given professional society will tend to focus on a common domain of inquiry, share a common mission, and employ a similar range of techniques. Network analyses will likewise tend to track scientists who share many of these common characteristics. In the analysis of cell biology in this book, I will be focusing on three of the features of disciplines discussed above – the mission, research methods, and institutions. Together, these resulted in a discipline devoted to the domain of cellular mechanisms.

In the discussion so far I have not considered the generality or specificity of a discipline. Each of the criteria, however, can be applied so as to pick out units that vary considerably in specificity or scope. Focusing just on professional organizations, some (such as the American Association for the Advancement of Science or The Royal Society) attempt to serve the whole of science, others (such as the Biochemical Society and the American Society for Cell Biology) have a more limited scope, and still others focus on specific phenomena (the Protein Society or the RNA Society) or techniques (The Tissue Culture Association or the Electron Microscope Society of America). Though terminology for units at different levels of generality involves very fuzzy, overlapping boundaries, the range from most to least general might be taken to include (1) molar disciplines (e.g., biology, chemistry); (2) operational units called either disciplines or fields[5] (e.g., ecology, cell biology, biochemistry); (3) research areas[6] that focus on particular phenomena within the scope of a discipline (these can be rather slippery; e.g., the research areas of intermediary metabolism and cellular respiration have overlapped in different ways at different times); (4) smaller units that can be delimited by time, space, subdomain, or affinity (e.g., the Institute for Enzyme Research at the University of Wisconsin; (5) finally, the individual investigator or research team.

Several levels in this hierarchy are relevant for understanding the creation of cell biology. At various points I will be focusing on individual scientists and laboratories and will discuss several research areas. But cell biology

---

[5] In definitions and in actual use, these terms overlap substantially but not completely. For example, disciplines may tend toward greater size or tighter ties to institutional structures than fields. Any such differences need not be adjudicated here.

[6] They are called *research areas* to help distinguish them from the larger fields to which they belong, but are similar in meaning to *research fields* or *scientific specialties*. Historian of science Larry Holmes offered this discerning characterization: "A research field is more than a network of communication and ties of professional interest. It is also an ongoing investigative stream composed of the intersecting investigative pathways of each of the individual scientists (or integrated local research groups) who participate actively at its creative forefront" (1992, p. 7).

itself is a discipline or field that incorporated a number of research areas. The institutions developed to serve it, especially the American Society for Cell Biology and the *Journal of Cell Biology*, took as their focus investigations into the various organelles of the cell and their functioning. Even when they worked in different research areas, the investigators who affiliated with these organizations shared a common mission of understanding the structure and function of the mechanisms comprising cells. They also shared a number of research instruments and techniques, especially electron microscopy and cell fractionation.

Accounts of disciplines sometimes treat them as eternal, non-temporal entities. My reference to the creation of cell biology beginning in the 1940s highlights the fact that they are historical entities. Starting in the nineteenth century new disciplines have repeatedly arisen. New disciplines or research areas can emerge for any one of a number of reasons. Sometimes a particular investigative strategy and a mission and domain adapted to it arise within a discipline and the divergence is not comfortably accommodated (e.g., molecular biology arose in part from biochemistry but separated, producing an uneasy relationship for many years).[7] Sometimes a genuinely new domain comes into existence along with people, methods, and a mission (e.g., the rise of computer science in the mid-twentieth century). A very common pattern, though, is for new disciplines to emerge in an unoccupied domain lying between existing disciplines for which no existing discipline possesses the needed research tools.

Research in contemporary science is often interdisciplinary in character, and interdisciplinarity is often touted as a virtue. But there has been little discussion of the desired outcome of interdisciplinary research. The best account remains Darden and Maull's (1977) account of what they called *interfield theories*. They described an interfield theory as "a different type of theory . . . which sets out and explains the relations between fields" (p. 48). They suggested and discussed several types of interfield relation: (a) structure-function, e.g., physical chemistry targets the structure of molecules while biochemistry describes their function; (b) physical location of a postulated entity or process, e.g., the chromosomes identified in cells by cytologists provide the physical location of the genes postulated by geneticists (a case that also exemplifies structure–function and part–whole relations); (c) physical nature

---

[7] There are numerous books and papers on this case, but see especially the Fall 1996 special issue of *Journal of the History of Biology*, 29 (3), on "The tools of the discipline: Biochemists and molecular biologists." Rheinberger, Chadarevian, Gaudillière, and Burian emphasized methods and Creager, Gaudillieère, and Kay attended particularly to individuals and institutions.

of a postulated entity or process, e.g., biochemistry specifies the physical realization of entities postulated by the operon theory in genetics; (d) cause–effect; e.g., biochemical interactions are a cause of heritable patterns of gene expression. More recently Darden (Darden & Craver, 2002; Darden, 2005) has emphasized mechanisms as a frequent focus of interfield theories, and indeed, all four of these relationships figure in accounts of mechanisms.

My interest here is not in interdisciplinary research generally, but rather in interdisciplinary research that gives rise to a new discipline. New disciplines do not always result even if the interdisciplinary engagement is maintained over a long period of time. In Bechtel (1986a) I contrasted cases such as cognitive science, in which scientists maintained their primary allegiances to the contributing disciplines (e.g., psychology, artificial intelligence, linguistics), with cases such as biochemistry, in which a new discipline developed that secured the primary commitment of future practitioners. I characterized the former as *interdisciplinary research clusters.* Although they develop similar institutional structures as disciplines (professional societies and journals) and are interested in a common domain of phenomena, clusters do not employ distinctive research techniques. Rather, collaborators draw upon the techniques and employ the standards of successful explanation from their home disciplines. Collaboration is motivated by the advantages of coordinating the results of multiple disciplines in understanding a phenomenon of common interest.

Interdisciplinary research gives rise to new disciplines when research addresses phenomena that are not the focus of any of the contributing disciplines and involves new research tools and techniques that facilitate the new inquiry. These also enable researchers to pursue explanatory goals that are not those of any of the existing disciplines. When new institutional arrangements are developed, they tend to become the primary institutional home of the practitioners and they will characterize themselves using the name of the new discipline. This, I will argue, is what happened in the case of cell biology.

3. THE NEW DISCIPLINE OF CELL BIOLOGY

By the late 1960s cell biology had acquired the characteristics I identified for disciplines in the preceding section. Cell biologists pursued a distinctive domain (cell organelles and their functions), employed a new set of tools (especially electron microscopy and cell fractionation), pursued a distinctive mission (determining the mechanisms that enabled organelles to perform their functions), and created new professional institutions (especially journals and professional societies). This process began in the 1940s, and thus I have chosen the period 1940 to 1970 as the scope of my inquiry. In many respects, though, the endeavors of cell biologists were continuous with those of earlier scientists: Cells and their internal structure had been the focus of extensive investigation in nineteenth-century cytology. Yet, the name *cell biology* was deliberately chosen to demarcate the new discipline from cytology. Another discipline, biochemistry, had established itself at the beginning of the twentieth century to study the chemical activities within cells, with a particular focus on enzymes, and achieved remarkable success in its first four decades. Chapter 3 reviews these accomplishments, showing that each of these disciplines already was addressing aspects of what was to become the domain of cell biology: cell structures in cytology and chemical operations associated with cell structures in biochemistry. However, proceeding on their separate paths, they failed to integrate findings on cell structure and function into a unified account. Moreover, as noted above, a discipline is more than a domain. What most clearly differentiated early cell biology from both cytology and biochemistry was the nature of the investigations cell biologists pursued and the tools they employed.

The investigators who created cell biology were strongly committed to inquiry that related knowledge of the parts of the cell (organelles) and the chemical operations that took place in those parts. The methodologies that had been so successful in the early decades of biochemistry involved extracting the responsible enzymes and substrates from cells and studying reactions that did not depend on the specifics of cell structure. So for many biochemists, cells were unimportant. A number of cytologists, on the other hand, were eager to relate cell parts to chemical operations, but for the most part they lacked the tools to make the connections. Thus, despite the focus on cell structure and function by cytology and biochemistry, Chapter 3 shows that in 1940 there remained a terra incognita between these two fields.

Integration of cytological and biochemical approaches were critical for investigators to understand many of the mechanisms responsible for cell activities. However, there is one conspicuous instance of a cell mechanism for which cytologists figured out the basic mechanism schema without reliance on biochemistry – cell division. With the introduction of new dyes in the late nineteenth century, researchers such as Edouard van Beneden, Hermann Fol, and Walther Flemming were able to identify chromosomes in the cell nucleus and determine the sequence of operations in which they figured in cell division. Further, August Weisman and others recognized linkages between chromosomes and hereditary material, and Carl Correns determined how the events of cell division ensured the transmission of hereditary factors to daughter cells. To discover that DNA was the hereditary material and determine the mechanism by which it replicated, new tools had to be developed. This was done by molecular biologists in the 1940s and 1950s, but from the point of view of cell mechanisms, those discoveries filled in the schema developed much earlier (Darden, 2005). I will briefly analyze this case in Chapter 3, but otherwise my focus will be on mechanisms situated in the cytoplasm of the cell, that is, all of the cellular materials enclosed within the cell membrane excepting the nucleus.

The cytoplasm is the locus of a host of cellular mechanisms that are responsible for taking in material from the cell's environment, breaking down this material as well as worn out pieces of its own structure, and synthesizing new components of its own structure or materials for export out of the cell. All of these mechanisms require energy; accordingly, several additional mechanisms are devoted to procuring and making energy available to them. To understand any of these mechanisms, it was not enough to have ideas about the phenomena they produced; scientists needed means of investigation. With the development of stains in the late nineteenth century, cytologists began to secure evidence of the occurrence of structures, called *organelles*, in what until then had appeared as formless cytoplasm. However, the evidence was highly contested and would be until techniques became available that went beyond those of light microscopy. Biochemists, on the other hand, routinely destroyed cell structure because it provided the best means of securing preparations in which they could study chemical reactions. To link reactions to the particular organelle in which they occurred (and in some cases to study them at all) required more refined preparations. Two instruments developed by physicists and chemists in the 1920s and 1930s offered promise for entering these unexplored territories – the electron microscope and the ultracentrifuge. Biologists, however, confronted significant challenges in developing the techniques needed to deploy these instruments to study mechanisms operative in cells. These challenges and the techniques that were created to surmount them will be examined in Chapter 4.

New research techniques pose both an engineering and an epistemic challenge. As in prototypical engineering tasks, in order to use the ultracentrifuge or the electron microscope, researchers had to figure out ways to accomplish new tasks – for example, to release the contents of cells from the cell membrane without disrupting internal structures and to stain cell components so that they would differentially diffract electrons. Engineering tasks like these give rise to the epistemic challenge of showing that the results reflect the phenomenon of interest and are not artifactual. Because the evaluation of engineering solutions is frequently grounded in empirical exploration, rather than in detailed knowledge of how the techniques work, scientists use indirect measures to answer the epistemic challenge. Chapter 4 will focus on three such indirect measures used by cell biologists – the determinateness and repeatability of the resulting evidence, consilience of evidence from multiple techniques, and the coherence of the evidence with plausible accounts of the mechanism under study.

In Chapters 5 and 6 I turn to how these new techniques were deployed to better understand cell mechanisms.[8] Chapter 5 concentrates on one laboratory located at the Rockefeller Institute that conducted many of the pioneering inquiries. The initial focus of this laboratory was cancer. Albert Claude centrifuged tumor cells in an attempt to find within them particles involved in the transmission of cancer. When the particles he discovered turned out not to be unique to cancer cells, he shifted his attention to studying those isolated from normal cells. Collaborating with chemists, he identified the chemical reactions associated with centrifuged fractions containing those particles. In collaboration with Keith Porter, he also began employing electron microscopy to compare the particles in the fractions with components of intact cells to determine their origins. A major accomplishment of this research was to demonstrate that a fraction that was highly specific for a certain cell organelle – the mitochondrion – contained the critical enzymes involved in cellular respiration. Mitochondria were dubbed the power plants of the cell, and they became a focus of the analytic powers of cell biologists.

Chapter 6 describes how in the 1950s and 1960s a growing cadre of researchers adopted the techniques of electron microscopy and cell fractionation and pursued the project of explicating the mechanisms operative in the cell cytoplasm. Figure 1.1 is a drawing portraying the organelles known to populate the cell that is representative of those offered by cell biologists based on the electron micrographs produced in the 1950s. Cell biologists and biochemists both contributed to increasingly fine-grained accounts of the structures within mitochondria and their roles in biochemical operations. A crowning achievement was the discovery that certain key enzymes were embedded in the inner membrane of the mitochondrion so as to spatially and temporally organize the reactions involved in electron transport and oxidative phosphorylation. The work with mitochondria provided an exemplar (in Kuhn's sense) of how to link functions with organelles that became a model

---

[8] Treating the development of new instruments and research techniques separately from their utilization in developing new knowledge is artificial because the investigators were putting their techniques to use to secure information about cell mechanisms as they were developing them. I have discussed them separately so as to do justice to the different epistemic issues involved in developing the techniques and using them to understand mechanisms.

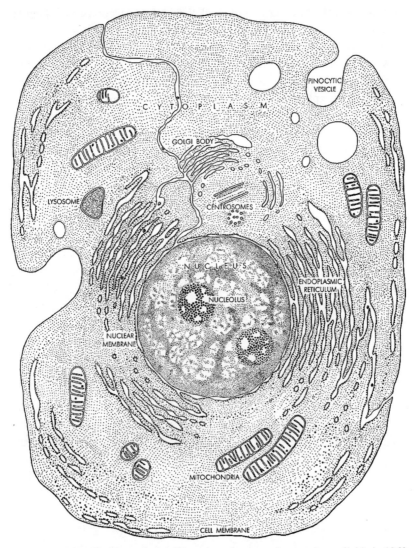

Figure 1.1. Diagram of a typical cell based on electron micrographs available in 1960. The four organelles that are the focus of the analysis in subsequent chapters – the mitochondrion, endoplasmic reticulum, lysosomes, and Golgi body (Golgi apparatus) – are clearly delineated. Reprinted from Jean Brachet (1961), The living cell, *Scientific American*, *205* (3), p. 7, with permission of Donald Garber, Executor of the Estate of Bunji Tagawa.

for investigations of other cell organelles and their functions, including the role of the ribosomes and endoplasmic reticulum in protein synthesis, the Golgi apparatus in cell secretion, and the lysosome in recycling cell waste.

As I discussed in the previous section, an important aspect of scientific disciplines is the creation of institutional structures. In the 1950s, research in the new discipline of cell biology was becoming highly specialized. It also fell sufficiently outside the scope of existing disciplines that numerous scientists felt the need to create new journals. The *Journal of Biophysical and Biological Cytology* was established in 1956 and rapidly established itself as the premier journal for the new community. In 1962 it changed its name to the *Journal of Cell Biology*, thereby recognizing the name increasingly favored for this discipline. The growing community also felt the need for regular scientific meetings; in 1960 they founded the American Society for Cell Biology. Creating a journal and a society were not small undertakings but required scientists to take substantial amounts of time away from research. The character of the journal and the society, moreover, were not foreordained, but resulted from explicit deliberation by these pioneers in cell biology. Chapter 7 will examine the processes involved in creating and nurturing the new journal and new society.

2

Explaining Cellular Phenomena
through Mechanisms

I do not in the least mean by this that our faith in mechanistic methods
and conceptions is shaken. It is by following precisely these methods
and conceptions that observation and experiment are every day enlarging
our knowledge of colloidal systems, lifeless and living. Who will set a
limit to their future progress? But I am not speaking of tomorrow but
of today; and the mechanist should not deceive himself in regard to the
magnitude of the task that still lies before him. Perhaps, indeed, a day
may come (and here I use the words of Professor Troland) when we
may be able 'to show how in accordance with recognized principles
of physics a complex of specific, autocatalytic, colloidal particles in
the germ cell can engineer the construction of a vertebrate organism';
but assuredly that day is not yet within sight of our most powerful
telescopes. Shall we then join hands with the neovitalists in referring
the unifying and regulatory principle to the operation of an unknown
power, a directive force, an archaeus, an entelechy, or a soul? Yes, if we
are ready to abandon the problem and have done with it once and for all.
No, a thousand times, if we hope really to advance our understanding
of the living organism.

(Wilson, 1923, p. 46)

The focus of this book is creation of cell biology in the mid-twentieth century
as a distinct field of biology devoted to discovering and understanding the
mechanisms that account for the ability of cells to live. The notion of mechanism has a long history in philosophy (a brief sketch follows), but it was largely
eclipsed in twentieth-century philosophy of science, which emphasized laws
and, not coincidentally, physics rather than biology. Mechanisms have begun
again to be the focus of discussion in very recent philosophy of science. I will
devote a major portion of this chapter to discussing the central features of
this newly emerging conception of mechanism and mechanistic explanation.
While many of the philosophers advancing this notion of mechanism have

19

been principally concerned with biology, the notion of mechanism on which consensus has settled is insufficiently biological. Accordingly, a further step in my discussion will be to develop a more biologically adequate notion of mechanism, one that is itself inspired by the conception of a cell as the basic living unit. Because actual scientific investigation, such as that pursued in cell biology, is not concerned with the abstract character of a mechanism, but with investigating the details of actual mechanisms, I will conclude by examining how mechanisms are studied, and the challenges cell biology faced in discovering the mechanisms in cells that account for the activities of life.

1. HISTORICAL CONCEPTIONS OF MECHANISM

In western thought, the notion of mechanism originated with the design by the ancient Greeks of machines, such as the wedge, to perform work. Greek thought, however, usually contrasted mechanics with nature – machines facilitated accomplishing work that opposed natural forces. This is clear in the pseudo-Aristotelian text *Mechanica*:

> Nature often operates contrary to human interest, for she always follows the same course without deviation, whereas human interest is always changing. When, therefore, we have to do something contrary to nature, the difficulty of it causes us perplexity and art has to be called to our aid. The kind of art which helps us in such perplexities we call Mechanical Skill. (*Mechanica* 847a14f)

Accordingly, nature itself was not conceived as operating mechanically. Aristotelian philosophy in particular advanced an anti-mechanistic conception of nature. It emphasized *telos*, the end state to be achieved by entities of nature, and the form, which resided in bodies and determined their nature and what they did. Thus, Aristotle stressed the distinctive forms (souls) of living things and the activities they made possible (nutrition and reproduction for plants; also sensation and locomotion for animals; and all of these plus reason for humans).

The Aristotelian perspective was still dominant when, in the seventeenth century, numerous natural philosophers rebelled and developed a program of explaining natural phenomena mechanically. Unlike the Greek mechanics, the seventeenth-century mechanists identified natural phenomena as themselves mechanical and offered mechanical explanations of naturally occurring activities.

Although Aristotelian mechanics set nature in opposition to mechanism, the Greek atomist tradition advanced a view closer in spirit to the later mechanical philosophy. Such philosophers as Leucippus (ca. 480–ca. 420 BCE) and Democritus (ca. 470–ca. 380 BCE) were called *atomists* because they sought to explain phenomena in nature by appealing to their constituent elementary particles, or atoms. They characterized these particles in terms of their shape and size, and then appealed to these properties to explain the characteristics of the macroscopic objects they comprised. Thus, Democritus proposed that hot bodies were hot because they were composed of elements of fire, which were small and round, whereas cold bodies were cold because they were composed of larger particles with sharp points. The early modern mechanists followed the atomists in appealing to the shapes of the hypothetical minute components of material objects to explain the properties of macroscopic objects.

One of the most prominent of the early modern mechanists was Galileo Galilei (1564–1642). He is celebrated for developing a mechanics of simple moving bodies, an account he extended to explain celestial phenomena such the movement of the earth around the sun. He also developed hypothetical mechanisms to account for mundane phenomena, and here the inspiration of the atomists is most clear. For example, he offered the following explanation of the power of heat to liquefy bodies:

> The extremely fine particles of fire, penetrating the slender pores of the metal . . . fill the small intervening vacua, and . . . set free these small particles from the attraction which these same vacua exert upon them and which prevents their separation. Thus the particles are able to move freely so that the mass becomes fluid and remains so long as the particles of fire remain inside; but if they depart and leave the former vacua, then the original attraction returns and the parts are again cemented together. (Galilei, 1638/1914, p. 19)

Rene Descartes (1596–1650) provided perhaps the fullest development of the mechanical philosophy in the early modern period. As he said in *Principia*, "I have described this earth and indeed the whole universe as if it were a machine: I have considered only the various shapes and movements of its parts" (Descartes, *Principia* IV, section 188). In appealing to the motions of the parts as well as their shapes, Descartes added to the resources of the ancient atomists. A key component of his mechanistic conception was that the movement of one object would cause movement in another. In particular, because Descartes did not allow for empty space, any movement of one object required the movement of other objects.

Having emphasized the change caused by the contact of one body with another, Descartes, in contrast to both the ancient mechanics and to the Aristotelians who had seen mechanics as opposed to nature, argued for treating the natural world as a mechanism and all events in it as the result of the operation of a mechanism. In so doing he denied any ontological difference between natural phenomena and the workings of human made mechanisms:

> I do not recognize any difference between artifacts and natural bodies except that the operations of artifacts are for the most part performed by mechanisms which are large enough to be easily perceivable by the senses – as indeed must be the case if they are to be capable of being manufactured by human beings. The effects produced by nature, by contrast, almost always depend on structures which are so minute that they completely elude our senses. (Descartes, *Principia* IV, section 203)

Descartes' conception of mechanism extended as well to the biological world, including the functioning of human bodies. In this he appealed frequently to metaphors with human-made machines, using them to suggest explanations of features of biological systems that initially appear to set them apart. For example, to explain the ability of animals to initiate motion, he appealed to such artifacts as clocks and mills that succeeded in moving on their own. Although William Harvey had already offered his own mechanical pump model for the circulation of blood, Descartes proposed to explain circulation as resulting from heating that caused the expansion of droplets of blood, which then forced their way through the arteries. He further appealed to the hydraulically moved statuary in the Royal Gardens to provide a model of the ability of the nervous system to transmit sensory and motor signals from the brain. He proposed that in animals the brain directed activity by altering the flow of very fine matter, the animal spirits, through the nerves (see Figure 2.1).

A generation after Descartes, the British chemist Robert Boyle (1627–91) coined the term *mechanical philosophy*. Boyle directed part of his empirical work to developing the air pump as a device for creating vacuums. In allowing for a vacuum, that is, space unoccupied by objects, Boyle developed a different perspective on a mechanistic conception of nature than Descartes had endorsed. Furthermore, his interest in the air pump was not just that of an engineer: Boyle sought to explain the ability of air to exert pressure, and proposed a mechanical model of air molecules as small springs that could be compressed and later expand. As exemplified in his account of air, Boyle's general strategy was to appeal to the shape and motion of hypothetical small particles to explain the properties of different chemical substances.

Although the notion of mechanism was a major feature of early modern philosophy, another tradition developed in the same period ultimately superseded it in philosophical accounts of science. Instead of explaining the

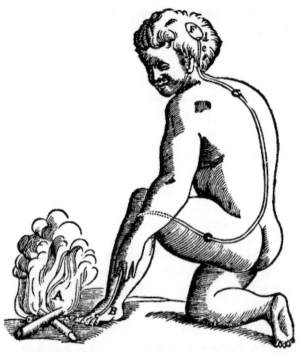

Figure 2.1. Descartes' representation of the nerve transmission in reflex action. Animal spirits traverse the path from the sense organs to the brain and back, resulting in motion. Reprinted from René Descartes (1664), *L'Homme*. Paris: Charles Angot.

activities of the natural world as workings of a mechanism, this tradition placed principal emphasis on *laws* of nature. Ronald Giere (1999), among others, has proposed that this view was a product of the theological perspective of certain early modern philosophers – those who conceived of the principles governing the natural world on the model of civil laws, albeit with God, not humans, as the lawgiver. Appeal to a lawgiver continued as a feature of this tradition, but what gave the law-based tradition traction was that scientists learned to describe regularities in the universe economically and precisely in laws that were appealed to in explaining particular phenomena. The perspective is clearly manifest in Newton, who posited forces in nature to explain phenomena from the terrestrial to the celestial. His well-known three laws of motion are an exemplar of this approach, but he pursued the approach more generally. To explain the reflection of light, for example, he posited a repulsive force between a light ray emitted and its source as soon as a ray left the area of attraction. Vaporization was similarly attributed to a repulsive force. Moreover, Newton suggested a systematic relationship between repulsive and attractive forces in general: "As in Algebra, where affirmative Quantities vanish and cease, there negative ones begin; so in Mechanicks, where Attraction ceases, there a repulsive Virtue ought to succeed" (Opticks, Query 31).

What disturbed many mechanists about Newton's approach was that he envisaged forces as operating over a distance, whereas the mechanical philosophy, especially as developed by Descartes, insisted on contact in any causal interaction. Newton himself, in the preface to *Principia*, tried to downplay this difference, referring to forces as mechanical principles:

> I wish we could derive the rest of the phenomena of Nature by the same kind of reasoning from mechanical principles, for I am induced by many reasons to suspect that they may all depend upon certain forces by which the particles of bodies, by some causes hitherto unknown, are each mutually impelled towards one another, and cohere in regular figures, or are repelled and recede from one another. These forces being unknown, philosophers have hitherto attempted the search of Nature in vain; but I hope the principles here laid down will afford some light either to this or some truer method of philosophy. (1687, Preface to the First Edition)

As Boas (1952) discussed in her classic history of the mechanical philosophy, two schools of Newtonians developed. One school accepted forces operating without contact while the other insisted on maintaining some version of mechanical contact as involved in all actions of forces:

> One, comprising those whom we call Newtonians today, accepted the concept of action at a distance; rejected mechanical explanations, denying their necessity; converted the ether into magnetic and electric fluids where necessary; and generally followed the theories enunciated in the *Principia* rather than the more speculative hypotheses of the *Opticks*. But there were other physicists of the eighteenth century who followed the more mechanical aspects of Newton's theory and used his ether to form a highly complex system. (Boas, pp. 519–20)

2. TWENTIETH-CENTURY CONCEPTIONS OF MECHANISM

The dominant twentieth-century philosophy of science, developed by philosophers and scientists who characterized themselves as *logical empiricists*, adopted a non-mechanical emphasis on explanation in terms of laws. Part of the reason for this was the logical empiricists' generally Humean, anti-metaphysical perspective, which rejected positing hidden causal relations between events. According to Hume, we observe sequences of events but do not observe anything connecting one event to the next. Hume favored a

24

minimalist concept of cause and effect, in which an event that regularly precedes another type of event can be denoted as its cause; Hume thought no further explanation should be sought. Being generally skeptical of the human capacity to satisfactorily address metaphysical issues, the logical empiricists went a step further and tried to avoid reference to causes by focusing instead on the linguistic statement of laws. They developed a conception of explanation, known as the *deductive-nomological model* of explanation, that involved derivation of an observation statement from statements of laws along with statements specifying initial conditions (Hempel & Oppenheim, 1948).

In contrast to the logical empiricists, Wesley Salmon (1984) developed an account of scientific explanation that made the notion of cause itself central. For Salmon, the way to answer the question of why something happened was to identify what caused it. Salmon tried to formulate a non-problematic account of what a cause is, but his proposals have in turn been the focus of critical objections.[1] I will not pursue this debate because my interest in Salmon is not in his analysis of causation, but his characterization of his account as "causal-mechanical." Salmon was one of the first twentieth-century philosophers of science to revive the interest in mechanistic explanation. It is not clear, however, what Salmon intended by invoking the term *mechanical*. As Glennan (2002) noted, Salmon never explicated the notion of a mechanism, settling instead for such comments as "explanations reveal the mechanisms, causal or other, that produce the facts we are trying to explain" (Salmon, 1989, p. 121).

Before advancing an account of what a mechanism is, though, let me note an important respect in which Salmon's account offers a significant advance over seventeenth- and eighteenth-century mechanical philosophy. Whereas the early modern mechanists allowed only such properties as the shape and motion of particles to figure in accounts of mechanisms, Salmon (1984, p. 241) broadened the category to include such constructs as force fields: "We have to change our mechanistic view from the crude atomism that recognizes only the motions of material particles in the void to a conception that admits such non-material entities as fields, but for all of that, it is still a mechanistic world view. Materialism is untenable, but the mechanical philosophy, I believe, remains viable." This expansion repairs the breach in the mechanical philosophy introduced by Newton's appeal to forces. Mechanical interactions no longer require contact between causes and their effects, but can accommodate more distal actions such as gravitational and magnetic attraction. (This expansion of the

---

[1] See Salmon (1984) for his initial proposal. For a critical response, see Kitcher (1989). Salmon (1994) proposed an alternative account, which in turn has been criticized by Dowe (1995).

conception of mechanical activity had in practice long become commonplace. For example, nineteenth-century mechanistic accounts of physiological processes incorporated energy-intensive operations such as chemical bonding and catalysis.)

3. CURRENT CONCEPTIONS OF MECHANISM

With increased attention given to the biological sciences in the last two decades, a number of philosophers of science have begun attending to mechanistic explanation. They have addressed the absence of an appropriate framework by offering initial proposals that overlap in some important respects but also vary in terminology, scope, and emphasis (see, for example, Bechtel & Richardson, 1993; Glennan, 1996; Machamer, Darden, & Craver, 2000).[2] I first provide a basic characterization of mechanisms as found in nature and then elaborate it into a framework for mechanistic explanation:

> A mechanism is a structure performing a function in virtue of its components parts, component operations, and their organization. The orchestrated functioning of the mechanism is responsible for one or more phenomena.

Moreover:

- The component parts of the mechanism are those that figure in producing a phenomenon of interest.

[2] With Robert Richardson, I characterized mechanistic explanations as accounting "for the behavior of a system in terms of the functions performed by the parts and the interactions of these parts . . . A mechanistic explanation identifies these parts and their organization, showing how the behavior of the machine is a consequence of the parts and their organization" (Bechtel & Richardson, 1993, p. 17). Whereas we focused on the "functions" (operations) that parts perform, Glennan emphasized the properties of parts in stating what he originally (1996) called laws and now (2002) calls "invariant change-relating generalizations." These are instantiated in "interactions" in which "a change in a property of one part brings about a change in a property of another part" (2002, p. S344). Machamer, Darden, and Craver characterize mechanisms as "entities and activities organized such that they are productive of regular changes from start or set-up to finish or termination conditions" (2000, p. 3), emphasizing the distinct metaphysical status of "entities" (parts) and "activities" (operations). Tabery (2004) has proposed a partial synthesis in which activity and property changes are seen as complementary. I discuss the use of the term *operation* rather than *activity* in note 7 below. Finally, Machamer et al. (p. 3) include a characterization of mechanisms as "productive of regular changes from start or set-up to finish or termination conditions." I am concerned that such an emphasis helps to retain a focus on linear processes whereas mechanisms, when they are embedded in larger mechanisms, are continuously responsive to conditions in the larger mechanism, for tractability scientists tend to focus on the conditions in which an operation is elicited and on its contribution to the behavior of the overall mechanism. However, they often have to counter this analytical perspective to appreciate the dynamics at work in the system.

- Each component operation involves at least one component part. Typically there is an active part that initiates or maintains the operation (and may be changed by it) and at least one passive part that is changed by the operation. The change may be to the location or other properties of a part, or it may transform it into another kind of part.
- Mechanisms may involve multiple levels of organization.
- Operations can be organized simply by temporal sequence, but those in biological mechanisms tend to exhibit more complex forms of organization.
- Mechanisms can be dynamic and can change both ontogenetically and phylogenetically.

Several features of this characterization of a mechanism require elaboration.

Mechanisms Explain Phenomena

The conception of a mechanism is intimately tied to the context of explanation: Mechanisms are identified in terms of a phenomenon for which explanation is sought. In logical empiricist philosophy of science there was a tendency to construe explanations as explaining observation statements, which were taken to be theory-neutral characterizations of events. This view became problematic with the contention, stemming from Hanson (1948) and Kuhn (1962/1970), that observations are theory-laden in that what scientists observe is influenced by the theories they hold. This seemed to threaten circularity with theories being tested by observations that are already shaped by the theories being tested. Although there are simpler ways of showing that this circularity is not vicious,[3] Bogen and Woodward challenged the very idea that observations are what scientists explain. They contrasted observations and phenomena. Observations provide data but – except when the data does not turn out as expected and the investigator seeks to explain why – it is not data that scientists explain. Rather, they explain *phenomena* – occurrences in the world about which data can be procured.[4] Although there can be singular phenomena (the big bang or the birth of a particular organism), the

---

[3] For example, it is sufficient to note that even if the way in which we observe something is affected by our theories and other beliefs, this does not entail that we can observe whatever the theory predicts. Holding a theory that grass turns red in normal daylight will not make it the case that I will see red grass.

[4] Some purported phenomena do not exist. Bogen and Woodward cited the example of N-rays, a phenomenon that French physicists at the beginning of the twentieth century posited on the basis of several observations; ultimately, more careful data collection demonstrated that such a phenomenon did not exist.

phenomena of interest in science can be captured in generalizations involving, for example, a functional relationship between variables, or the fact that a certain kind of event regularly occurs only if a certain other type of event has just occurred. Data play an important role in identifying and providing evidence for phenomena, but it is the phenomena so identified that are the objects of explanation.

Bogen and Woodward offered as examples of phenomena "weak neutral currents, the decay of the proton and chunking and recency effects in human memory" (1988, p. 306). In biology, DNA replication or alcoholic fermentation would be comparable examples. It is often possible to characterize phenomena quantitatively. Bogen and Woodward consider the example of lead melting at 327° C. Galileo established that the distance traveled by an object falling freely near the surface of the earth is sixteen times the square of the number of seconds it falls. A biological example of a quantitative phenomenon is that the maximum number of molecules of adenosine triphosphate (ATP) formed in normal cells via oxidative phosphorylation per oxygen molecule consumed is three. Phenomena may also be characterized with different degrees of specificity. An individual scientist might, for example, identify the phenomenon for her investigation as the synthesis of a particular protein in a specific type of cell occurring in a particular species living under specified conditions. The author of a review paper might address the phenomenon of synthesis of that protein in a variety of cell types and species. At the most general level, a textbook author might write a few pages simply on "protein synthesis."

Identifying and characterizing phenomena is a challenging scientific activity that consumes considerable resources of time, money, and ingenuity. A purported phenomenon must be shown to be genuine and its generality determined. Some purported phenomena do not stand up under scrutiny and must be discarded. Although I will emphasize the importance of a specification of the phenomenon for the development of mechanistic explanations, it is important to note at the outset that scientists often revise their characterizations of phenomena in the course of investigating the mechanisms they take to be responsible for them. In *Discovering Complexity*, Richardson and I referred to such revisions as *reconstituting the phenomenon* and we offered the example that, in the course of investigating the mechanism of gene expression, researchers repeatedly revised the conception of what genes code for. In the 1860s Gregor Mendel spoke of factors for traits. In the 1910s Thomas Hunt Morgan and his collaborators sought to localize genes for such traits as eye color, but in the 1940s Beadle and Tatum's inquiry with mutations in *Neurospora* led them to link genes instead to individual enzymes.

If the revision to the conception of the phenomenon is so major that little of what the mechanism was thought to be doing is still recognized as occurring, the work toward articulating the mechanism for the phenomenon as originally characterized may prove to have been in vain. Most often, though, the change in the characterization of the phenomenon takes the form of a revision, not a wholesale replacement of the old conception, and the changes in the account of the mechanism are accordingly more restricted. For example, early investigators construed fermentation as a catabolic activity breaking down sugar and yielding alcohol (with heat as a by-product). Once researchers recognized that the energy released was captured in high-energy phosphate bonds that were used as energy sources for other cell activities, the conception of the phenomenon to be explained was revised. It was now a mechanism for converting the energy of foodstuffs into a form useful in such other activities as cell division. Yet, because the catabolic breakdown of sugar to alcohol remains part of this process, much that had been learned about the mechanism of fermentation still applied after the phenomenon had been reconceptualized.

It bears emphasizing that the project of providing explanations, including mechanistic explanations, starts with the identification of a phenomenon. This is where the functioning structure gets determined, constraining what will count as a successful identification of relevant parts and operations and their organization (Kauffman, 1971). If the operation of some entity does not contribute to the production of a given phenomenon, it is not part of the mechanism responsible for that phenomenon. On this construal, different mechanisms may be instantiated in the same substance in the same spatial-temporal region and may share many component parts and operations. What unites one set of parts and operations into a given mechanism is their organization and their orchestrated functioning in producing a particular phenomenon.

I will conclude this discussion of phenomena by introducing an illustration I will continue to use to characterize various features of a mechanism. After the investigations of Harvey, pumping of blood through the circulatory system was a well delineated phenomenon. Although we now take the phenomenon to be obvious, until Harvey established the more general phenomenon of the circulation of the blood, the phenomenon of pumping blood was not recognized. Rather than conceiving of circulation, investigators assumed that both the arteries and veins transported material to the bodily tissues and that this phenomenon was readily accounted for as a result of newer material pushing older material along. Once Harvey established that the blood circulated, the need for a pump to move blood was recognized and the functioning heart was identified as the mechanism responsible for this phenomenon.[5] The importance of specifying the phenomenon to be explained is illustrated with this example: Until the heart was recognized as performing the function of pumping blood, there was no interest in understanding the way in which this occurred. Moreover, hearts also do other things – they make sounds and someone might want to explain that phenomenon. That, however, is a different phenomenon that involves a different mechanism – a system of parts and operations that presumably shares some components with the mechanism for pumping blood but is not identical to it.

Component Parts and Component Operations

The next aspect of mechanisms to emphasize is that mechanisms consists of component parts and component operations.[6] Figure 2.2 illustrates key components of the heart viewed as a mechanism for pumping blood. Component parts of the heart include the atria and ventricles, the valves between the atria and ventricles and between the ventricles and arteries, and the blood itself. Component operations include the contraction and relaxation of the atria and ventricles and the opening and closing of the valves. Blood is forced out of the atria and ventricles as they contract, and prevented from flowing back by the closing of the valves afterward.[7] The blood here is a part of the mechanism, but one that is operated on rather than itself performing operations (in the context of this phenomenon). Although in this case the parts that perform

---

[5] In many cases, however, the entity or system responsible for the phenomenon is not immediately obvious and must be discovered. For example, in Bechtel and Richardson (1993), we discussed the extended controversy in the nineteenth century of the locus of control for respiration until Pflüger (1875) established that it occurred within individual cells.

[6] When I am emphasizing the thing performing the operation, I use the term *part* or *component part* and when I am emphasizing what the part does, I speak of *operation* or *component operation*. I sometimes use the term *component* alone where it is not important to be specific, or to designate jointly the part and the operation it performs.

[7] Machamer, Darden, and Craver employ the term *activity* to draw attention to the fact that components of mechanisms are active. The mechanical philosophy of the seventeenth century departed from Aristotelian philosophy in construing elements of nature as passive, doing something only if acted upon. In some accounts, the giant clock-like machine of nature was wound up at the outset and motion in the world is the playing out of that initial state. The term *activity*, however, does not readily capture the fact that in most operations there is also something acted upon. This is the reason I have preferred the term *operation*. Typical of the operations I have in mind are the reactions of chemistry which prototypically involve a catalyst, a reactant, a product, and often a cofactor. In some reactions there is no need for a separate catalyst as the energetic factors are such that the reaction will occur spontaneously. In autocatalytic reactions, which are highly relevant in living systems, the product of the reaction also serves as a condition for more iterations of the reaction.

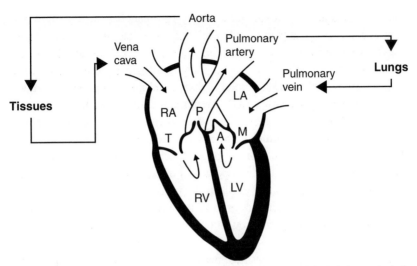

Figure 2.2. An example of a mechanism: the heart pumping blood. Labeled parts include RA: right atrium; LA: left atrium; RV: right ventricle; LV, left ventricle; T: tricuspid valve; M: mitral valve; P: pulmonary valve; A: aortic valve.

operations are separate from those operated on, in other cases the parts that perform operations may also be affected by the parts on which they operate (as I will discuss in Section 6, such feedback provides a prime way in which mechanisms can be self-regulating).

The component parts and operations of a mechanism do not present themselves to the scientist neatly distinguished and labeled as in a textbook. The investigations resulting in an understanding of the mechanism require *decomposing* it (taking it apart), conceptually if not physically. Corresponding to the division of parts and operations, there are two types of decomposition. What I refer to as *structural decomposition* decomposes a structure into component parts while *functional decomposition* decomposes the function into component operations. Though sometimes coordinated, it is not uncommon for these two types of decomposition to be pursued independently of each other, by scientists in different fields employing different tools. Often one decomposition proceeds more rapidly and successfully than the other, with considerable time elapsing before the slower inquiry catches up.

One example of structural decomposition is the discovery via anatomical dissection that (1) the body has a heart; and (2) the heart has four chambers (RA, LA, RV, and LV) and at least four valves (T, M, P, and A). Another example is the discovery via microscopy that (1) tissues are composed of cells; (2) cells contain a plasma membrane, nucleus, and cytoplasm; (3) cytoplasm

31

contains an aqueous cytosol and a variety of organelles such as mitochondria and the Golgi apparatus; (4) each organelle has an internal structure (description and further levels varying by organelle). An example of functional decomposition is the discovery via physiological investigation that the overall function of pumping blood includes multiple component operations of contraction and relaxation (at different times in different chambers) and of opening and closing (of valves). The parentheticals indicate that it is generally difficult to specify operations without some indication of the parts involved. It is useful to have a separate notion of functional decomposition, though, because progress in identifying operations often can proceed when there is minimal knowledge of some or all of the parts involved. For example, biochemists in the early twentieth century decomposed the overall function of cellular respiration into numerous biochemical reactions (operations) while, structurally, they had basic knowledge of the substrates and products (passive parts) but little more than invented names for the enzymes (active parts) that were assumed to catalyze the reactions.

Ultimately, the full characterization of a mechanism requires mapping the operations into which the overall function of the mechanism is decomposed onto the parts into which the structure is decomposed. I use the term *localization* for such mappings and discuss them further below. Uncovering the organization of the parts and operations at a given level – not merely identifying them – is also crucial. Such a combined perspective is often required to understand fully how the mechanism generates its behavior, because frequently the spatial layout of the parts enables or facilitates the temporal organization of their operations. As a practical matter, moreover, structure and function frequently provide critical insights into the other. Learning the structural character of a part can provide insight into how it carries out its operation. Understanding the operation that is being performed often provides clues as to what sort of part is responsible. As we will see, many of the major contributions of modern cell biology to understanding the mechanisms responsible for the phenomena exhibited in living cells involved the ability to relate various organelles to the physiological operations they perform and, at a lower level, certain components of the organelles with particular biochemical operations.

Organization and Orchestration

A point just mentioned – that it is important to determine how parts and operations are organized – deserves elaboration. A mechanism is typically not just a collection of independent parts, each carrying out its operation in isolation. Rather, parts and operations are generally integrated into a cohesive, functioning system. In the heart, the veins, atria, ventricles, and arteries must be appropriately spatially related to each other. Further, the valves must be located at the right place and oriented so as to prevent blood from flowing backwards through the system. This exemplifies the fact that the parts of a mechanism are organized. Its operations also are organized. Each may occur in its turn – a simple temporal ordering – or there may be a good deal of overlap, interdependency, or other complications. As the timing becomes more complex, especially in mechanisms with many parts, we further say that the operations are *orchestrated* so as to produce the phenomenon of interest. For example, the phenomenon of the heart pumping blood depends on the spatial organization of the component parts, the temporal organization of the component operations (movements), and the fine-grained spatiotemporal orchestration of the moving parts in real time.

There are several reasons organization is so important. If one part is to operate on the product produced by the operation of another part, it needs to have reliable access to that product. One way to ensure this is for the two parts to be spatially contiguous. Another is to provide a mode of transport between them. The assembly lines used in human manufacturing adopt this mode of organization – components are brought to the line and added to the emerging product in sequence. Organization in biological systems, however, is seldom as simple as the sequential arrangement used in assembly lines. One of the key features of organization in biological mechanisms is the incorporation of feedback and other kinds of control systems that allow the behavior of some components of the mechanism to be regulated by other components of the mechanism. The role of more complex modes of organization is a major feature that renders the biological conception of mechanism different from that which suffices for non-biological mechanisms, as discussed further in Section 6.

4. REPRESENTING AND REASONING ABOUT MECHANISMS

Mechanisms are real systems in nature, which led Salmon (1984) to identify his causal-mechanical approach to explanation as *ontic*, as it appeals to the actual mechanism in nature. He contrasted it with an *epistemic* conception of explanation that appeals to laws and derivations from laws, which are clearly products of mental activities. Salmon's insight is important, but the ontic/epistemic distinction does not properly capture it. He is right that in mechanistic explanation a scientist appeals to causal relations and mechanisms operative in nature, which are taken to generate or produce the phenomena being explained. However, it is important to note that offering an explanation is still an epistemic activity and that the mechanism in nature does not directly perform any explanatory work.[8]

There are several ways to appreciate the epistemic character of mechanistic explanation. First, the mechanisms operative in our cells were operative long before cell biologists discovered and invoked them to explain cellular phenomena. The mechanisms are not themselves the explanations; it is the scientist's discovery and rendering of aspects of the mechanism that produces what counts as an explanation. Second, the difference between the mechanism and the mechanistic explanation is particularly obvious when considering incorrect mechanistic explanations – in such a case a scientist has still appealed to a mechanism, but not one operative in nature. Such a mechanism exists only in the representation offered by the scientist. It is thus the mechanism as represented, not the mechanism itself, that figures in explanation. (It is also the phenomenon as represented that scientists seek to explain.) Thus, scientists offer a mechanistic explanation by identifying and representing parts and operations regarded as key to producing the phenomenon and showing how, appropriately organized, they can do so.[9]

Mechanisms can be represented either in linguistic descriptions or in diagrams. Philosophical accounts of science have tended to privilege linguistic representations and regard diagrams as at best crutches for following the linguistic argument. When one considers the actual practice of scientists in reading papers, however, the tables seem to be turned. It is common for readers to scan the abstract and then jump to key figures. To the extent that crutches are involved, the figure captions that provide commentary on the figures play this role. Consider a paper in which a mechanistic explanation is proposed. The diagram provides a vehicle for representing the complex interactions among operations, while the commentary can only characterize these one at a time. The text of the paper then provides yet further commentary: about how the mechanism is expected to operate (introduction), how evidence as to its operation was procured (methods), what evidence was advanced (results), and the interpretation of how these results bear on the proposed mechanism

---

[8] This point is eloquently stated by Christian de Duve (1984, p. 18) at the outset of his masterful *Tour of the Living Cell*: "Every object, every site, every happening, every process, every mechanism that will be pointed out as though it were there to be seen is actually a product of individual human brains mulling and churning over collections of images and sets of figures, themselves the products of recordings made by intricate instruments on biological materials subject to complex experimental manipulations."

[9] I thank Cory Wright for impressing on me that a mechanism in nature does not itself explain anything.

(discussion). The detailed commentary is important, but it is the diagram that represents the mechanism. As just one example of the saliency of diagrams, Christian de Duve, whose role in discovering the lysosome will be discussed in detail in Chapter 5, recollects that his discovery of the lysosome was sparked by an unexpected failure in his biochemical investigation of a liver enzyme. "By some fortunate coincidence, my recent readings had included [two 1946 papers by Claude and] I immediately recalled Claude's diagrams showing the agglutination at *p*H 5 of both large and small granules, and concluded that our enzyme was likely to be firmly attached to some kind of subcellular structure" (de Duve, 1969, p. 5).

The importance scientists place on diagrams should lead us to question whether they are in fact superfluous. Are there reasons a scientist might prefer to represent certain information diagrammatically rather than propositionally? More importantly, are there different processes of reasoning with diagrams than with propositions such that an account of science that focused only on logical inference would fail to capture an important aspect of explanatory reasoning?

The motivation for using diagrams to represent mechanisms is obvious. Unlike linguistic representations (except those found in signed languages), diagrams make use of space to convey information. As the heart example revealed, spatial layout and organization is often critical to the operation of a mechanism itself. As in a factory, different operations occur at different locations. Sometimes this serves to keep operations separate from one another and sometimes it serves to place operations in association with one another. These spatial relationships can be readily shown in a diagram. Even when information about the specific spatial layout is lacking or not significant, one can use space in the diagram to relate or separate operations conceptually. Moreover, diagrams can take advantage of dimensions other than space that visual processing can access, including color and shape.[10]

Time is at least as important as space to the operation of a mechanism – one operation proceeds, follows, overlaps, or is simultaneous with another operation. This can be captured by using one of the spatial dimensions in a diagram to convey temporal order. This of course presents a problem: Most diagrams are two dimensional and this leaves just one dimension for everything other than time. One solution – as exemplified in the heart diagram – is

---

[10] These can either be iconic – representing the actual color or shape of the parts of a mechanism, or they can be symbolic. fMRI diagrams of brain activity are a well-known example of the symbolic use of color – colors scaled from hot to cold are used to represent such things as strong to weak activations or high to low statistical significance of the increase of the activation above some baseline.

to make strategic use of arrows to represent temporal relations, leaving both dimensions free to represent the mechanism's spatial or similarity relations. Another solution is to use techniques for projecting three dimensions onto a two-dimensional plane.

Whether the temporal order of operations is represented by means of a spatial dimension or by arrows, a diagram has clear advantages over linguistic description. The most obvious advantage – that all parts and operations are available for inspection simultaneously – probably is the weakest one. Due to processing limitations, people can take in only one or a few parts of the diagram at a time. Nonetheless, more so than when reading text, they have the freedom to move around it in any number of ways; and as the diagram becomes more familiar, more of it can be taken in at one time. A stronger advantage is that diagrams offer relatively direct, iconic resources for representation that can be invaluable. For example, it is immediately apparent in the heart diagram that blood is being pumped simultaneously from the two atrial chambers to the two ventricles and that these two parallel operations are in a sequential relationship to two other parallel operations (pumping from the two ventricular chambers).

The value of consulting a diagram in this way is even more apparent in mechanisms with feedback loops, through which an operation that is conceptually downstream (closer to producing what is taken to be the product of the mechanism) has effects that alter the execution of operations earlier in the stream at subsequent time steps. Multiple examples can be found within the cellular respiration. As biochemists discovered in the 1930s, it is composed of three connected submechanisms (as illustrated in the next chapter in Figure 3.16). When these are further unpacked, they are seen to involve coordinated biochemical operations, including feedback operations. Figure 2.3 shows an important feedback loop that operates at the interface between the first two submechanisms – glycolysis and the citric acid cycle. The diagram aids understanding by spatially laying out the parts of the system (compounds such as pyruvate) and by using the vertical dimension as well as arrows to indicate the sequence of operations (solid arrows for the reactions and a dotted arrow for the feedback loop).

An important principle recognized by cognitive scientists engaged in modeling reasoning computationally is that it is essential to coordinate the modes of representation and procedures of inference. If diagrams are an important vehicle for representing mechanisms, then it is necessary to consider how people reason about diagrams. Philosophers since Aristotle have often assumed that procedures of logic, especially natural deduction, describe our reasoning. But logic operates only on linguistic representations, so if scientists reason

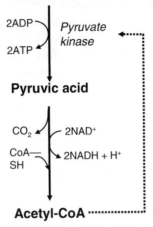

Phosphoenolpyruvate

Figure 2.3. Feedback loop in the linkage between glycolysis and the citric acid cycle. In the final reaction of glycolysis, phosphoenolpyruvate produces pyruvic acid. Pyruvic acid then produces acetyl-CoA, some amount of which is needed to continuously replenish the citric acid cycle (not shown). If more acetyl-CoA is produced than can be used in the citric acid cycle, it accumulates and feeds back (dotted arrow) to inhibit pyruvate kinase, the enzyme responsible for the first step in the reaction. This in turn will stop glucose from entering the glycolytic pathway.

with diagrams, the operations of reasoning must be different. To understand how scientists reason with diagrams it is helpful to keep in focus the fact that mechanisms generate the phenomenon in virtue of their component parts performing their operations in a coordinated manner. The kind of reasoning that is needed is reasoning that captures the actual operation of the mechanism, including both the operations the components are performing and the way these operations relate to one another.

One limitation of diagrams when it comes to understanding mechanisms is that they are static. Even if they incorporate arrows to characterize the dynamics of the mechanism, the diagram itself doesn't do anything. Thus, it cannot capture the relation of the operation of the parts to the behavior of the whole mechanism. Accordingly, the connection together must be provided by the cognitive agent. The cognizer must imagine the different operations being performed, thereby turning a static representation into something dynamic.[11]

---

[11] Animated diagrams relieve people of this difficult task and are often far more instructive to novices. Thomas M. Terry of University of Connecticut has produced some excellent ones that make clear how the many operations in cellular metabolism are related. He has them posted at http://www.sp.ucon.edu/~terry/images/anim/ETS.html. Another good site for such diagrams,

Mary Hegarty (1992) called the activity of inferring "the state of one component of the system given information about the states of the other system components, and the relations between the components" *mental animation* and emphasized its importance to the activities of designing, troubleshooting, and operating mechanical devices (p. 1084). Obtaining reaction time and eye movement data while people solved problems about relatively simple pulley systems, she investigated the extent to which inference processes are isomorphic to the operation of the physical systems. One way they were not isomorphic is that the participants made inferences about different components of the system (i.e., individual pulleys) separately and sequentially even though in the physical system the components operated simultaneously. The participants found it considerably harder, however, to make inferences that required them to reason backward through the system rather than forward, suggesting that they animated the system sequentially from what they represented as the first operation, in this respect preserving isomorphism with the actual system.

Accepting the claim that people, including scientists, understand diagrams of mechanisms by animating them, a natural follow-up question concerns how they do this. A plausible initial proposal is that they create and transform an image of the mechanism so as to represent the different components each carrying out their operations. In perception we have experience of parts of the system changing over time, and so the proposal is that in imagination we animate these components by invoking the same processes that would arise if we were to watch an animated diagram. This proposal needs to be construed carefully, as a potential misunderstanding looms. Reference to a mental image should not be construed as reference to a mental object such as a picture in the head. Recent cognitive neuroscience research indicates that when people form images they utilize many of the same neural resources that they do in perception (Farah, 1988; Kosslyn, 1994).[12] Thus, what occurs in the head in forming an image is activity comparable to that which would occur when seeing an actual image. Barsalou (1999) speaks of this neural activity as a *perceptual symbol.*

---

which also provides links to Terry's diagrams, is http://www.people.virginia.edu/~rjh9u/atpyield.html.

[12] Within cognitive science there has been a heated controversy over whether the representations formed in the cognitive system are really image-like (Kosslyn, 1981; Kosslyn, 1994; Pylyshyn, 1981; Pylyshyn, 2003). This discussion can remain neutral on this issue since the fundamental issue is not how the cognitive system encodes its representations but what it represents something as. The visual system represents objects as extended in space and changing through time. What is important here is that scientists can represent mechanisms mentally in much the way that they represent diagrams that they encounter (albeit with less detail than when actually looking at the diagram).

Thinking with perceptual symbols then involves the brain initiating sequences of operations that correspond to what it would undergo if confronted with actual input from visual objects behaving in a particular manner. Barsalou refers to this as *simulation*. Moreover, simulation is not restricted to repeating those sequences of neural processes that occurred in previous experience. Just as we can imagine objects that we have never seen by recombining components of things we have seen, we can imagine sequences of changes that depart from those that we have actually encountered.

Although humans are relatively good at forming and manipulating images of rather simple systems, if what we are imagining is the working of a rather complex mechanism that has multiple components interacting with and changing each other, we often go astray. We fail to keep track of all the changes that would occur in other components of the system in response to the changes we do imagine. Thus, the usefulness of mental animation for understanding a mechanism does reach a limit. Ordinary people may simply stop trying at this point, but scientists and engineers often find it important to do better and hence have created tools that supplement human abilities to imagine a system in action. One tool involves building a scale model (or otherwise simplified version) of a system and using it to determine how the actual system would behave. The behavior of the scale model *simulates* that of the actual system. For example, the behavior of objects in wind tunnels can be used to simulate phenomena involving turbulence in natural environments. If instead an investigator can devise equations that accurately characterize the changes in a system over time, the investigator can often determine how the system will behave by solving the equations without actually building a scale model. In this case the simulation is done with a mathematical model rather than a physical model. The advent of the computer provided both a means of solving the equations of a mathematical model and an additional means of simulating systems. Higher level computer languages are designed to represent complex structures and their interactions, and by using these resources, one can often create a computer simulation of the interactions in a complex system (Jonker, Treur, & Wijngaards, 2002).

These different modes of simulating a system all provide an important advantage when a mechanism is complex with multiple operations occurring simultaneously – they do not lose track of some of the interactions as a human imagining the operation of the mechanism often does. But even when it is a human who is doing the imagining, what he or she is doing can also be characterized as simulating the mechanism.

Although a mechanism can be represented by means of a diagram, it can also be described linguistically. Is there any fundamental difference between linguistic and diagrammatic representations? Larkin and Simon (1987) considered diagrams and linguistic representations that are informationally equivalent and analyzed how they can nonetheless differ with respect to ease of search, pattern recognition, and the inference procedures that can be applied to them. In part these differences stem from the fact that information that may be only implicit in a linguistic representation may be made explicit, and hence easier to invoke in reasoning, in a diagram (Larkin & Simon, 1987, p. 65).[13] More recently, Stenning and Lemon (2001) suggested that diagrams are more constrained in expressive power than propositions and accordingly are more tractable. They also argued that the advantage provided by these constraints is dependent upon the subject supplying an interpretation that makes them available.

5. LEVELS OF ORGANIZATION AND REDUCTION

The part–whole relationship between a mechanism's component parts and its structure can be understood as falling within the type of hierarchical, mereological framework that systematic biologists and others have long used to bring orderliness to types of entities at different levels. The relationship between a mechanism's component operations and its overall function have roughly the same character, though less attention has been paid to systematizing this kind of relationship. What is important here is that both kinds of components (the parts and their operations) can be regarded as occupying a lower level than the mechanism itself (a structure with a function). Because of this difference in levels, mechanistic explanations are commonly characterized as *reductionistic*.[14] The notion of reduction that arises with mechanistic explanation, however, is very different from that which has figured either in popular discussions or in recent philosophy of science, and its consequences are quite different. In these discussions, appeals to lower levels are thought to deny the efficacy of higher levels. While the functioning of a mechanism

---

[13] Larkin and Simon commented, "In the representations we call diagrammatic, information is organized by location, and often much of the information needed to make an inference is present and explicit at a single location. In addition, cues to the next logical step in the problem may be present at an adjacent location. Therefore problem solving can proceed through a smooth traversal of the diagram, and may require very little search or computation of elements that had been implicit" (1987, p. 65).

[14] Wimsatt (1976) was one of the first philosophers to recognize the relation between mechanism and what most scientists mean by *reduction*: "At least in biology, most scientists see their works as explaining types of phenomena by discovering mechanisms, rather than explaining theories by deriving them or reducing them to other theories, and *this* is seen as reduction, or as integrally tied to it" (p. 671).

depends upon its constitution, it also depends on its context, including its incorporation within systems at yet higher levels of organization. Mechanistic reductionism neither denies the importance of context or of higher levels of organization nor appeals exclusively to the components of a mechanism in explaining what the mechanism does. The appeal to components in fact serves a very restricted purpose of explaining how, in a given context, the mechanism is able to generate a particular phenomenon.

Before explicating further what mechanistic reduction involves, I need to clarify the notion of level. The notion of level is widespread in both philosophical and scientific discussions (see Churchland & Sejnowski, 1992) and it is often assumed that levels cut across all of nature so that there are levels of subatomic particles, atoms, molecules, and so forth. Extending this to the living world, there are levels of cellular organelles, cells, tissues, organs, organisms, societies, etc. There are problems in fitting such a conception of levels together with the scientific practice of explaining phenomena because often the crucial operations in nature cross these levels – electrons interact with molecules, ions with membranes, and single-celled organisms with organisms containing multiple organs. Accordingly, it seems wise to abandon the attempt to demarcate levels that transect nature; rather, I restrict the identification of levels to local contexts in which mechanistic explanations are offered for particular phenomena.

The notion of level enters into discussions of mechanisms in virtue of the fact that a mechanistic explanation decomposes a mechanism into its component parts and operations. Thus, an investigator begins with a phenomenon – for example, the phenomenon of organisms taking in oxygen and releasing carbon dioxide and water. In the attempt to explain this phenomenon the investigator decomposes the structure (the organism's body) into component parts, decomposes its function (respiration) into component operations, and determines how they are organized and orchestrated to produce the phenomenon. With respect to this phenomenon, the operative parts of the mechanism – the lungs, blood, tissues, etc. – constitute entities at a lower level of organization than the respiring organism. Another investigator may be interested in how tissues perform their role in this mechanism and decompose them into cells and then, within cells, into different organelles involved in cellular respiration. These components are then at a lower level than those identified in the first decomposition. In principle this process can continue indefinitely, but in practice it stops after two or three rounds of decomposition within a given research area or line of investigation.

The fact that components are contained within mechanisms ensures that components are of a smaller size than the mechanism itself, but, as the

41

examples in the last paragraph reveal, not all entities at a given level will be of the same order of magnitude. The level at which a particular entity resides depends on the role it is playing in the mechanism. A heart is a critical part of the circulatory system of a surgeon, which in turn plays a critical role in the ability of the surgeon to perform surgery. But when the surgeon is performing heart surgery, she is interacting with the patient's heart – holding it, reconnecting it, etc. In that interaction the patient's heart is at the same level as the surgeon performing the operation, whereas the surgeon's own heart is at a lower level. Likewise, in oxidative phosphorylation, there are protons in the atoms that comprise the lipid molecules of the inner mitochondrial membrane as well as protons being pumped across the membrane. The protons pumped across the membrane interact with it (as illustrated later in Figure 6.9), but protons in the lipid molecules are at a lower level than the protons being pumped.

So far I have emphasized that in a mechanistic account of levels it is the components of a mechanism that are the denizens of a lower level. In contrast, it is the mechanism as a whole and the things with which it interacts that inhabit the higher level. These interactions with other things may greatly influence the behavior of the mechanism and many of these things may themselves be fruitfully construed as mechanisms. In some cases a mechanism may be part of an organized system in which its behavior is coordinated with that of other entities to perform yet another function. In that case, the first mechanism is a component of the yet higher-level mechanism. If the organization in the higher-level mechanism results in the imposition of constraints on the behavior of the first mechanism, such embedding of a mechanism into a higher-level mechanism can be highly relevant for the scientist trying to explain the operation of the initial mechanism. Moreover, just as one can go up from a given mechanism to a higher-level mechanism, one can proceed down from a component of a mechanism to yet lower-level components. A mechanistic account thus gives rise to a cascade of levels.

Given the differences in the way levels are characterized, it is not surprising that the mechanistic account of reduction looks very different from traditional philosophical accounts of reduction which hold that higher-level theories are reduced by logically deriving them from lower-level theories (Nagel, 1961; Causey, 1977). For such a derivation to succeed, all the information required to generate the higher-level must be contained in the lower-level theory. If this is the case, a successful theory reduction renders the derived theory superfluous – everything is explained equally well with the lower-level theory. With mechanistic explanations, accounts of the lower level do not offer a complete theory. None of the components, alone, generates the phenomenon.

Even the whole collection of components does not produce the phenomenon except when appropriately organized. But it is precisely by organizing the parts and their operations that a higher level is constituted, thus providing a bridge linking the level of the components to the level of the mechanism.

It is important to emphasize that a mechanistic account, while appealing to parts and their operations, also appeals to the functioning of the mechanism as a whole and how, in virtue of its function, it interacts with entities in its environment. There is, therefore, a critical difference in what is investigated at the lower and higher levels. At the higher level investigators characterize the functioning of the mechanism as it is situated in its environment. Pasteur, for example, characterized yeast cells switching from aerobic to anaerobic metabolism depending on the presence of oxygen in their environment. Bio-chemical studies, which were not available to Pasteur, could open up the mechanism and reveal the components and their operations that made yeast capable of this switch. The biochemical processes, however, only suffice to explain how the function is performed *in the particular context*. Even after the component parts and operations are identified, it remains the case that the mechanism as a whole in its context performs a function: Switching from aerobic to anaerobic metabolism remains something done by yeast cells (or at least the whole glycolytic system in them). The upper level is far from superfluous – explanation requires both the account of what is happening at the higher level and the account of the components of the mechanism. In this respect, the kind of reduction that arises with mechanistic explanations is also compatible with recognizing the autonomy of higher levels of organization. The operations at a level are unique to the level and must be investigated with the appropriate tools.

The role of organization in generating higher levels is made clear by considering the point of view of an engineer. When an engineer faces a task, what she must do is draw upon the operation of components she already has available and organize them in a new way to accomplish the task. (She may have to iterate this strategy by decomposing what she is trying to accomplish into yet finer-grained operations, some of which can be performed by existing components, but others requiring another level of decomposition so as to design components that can perform those subsidiary tasks.) When she has finished, she has built something new, perhaps something for which she could secure a patent. We would not expect the patent office to deny her a patent because all of the components were already known to her – they were also known to the others who failed to have the insight needed to develop the new mechanism. What she has done is to build a mechanism that performs a new function.

This appeal to engineering highlights the fact that mechanisms perform different functions than their components and emphasizes the importance of organization of parts and operations in accomplishing the new function. If the engineer relied on off-the-shelf components in building her mechanism, then the only thing she added was organization. It is the only thing she added, but it is far from trivial. It is for discovering the way to organize parts and operations that the engineer earns her patent. But there is a further factor that successful engineers take into account – the context in which the mechanism performs its function. One common way in which engineered products fail is when they are put into operation in contexts other than those for which they were designed. The wrong context can hinder a mechanism's operation whereas the appropriate context may provide things which are co-opted in the mechanism's operation. This is even more true of biologically evolved mechanisms than humanly engineered ones. Evolution is an opportunist, and if something can be relied upon in the mechanism's environment, then it doesn't have to be generated by the mechanism. Vitamins provide just one well-known example. Because our ancestors could generally count on the availability of vitamins in their foods, there was no evolutionary pressure for us to retain the ability to synthesize them. Insofar, however, as such environmental factors are necessary for the functioning of the mechanism, mechanistic explanations need to focus on the mechanism's context, not just its internal configuration.

6. ORGANIZATION: FROM CARTESIAN TO BIOLOGICAL MECHANISMS

I described previously how, in reviving the mechanical philosophy, Salmon significantly expanded the toolbox of features in terms of which mechanisms could be understood. Instead of just shape and motion (Descartes' features), modern mechanists can appeal to such things as gravitational and electromagnetic fields. In biology the comparable expansion brings in chemical bonds and catalysis. But biological mechanisms require expanding the toolbox Descartes supplied in yet a different way, one that focuses on organization of parts and operations. The appeal to engineering showed how organization enables a mechanism to perform functions which its parts alone cannot, but the organization of biological systems is distinctively different from the organization typically exhibited in humanly engineered systems. When humans think of putting components together we usually think of linking them together in series so that each component operation sends its product to the next component operation in the series. From biological systems, however, we have learned of more complex modes of organization that achieve rather surprising results.

The significance of organization for biological mechanisms was brought home in the nineteenth century by challenges from biologists who denied that mechanisms could account for the phenomena of life. These biologists, known as *vitalists*, highlighted ways in which biological systems function differently than non-biological systems. Xavier Bichat (1805) is an important example. In many respects, Bichat was pursuing a program of mechanistic explanation. He attempted to explicate the behavior of different organs of the body in terms of the tissues out of which they were constructed. He decomposed these organs into different types of tissues that varied in their operations and appealed to the operations of different tissue types to explain what different organs did. But when Bichat reached the level of tissues, he abandoned the mechanist program. This was because tissues exhibited two features which he thought defied mechanistic explanation. First, tissues are indeterministic in their response to external stimuli. In contrast, machines as he conceived them always respond the same when presented with the same stimulus. Second, they seem to resist those environmental forces that threaten them. It is relatively easy to take apart or interrupt a machine and stop its operation, but living tissues are often difficult to thwart or kill. These differences, Bichat thought, undermined any hope of providing a fully mechanist account of living tissues.

Several decades later Claude Bernard (1865) sketched a mechanistic answer to Bichat. It involved identifying principles of organization in living systems that could account for the features to which Bichat pointed. Bernard presented his answer by distinguishing two environments. The term *environment* is typically used only for the first – the external environment in which the organism as a whole lives. Bernard proposed that biologists also need to consider the local environment of each organ (mechanism) within a living organism. This he termed the *internal environment*. Component mechanisms in the living organism interact directly with this internal environment, not with the external environment. This internal environment provides a buffer between the conditions in the external environment and the local mechanisms. The various organs respond to the conditions of the internal environment, and these responses might be quite deterministic – when conditions in the internal environment differ, the organs predictably behaved differently. For example, decreased glucose levels in the blood could lead to lowered metabolic activity in somatic tissues or reliance on a different metabolite.

How did the internal environment serve as a buffer? Bernard proposed that each mechanism within the organism monitored an aspect of the internal environment and operated to maintain that feature of the internal environment in a constant condition.[15] The joint consequence of different internal mechanisms each operating to maintain a given feature of the internal environment is that each mechanism enjoys a stable environment in which to operate. Hence, each is buffered from conditions outside the organism – as external conditions begin to change conditions inside the organism, the appropriate mechanism registers the change in the internal environment and acts to restore it to its normal condition. Because of this internal buffering, the various other mechanisms, and the organism as a whole, do not show regular causal responses to perturbations in the external environment and so are indeterministic in the manner Bichat noted. Moreover, insofar as each component mechanism is successful in performing its activity as needed to keep the internal environment constant, the organism continues to maintain itself in the face of conditions that would seem to have the potential for destroying it. Thus, he can explain why organisms seem to resist external factors that would seem deadly to them.

Bernard's account leaves some important features of biological mechanisms unexplained. Perhaps the most important feature left unexplained is *how* it is that component mechanisms maintain "the constancy of the internal environment" (Bernard, 1878a, p. 113). In order to do this, though, each component mechanism must itself have a complex structure, including components that serve to detect when the operation is needed and a means to activate the mechanism so that the operation is performed when and only when it is needed. Walter Cannon (1929) introduced the term *homeostasis* (from the Greek words for *same* and *state*) for the capacity of living systems to maintain a relatively constant internal environment. He also sketched a taxonomy of strategies through which animals are capable of maintaining homeostasis. The simplest involve storing surplus supplies in time of plenty, either by simple accumulation in selected tissues (e.g., water in muscle or skin), or by conversion to a different form (e.g., glucose into glycogen) from which reconversion in time of need is possible. Cannon noted that in most cases such conversions are under neural control. A second kind of homeostasis involves altering the rate of continuous processes (e.g., changing the rate of blood flow by modifying the size of peripheral blood vesicles to maintain uniform temperature).

Cannon's particular interest was the role of the autonomic nervous system in regulating supplies or processes, but during the same period biochemists

---

[15] Bernard, for example, says, "all the vital mechanisms, however varied they may be, have only one object, that of preserving constant the conditions of life in the internal environment" (1878a, p. 121, translated in Cannon, 1929, p. 400).

were discovering the ubiquity of cyclic organization of biochemical processes and the capacity of such cycles to provide auto-regulation of these processes. One of the first cycles proposed was by Otto Meyerhof, drawing upon the research of Archibald Hill on two periods of heat generation in muscle contraction, one accompanying formation of lactic acid during muscle contraction itself and the second accompanying the subsequent disappearance of lactic acid during a recovery phase (Hill, 1910; Hill, 1913). Meyerhof (1920; 1924) proposed a cycle, which he termed the *lactic acid cycle*, in which approximately three-quarters of the lactic acid formed during the anaerobic contraction phase was resynthesized to glycogen at the expense of the remaining molecules of lactic acid, which were further oxidized to carbon dioxide and water. As I will discuss in the next chapter, other proposals for cycles soon followed and as their ubiquity became clear, the question of why they occur so frequently arose. One reason is that they provide a means of regulation via such procedures as feedback control.[16] We saw an example of this in Figure 2.3. As another example, the cycle in which ATP is broken down to ADP and P_i in the course of muscle work and resynthesized during energy metabolism provides a means by which energy metabolism can be regulated to proceed only when ATP is needed for work.

Systems organized to regulate themselves are not unique to biology. Maintaining constant conditions is also important in many human artifacts and engineers have invented ways for doing so. Water clocks, for example, required that the water-supply tank be maintained at a constant level so that water ran out at a constant rate, and a feedback control system for such clocks was developed by Ktesibios in approximately 270 BCE. Windmills need to be pointed into the wind, and British blacksmith E. Lee developed the fantail as a feedback system to keep the windmill properly oriented. A temperature regulator for furnaces was developed by Cornelis Drebbel around 1624. Although each of these used a version of negative feedback, they were isolated developments limited to the machines in which they were utilized.[17] The governor developed by James Watt for his steam engine attracted much

---

[16] Mercer described the general consequence of such organization, which he attributed to James Danielli's "generalized cell theory": "The whole complex of enzymatically controlled reactions – in which the product of one reaction forms the substrate for the next, and often a final product is returned (or fed-back) to re-enter a cycle of reactions at another point – constitutes a system in dynamic equilibrium buffered against change, so long as material and energy is fed into it" (1962, p. 50).

[17] Harold Black's development in 1927 of negative feedback as a means of controlling feedback distortion in amplifiers such as those used in telephones – by feeding back the signal from the amplifier so as to compare it with the input signal – illustrates how the principle had to be rediscovered in each individual case.

more attention, in large part because it became the focus of James Clerk Maxwell's (1868) mathematical analysis of control systems using differential equations. Watt's challenge was to maintain the steam engine at a constant speed despite the fact that the loads upon it varied (as, for instance, different sewing machines attached to it in a textile factory would come on and off line). His solution was to attach a spindle to the drive shaft and then attach moveable arms to the spindle (see Figure 2.4). Centrifugal force would cause these arms to open up more the faster the engine was running. To these angle arms Watt attached a linkage mechanism that would reduce the opening of the steam valve the more the arms opened and increase the opening the more they dropped. Whenever the engine ran too fast, the arms would rise and that would cause the valve to close, slowing the engine down. Whenever it ran too slowly, the arms would drop, and that would cause the valve to open, speeding the engine up.

As technological systems developed for which control was critical, negative feedback came to be recognized as a powerful tool. It provided the foundational idea for the cybernetics movement,[18] of which Norbert Wiener was a driving force. Wiener had interest and training in biology before earning his Ph.D. in mathematical logic and making important contributions to pure mathematics. Because he often collaborated with biologists, his inspiration for emphasizing cyclic organization probably lay as much in biology as in mathematics and engineering. During World War II he and Julian Bigelow took on the challenge of developing a control system for antiaircraft fire (the challenge stemmed from the fact that the airplane moves a considerable distance in the approximately one-minute interval that it takes a projectile to reach it). They recognized that the problem was not essentially different from the problem an animal faces in moving its limbs and were inspired to employ negative feedback. Thus, their strategy was to predict from radar information the future location of the plane but then correct the prediction based on the difference between the actual and predicted location at the next timestep. They found, however, that if the feedback signal was at all noisy and the system responded too quickly, feedback caused it to go into uncontrollable oscillations. Wiener and Bigelow consulted with Mexican physiologist Arturo Rosenblueth, who reported similar behavior in human patients with damage to the cerebellum. In both cases, they concluded, averaging techniques had to be invoked to dampen the response.

---

[18] Indeed, Wiener was led to the term cybernetics while thinking about Watt's governor. Looking into the etymology of the word *governor*, Wiener followed the path to the Greek word for governor, *kybernan*, and encountered the Greek word *kybernetes*, steersman.

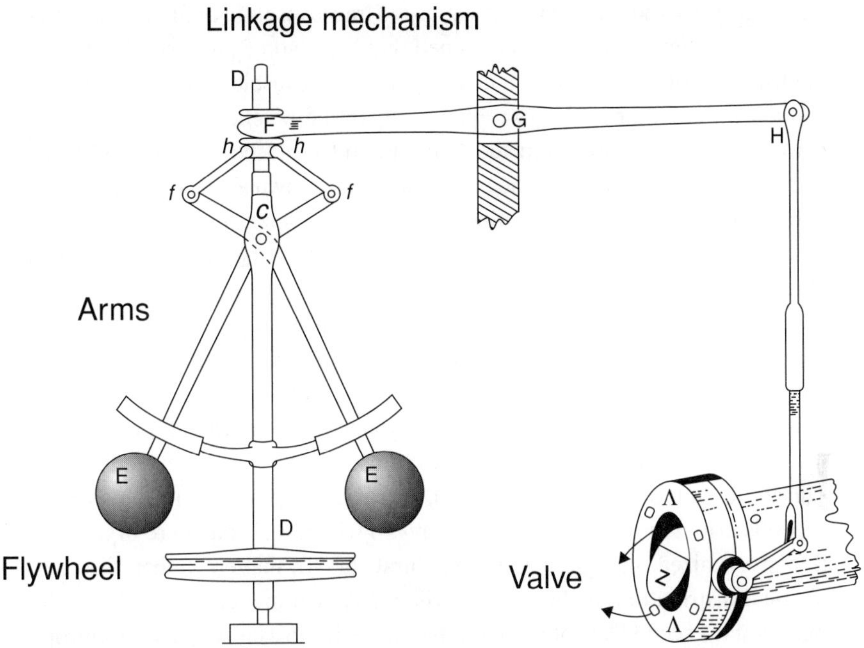

Linkage mechanism

Figure 2.4. The governor James Watt designed for his steam engine. An upright spindle is attached to the flywheel. Connected to it are arms with balls which, by centrifugal force, will move further out the faster the flywheel turns. There is a linkage mechanism which ensures that the valve closes as the balls move further out, and opens as they move closer in. This ensures that steam will flow at the rate needed to keep the flywheel moving at the desired speed. Drawing reproduced from J. Farley (1827), *A treatise on the steam engine: Historical, practical, and descriptive.* London: Longman, Rees, Orme, Brown, and Green, p. 436.

Wiener and his colleagues did not view their contribution only as an advance in technological design (in fact, Wiener's design failed due to the limitations of available hardware in which to implement it, and his contract with the Department of Defense was terminated). Rather, they thought they had articulated a basic principle of purposive and teleological behavior, whether in animals or machines. Accordingly they published their results in the journal *Philosophy of Science*.[19] They acknowledged that the concept of teleology

---

[19] On their account, *purposeful* is more general than *teleological.* "The term purposeful is meant to denote that the act or behavior may be interpreted as directed to the attainment of a goal – i.e., to a final condition in which the behaving object reaches a definite correlation in time or in space with respect to another object or event" (Rosenblueth, Wiener, & Bigelow, 1943, p. 18). They then invoked feedback to differentiate teleological from non-teleological behavior: "Purposeful active behavior may be subdivided into two classes: 'feed-back' (or 'teleological') and 'non-feed-back' (or 'non-teleological')" (p. 19).

was largely discredited because it suggested final causation or future outcomes directing earlier events, but maintained that the kind of purposeful behavior guided by feedback was sufficiently important to resuscitate the term. Wiener and his collaborators were so impressed with the potency of negative feedback that, with support from the Macy Foundation, they established twice yearly conferences. Initially the conference was known as the Conference for Circular Causal and Feedback Mechanisms in Biological and Social Systems, but after Wiener (1948) coined the term *cybernetics*, the name was changed to the Conference on Cybernetics.

Teleology is an important feature of biological systems, and cyclic organization is important for designing systems that can achieve ends on their own – whether the system is biological or an artifact designed by humans. There is, however, another feature of biological systems that is critical, but that has not received much attention to date by philosophers focusing on mechanism. The ability of mechanisms to function, and especially to regulate themselves, depends critically on the particular ways in which their parts are organized. Being organized, though, is not the natural state of matter. Rather, disorder, or equal distribution, is the state to which physical matter tends. This is the import of the second law of thermodynamics – that in a closed system, entropy (disorder) increases. The only truly stable state, referred to in thermodynamics as the state of maximum entropy, is one in which the components are equally and randomly distributed. When this obtains, the system is at equilibrium.

Given the second law of thermodynamics, biological mechanisms pose a puzzle. As highly organized systems, they are far from thermodynamic equilibrium. According to the second law, this organization should break down and such mechanisms should approach equilibrium. How is it possible to keep a system far from thermodynamic equilibrium – that is, to keep it organized? Part of the answer is that biological systems and their environment are not closed systems; energy is always entering from the outside. However, this energy must be appropriately directed to maintain organization. With human-made mechanisms, the most common means of maintaining organization is to rely on an external repair system, often a human being. Much like the original builder of a mechanism, the repair person utilizes energy, originating outside the system, to reorganize the components when the order between them breaks down and to replace components when they internally break down (become disorganized). In general, however, biological systems do not have external repair people to come in and expend energy to rebuild them. They must do it themselves. How is this possible?

A variety of twentieth-century theorists made important theoretical contributions that provide in basic outline an account of how biological systems maintain themselves far from thermodynamic equilibrium. One contribution is the recognition of self-organizing chemical systems such as the Belousov–Zhabotinskii (B–Z) reaction in which simple component reactions together give rise to complex patterns. Ilya Prigogine coined the term *dissipative structures* for constantly changing, highly organized systems (Nicolis & Prigogine, 1977). Such order depends on a constant flow of energy through the system, which is employed to create order but then dissipates as heat. Dissipative structures such as the B–Z reaction, however, can neither procure for themselves the needed energy flow to maintain themselves far from equilibrium nor the chemicals needed to continue the reactions. Hence, they soon terminate. In the early 1970s, Francisco Varela and Humberto Maturana introduced the term *autopoietic machine* for systems such as living organisms that maintain themselves by producing their own components. They emphasized that the components of such systems "1) through their interactions and transformations continuously regenerate and realize the network of processes (relations) that produced them; and 2) constitute it (the machine) as a concrete unity in the space in which they exist by specifying the topological domain of its realization as such a network" (Varela, 1979, p. 13).

Both characteristics of autopoietic machines deserve comment. First, to regenerate themselves, such machines require machinery that is directed toward making their own components. Such machinery requires a source of energy and raw ingredients for constructing itself. Hence, the system must be an open system situated between an energy source (a more structured, less equally distributed, state of matter) and an energy sink (which is less structured). But crucially it must capture this energy in a useable format (e.g., gradients, chemical bonds) so that it is available for energy demanding operations. Second, the smooth operation of the system depends upon the organization of the components so that different operations are orchestrated (e.g., energy and raw materials are delivered to the location where new parts are synthesized). Minimally, this requires some way to keep needed constituents together and separated from the environment in which the system is situated. That is, the system requires a partially porous boundary (e.g., a semi-permeable membrane) that allows selected materials to cross. Moreover, the system itself must be able to control what is admitted and what is expelled from the system. Ultimately, because the concern is with a system that remakes itself, this boundary needs to be one of the system's own making.[20]

---

[20] For a stimulating and provocative development of how a metabolic and a mebrane system, together with a control system, can result in a chemical system capable of exhibiting the basic features of living systems, see Gánti (2003).

Varela uses the term *autonomy* for systems that define and maintain their own boundaries. Ruiz-Mirazo, Peretó, and Moreno provide a useful characterization of an autonomous system as

> a far-from-equilibrium system that constitutes and maintains itself establishing an organizational identity of its own, a functionally integrated (homeostatic and active) unit based on a set of endergonic-exergonic couplings between internal self-constructing processes, as well as with other processes of interaction with its environment. (2004, p. 330)

Autonomy in this sense is a feature of any living system. Although each mechanism does not have to be autonomous, it must be part of a system that is.

The requirement of being built up through self-organizing processes also provides a means to address a key feature of Bichat's critique of mechanism with which I began this section. Bichat contended that living systems resist death. In fact, this generally takes the form of adaptive change over time so as to deal with novel environments. Understanding the adaptive capacities of biological organisms is challenging. Developmental theorizing, in both biology and psychology, has been driven by the polar positions of nature and nurture. Advocates of a nurture position maintain that the organization was not substantially prespecified, but rather resulted from interactions with the environment. But advocates of the nurture position have faced apparently insurmountable obstacles. The end-product of both biological and psychological development is a highly structured state. Ensuring the achievement of such structure seems extremely problematic if everything must be directed from the environment. The nature position seems improbable in its own right, requiring a vast amount and efficacy of innate information.

Within the sphere of cognitive development, the psychologist Jean Piaget proposed an alternative to the polarizing options of nature and nurture – a position he referred to as *constructivism*. He introduced this alternative as follows: "The essential functions of the mind consist in understanding and in inventing, in other words, in *building up structures* by structuring reality" (Piaget, 1971, p. 27). Piaget's attempt to find a middle path between nature and nurture, however, turned out to be bitterly contested. The philosopher of psychology Jerry Fodor objected, "It is *never* possible to learn a richer logic on the basis of a weaker logic, if what you mean by learning is hypothesis formation and confirmation" (1980, p. 148). Fodor's idea was that before one can test a hypothesis one must be able to formulate it, and that entails the capacity to represent it. Thus, the capacity to represent any hypotheses that one will ever test must be innate, making the native endowment very powerful. Fodor's claim is premised on the condition that learning consists of hypothesis formation and confirmation, which is itself contentious. To contest it, however, one must offer an alternative, and this has proven difficult. The notion that autonomous systems must be built ultimately out of self-organizing systems, however, perhaps offers the foundation for building an alternative to Fodor's assumption. Self-organizing systems are able to construct richer structures than they start with (see Elman et al., 1996, for a suggestive account of cognitive development).

This points us to yet another contrast with the notion of mechanism adequate for nonliving systems. The engineering strategy, as we have seen, is to build more complex systems out of simpler components by imposing organization upon existing components. This is a very productive strategy, but it depends upon the capacity of the engineer to propose new forms of organization. Biological systems, however, cannot rely on external engineers. (Given natural selection, they can of course rely on selective retention of chance variation, but the generation of new useful modes of organization through chance variation of already evolved systems is extremely uncommon.) Current modes of organization in artifacts are usually located near local peaks on adaptive landscapes so that small variations are likely to be detrimental. Fortunately, however, biological systems can utilize a different approach. If biological systems are self-organizing, they are also capable of self-reorganization. This means that they can do more than impose a new organization on existing components. Insofar as these self-organizing capacities remain active and not frozen within the system, the system can also generate from within new components capable of performing new activities.

At present we have only limited models of such self-organizing of new components with new capacities within living systems. These include cases of plasticity in nervous systems. Nervous systems wire themselves and components emerge and take on specific functions partly as a result of the neurons that synapse on them. As a result, tissue which in a normally developing individual will serve visual processing tasks will, in a brain in which there are no inputs from eyes, take on the processing of auditory or other sensory inputs (Pascual-Leone & Hamilton, 2001). And areas in the somatosensory cortex that process information from fingers will reorganize if the input changes if, for example, an investigator binds together two digits so that they can no longer respond differentially (Merzenich et al., 1990). Although the details are less clear, such processes provide the most plausible explanation for the existence of an area in the brains of literate humans that has apparently become specialized for visual letter patterns (Petersen et al., 1990). Such a processing area could not have been specified in a genome when that genome was selected in organisms that did not read written texts. It must be the product of self-reorganization of prior processing areas in organisms that had to adapt to and become adept at reading written characters.

7. DISCOVERING AND TESTING MODELS OF MECHANISMS

So far I have focused on articulating what are mechanisms and mechanistic explanations so as to make it clear what biologists, including cell biologists, are pursuing in their investigations. I have not said anything about the processes of discovering and evaluating claims about mechanisms. Because these are the prime activities of scientists, I turn next to the question of what can be said about these processes.[21]

The very conception of a mechanism sets the goals for its discovery – the investigator must identify the parts of the mechanism, determine what operations they perform, and figure out how they are organized to generate the phenomenon. Earlier I introduced the notion of *decomposition*, but emphasized the conceptual side of decomposition. To develop a model of a mechanism, a theorist decomposes it conceptually into parts and operations. But discovering a mechanism usually results from experimental engagement[22] with the mechanism. I differentiated two ways researchers decompose mechanisms – structurally or functionally – depending on whether they focus on component parts or component operations. As we turn to experiments, these two types of decomposition now correspond to two types of experimental

---

[21] Logical empiricists typically rejected the possibility of philosophical analysis contributing to understanding what Reichenbach (1966) called the context of discovery and instead focused on the context of justification, where it was thought that logic could characterize the relation of evidence to hypotheses. Interest in discovery was rekindled around 1980 (Nickles, 1980a; Nickles, 1980b). See in particular Darden (1991), who focuses on discovery in the context of theory revision. When mechanisms are the focus, it turns out that quite a bit can be said about scientific discovery.

[22] Another set of experiments aim simply to establish the phenomenon for which a mechanistic explanation is sought. Experimentation designed to assess the phenomenon itself generally does not try to take the mechanism apart, but rather seeks to determine regularities in its behavior by establishing relations between inputs to the mechanism, conditions of its operation, and output. This involves measuring the values of relevant variables and, often, the manipulation of some variables. Lavoisier and LaPlace (1780), for example, established the similarity of respiration by animals and ordinary combustion by placing respiring organisms or burning coal in a calorimeter and measuring the amount of carbon dioxide produced and of ice melted. This only involved setting up appropriate conditions for measuring the relevant variables. Pasteur (1861), on the other hand, determined that fermentation was an anaerobic phenomenon by manipulating the presence or absence of oxygen and showing that fermentation was suppressed in the presence of oxygen.

procedures – those directed at the identification of component parts and those focused on component operations.

Although the goal of discovering the component parts and operations of mechanisms is clear, actually succeeding often requires great ingenuity. Mechanisms generally do not directly reveal how they operate. All we observe is their functioning and often those observations require careful interpretation. Even when the internal operations involve changes in an identifiable substrate, the intermediate states of the substrate can be hidden because in a well-organized machine individual operations flow smoothly into each other without leaving any trace of their intermediate products. Think again of engineering, where the ultimate goal is to design a device of which the user is not even aware and which operates when needed without any intervention by the user. In the early days of a new technology this desiderata is often not realized. Drivers of early automobiles had to understand how they operated so as to be able to repair them after their frequent failures. Users of early computers had to understand their operation at the machine level, or at least at the level of the operating system, in order to get their programs to perform as desired. As computers matured such knowledge became unnecessary and the details of the operation were increasingly hidden from view (e.g., behind a smoothly functioning windowing system and higher-level programming languages).[23] An engineer who wants to emulate a competitor's system but lacks access to the design specifications often faces a serious challenge of reverse engineering – taking the system apart and trying to figure out how it was put together. This is the situation confronting investigators of biological systems who must rely on experimental manipulations to take apart and reveal the parts and operations within them.

Identifying Working Parts

Some experimental inquiries are designed simply to separate the parts of a mechanism. It is important to emphasize that the parts into which researchers seek to decompose a mechanism are ones which perform the operations that figure in the functional decomposition. The majority of ways of structurally decomposing a system will not result in parts that perform operations. As Craver (forthcoming) notes, one might dice any system into cubes, but these cubes do not individually perform operations in terms of which one can explain the phenomenon. To reflect this fact, I will, following Craver, refer to

---

[23] Herbert Simon (1996) made this point in the context of bridge design. When a bridge is functioning as intended, it reveals little of its design principles.

the relevant component parts as *working parts*. Although the goal is to find working parts, it is possible to decompose a system structurally independently of actually being able to determine the operations the various components perform. This involves, for example, appraising that component structures are likely to be distinct working parts on other grounds.

Let's begin with interventions that are designed to identify the parts of a mechanism. One way is to apply enough force that it breaks down into separate components that can then be further investigated as potential working parts. Deciding what counts as "enough" force is a challenge. Some ways of applying force will completely obliterate the components of interest – those that perform the operations within the mechanism. One could, for example, cut a brain into tiny cubes or homogenize a cell in a Waring blender. This is not problematic if the units of interest are at a sufficiently small scale (individual neurons or ganglia in the brain or individual enzymes in the cell) to survive intact. If the task of interest relies on larger units (brain regions or cell organelles), the intervention will have destroyed the working parts. There is a further challenge, though. Assuming that a researcher has arrived at a method that delivers the right amount of force to get parts of about the right size, how would she know whether these are the parts that divide the system at its *joints*? Consider performing an anatomical dissection. How does the anatomist know whether she has dissected out a working part (cut at the joints) or has cut within a part? Sometimes components (e.g., organs such as the heart, liver, and pancreas) exhibit clear boundaries and integrity. This does not entail that they are functionally relevant units, but it provides a good clue.

Physical separation is not the only possible strategy. In some fields researchers have a means of viewing internal components. Cytologists, for example, were able to distinguish cells as parts of organs, and nuclei as parts of cells, using the light microscope and smaller parts using the electron microscope. In Chapter 4 I will discuss the epistemic challenge in determining whether such visualizations are reliable. For now what is important to note is that visualization provided a partitioning based on appearance that was useful for many purposes, but needed to be corroborated by other methods to confirm correspondence to working parts. The history of attempts to identify working parts in the brain illustrates the challenge (Mundale, 1998). In highly convoluted brains such as the human brain, the existence of sulci and gyri seemed to provide natural boundaries between regions of cortex. Although these are still used as reference points in locating regions in the brain, it is now recognized that the process of folding that gives rise to sulci and gyri does not respect function as much as expected. Hence, other evidence is needed to identify the working parts. Brodmann (1909/1994) as well as several other investigators at the beginning of the twentieth century invoked other criteria such as the type of neuron found in an area and the thickness of different layers of cortex to demarcate regions. Although Brodmann lacked any means of linking neural operations with these areas, he clearly hoped that these criteria would differentiate working parts. Subsequently a variety of other indirect criteria, such as topographical mapping, have been used to refine such maps (van Essen & Gallant, 1994). The justification for using such indirect criteria is the assumption that they track features (e.g., morphologically distinct types of neurons) that should matter operationally.

Ultimately, it is the integration of structural decomposition with functional decomposition that provides real answers to the question of which parts are working parts. In subsequent chapters we will encounter a variety of methods employed for structural decomposition of cells and also the advantage attained when cell biologists integrated these results with those from functional decomposition.

Identifying Component Operations

Turning now to functional decomposition, here the strategy is to start with the overall function of the mechanism and figure out what lower-level operations contribute to achieving it. These operations are characterized differently in different sorts of mechanisms. In biochemistry, the typical operation is a chemical reaction catalyzed by an enzyme (active part) that transforms a substrate into a product (the passive parts). The biochemical system that performs glycolysis in cells, for example, catabolizes glucose to pyruvate via a series of such reactions, each of which may oxidize or reduce a substrate, add or remove phosphate groups, etc. Often it is possible to determine, at least to a first approximation, what the internal operations of a mechanism are without knowing what active parts perform these operations (the passive parts do need to be known, though, because the identity of the operation depends upon what is changed). In the case of glycolysis, biochemists were able to determine the chemical reactions that realized this function and to designate a responsible agent (enzyme) for each reaction without knowing the chemical structure of the agents (Bechtel, 1986b). Instead of direct access to the structures operative within the system, such decompositions rely on cleverly designed perturbations of the functioning of the system that provide clues to the component operations. In such cases, then, functional decomposition precedes structural decomposition.

Some experiments designed to identify internal operations involve manipulation and measurement of variables without explicitly going inside the mechanism. In such cases each manipulation should be precisely designed to target a specific internal operation, such that a distinctive effect on the behavior of the whole system will be obtained.[24] In most cases, though, experiments designed to determine the internal operations in a mechanism involve intervening or measuring changes within the mechanism. The strongest evidence may arise when an investigator can directly intervene on and measure effects in the operation of a component in the mechanism. Sometimes this can be approximated by separating a component from a mechanism, experimenting on it under realistic conditions to determine what it could do, and fitting that information into a coherent account of how the whole mechanism performed its function. Biochemists, for example, may attempt to isolate from cells what they take to be the relevant enzymes and cofactors for a given reaction and then conduct experiments on the isolated components. Frequently it is not possible to isolate the factors responsible for an operation within a mechanism in this manner, in which case researchers have to settle for more indirect routes for gaining information.

Regardless of whether it is structural or functional components that provide the initial target, two strategies for determining whether and how they figure in a mechanism are excitation and inhibition. The effects of these manipulations may then be registered in the overall behavior of the mechanism or more locally within the system. By inhibiting or removing a component, investigators force the system to operate without it. The hope is that from the altered behavior of the mechanism they can figure out what part or operation was involved. This is not always easy, as Richard Gregory argued:

> Although the effects of a particular type of ablation may be specific and repeatable, it does not follow that the causal connection is simple, or even that the region affected would, if we knew more, be regarded as functionally important for the output – such as memory or speech – which is observed to be upset. It could be the case that some important part of the mechanism subserving the behavior is upset by the damage although it is at most indirectly related, and it is just this which makes the discovery of a fault in a complex mechanism so difficult. (1961, p. 323)

Moreover, in many cases, the removal of a component part produces a whole cascade of effects, leading the mechanism to no longer carry out anything like

---

[24] Reaction time studies used in cognitive psychology provide an illustrative example. Different proposals for internal operations were linked with different timing expectations so that actual reaction time measures could decide between them.

the function normally associated with it. Removing a transistor from a radio, to use another of Gregory's examples (Gregory, 1968), may cause the radio simply to hum, perhaps giving the misleading suggestion that the transistor was a hum suppressor. Nonetheless, inhibiting or removing a part sometimes provides important clues as to what operation it performed. For example, if removing a particular enzyme inhibits a reaction and thereby an entire function (e.g., fermentation) and also leads to the build-up within the system of another substance, then it provides evidence both that the reaction contributes to that function and that the accumulating substance was an actual intermediate (or the product of a further reaction on that intermediate). Although single clues of this type are subject to multiple interpretations, the results of multiple different inhibitions within the system often provide insight into what operations different parts are performing.

The complement to inhibiting an operation is to stimulate it. For example, investigators can supply additional amounts of what they think is an intermediate substance in a chemical pathway and determine whether that results in a greater output from the pathway. This too can lead to anomalous results, either because the substance did not reach the right point in the pathway or because, without other components cooperating appropriately (perhaps via feedback loops), the operation was blocked. Often the stimulation technique can be fruitfully combined with an inhibition. For example, investigators might inhibit a reaction by removing the enzyme needed to obtain a particular product but also supply the product from outside. If the ultimate product of the reaction pathway continues to be produced normally, then the investigators can have high confidence that the enzyme they removed was part of the normal pathway.

Both inhibition and stimulation studies involve manipulating a component in the normal operation of the mechanism and detecting the consequences. Sometimes it is possible to observe the operation of the components directly. One example is single neuron recording in neuroscience, in which an electrode is inserted near a neuron and then various stimuli are presented to see which types produce increased (or decreased) spiking. Another example is radioactive tracer experiments, in which researchers label a substrate with a radioactive component and then detect where the radioactivity later appears within the mechanism. When such recordings are possible, they often provide the most compelling evidence of the sequence of internal processes. They do not, however, reveal the nature of the operation involved at each locus; hence, recording studies usually must still be combined with inhibition or stimulation studies to determine the operations.

Localizing Operations in Parts

I noted earlier that sometimes researchers are able to make greater progress in identifying parts than in specifying operations, or vice versa. One reason this happens in the history of science is that the techniques for carrying out one form of decomposition are developed in a different field and at a different time than those for carrying out the other form of decomposition. We will see in subsequent chapters that this was often the case with regard to mechanisms in the cell. Biochemists, focusing on functional decomposition into component operations, identified biochemical reactions. Cytologists, focusing on structural decomposition into component parts, identified those organelles that were within the resolving power of their microscopes. It was engagement between these fields that gave rise to cell biology. Increasingly, parts and operations in what previously had been a terra incognita were identified and aligned to provide new mechanistic accounts (e.g., it was found that one part of the mitochondrion – its inner membrane – performed the operation of maintaining a key energy gradient). Equally important, it was learned which of the newly discovered mid-level structures housed which lower-level biochemical reactions. For example, the cytosol was found to be the site of the reactions involved in glycolysis. This kind of alignment, crossing levels rather than remaining at a single level, at the very least provided a more complete account of the cell. Often there were further benefits, as important hints and constraints suggested by interlevel alignments led to improved accounts at one or more of the levels involved. At various points in the chapters that follow, I use the term *localization* to refer not only to alignments within a single mechanism but also to those that crossed levels in the cascade of mechanisms being uncovered by cell biologists.

The ability to link operations with parts often depends upon developing experimental procedures that can provide appropriate linking information. In the case of cell biology, the ability of electron microscopy to identify the structures found both in whole cells and in preparations of working parts isolated by cell fractionation provided the primary evidence relating biochemical operations with the organelles of the cell. As discussed in Chapter 4, an important contribution of such connections is that they provide a means of corroborating each decomposition. Linking a component operation with an independently identified component part provides evidence that both really figure in the mechanism. Failure to link operations with parts, on the other hand, can be grounds for doubting the existence of either the part or the operation. As we shall see, cell organelles were often claimed to be artifacts until they were shown to be the locus of important operations in the cell. Linking an operation to a structure also provides evidence that the hypothesized operation actually occurs.

Although I have stressed the possibility of either decomposition proceeding in advance of the other, it also often happens that some progress in localization accompanies decomposition. For example, in the preceding sections I often characterized experimental inquiries into operations as inhibiting either the component part or operation. If investigators have also hypothetically linked that operation with a component part, then they can test this by excising the part to inhibit the operation. And likewise investigators can use evidence about operations to confirm that the parts into which they have decomposed the system are indeed working parts. When, therefore, it is possible to coordinate hypotheses about structural and functional decompositions in the course of research, it is advantageous to do so.

Testing Models of Mechanisms

So far in this section I have focused on the discovery of mechanisms. However, science involves not just the advancement of hypotheses but also testing whether they are true in a given situation. While logical empiricists had little to say about discovery, they did attempt to articulate criteria for evaluating proposed laws. Essentially, this involved making predictions from the laws and evaluating the truth of these predictions. The challenge was to articulate a logic that would relate the truth or falsity of a prediction to the confirmation or falsification of the law. This is not the occasion to review the problems and solutions that have been proposed; I simply note the genesis of the problems for confirmation or falsification. Confirmation is challenging because there are always alternative possible laws from which one might make the same prediction (the problem of underdetermination). Falsification is challenging because a false prediction might be due to an error either in the proposed law or in one of the auxiliary hypotheses that figured in deriving the prediction (the problem of credit assignment).

Modulo the difference that scientists typically make predictions from mechanisms by simulating their operation rather than by making logical deduction from laws, the challenge for testing hypotheses is much the same for mechanisms as for laws. A researcher tests hypothesized mechanisms by inferring how the mechanism or its components will behave under specified conditions and uses the results of actually subjecting the system in nature to these conditions to evaluate the proposed mechanism. In principle the same challenges confront tests of mechanisms as tests of laws – (a) different mechanisms might generate the same predictions (underdetermination); and

(b) when a prediction fails, the problem might lie either with the model of the mechanism or with auxiliary hypotheses invoked in making the prediction (credit assignment). Although these problems are not eliminated, they are mitigated when researchers test accounts of mechanisms because their focus is typically not on the mechanism as a whole, but on specific components. Evidence is sought that a given component actually figures in the generation of the phenomenon in the way proposed. Accordingly, predictions are diagnostic – specifically targeted to the effects such a component would have (e.g., that if one intervened in a way known to incapacitate that component, there would be a specific change in the way the mechanism as a whole would behave).[25] If the part does not, for example, perform the operation hypothetically assigned to it, that is often evidence that unsuspected parts are involved and that the operation requires their orchestrated operation.

An important aspect of discovering and testing mechanisms is that inquiry does not simply consist of postulating and testing a mechanism. Rather, research typically begins with an oversimplified account in which only a few components and aspects of their organization are specified. Over time, it is repeatedly revised and filled in (Bechtel & Richardson, 1993). Machamer, Darden, and Craver (2000) refer to the simplified account as a *sketch* of a mechanism. Much of the discovery and testing involved in mechanistic explanation focuses on proposing components or forms of organization that are to be added to or used to revise parts of a sketch, and (often late in the process) localizing the worked-out component operations in the appropriate component part or parts. The entire process is typically a long-term endeavor (Bechtel, 2002a). The result looks much more like a research program (Lakatos, 1970) than like the classical account of theory-testing.

8. CONCLUSION

This chapter has focused on providing a general framework within which to understand cell biologists' mission of trying to understand cellular mechanisms. My goal has been to articulate what mechanisms are and how they figure in explanations. Accordingly, I have emphasized the role of diagrams as well as text in representing mechanisms and how simulation underlies reasoning about them. Because mechanisms inherently bridge multiple levels of organization (the components, the mechanism as a whole, and any larger

---

[25] See, for example, Darden (1990; 1991; 1992) on anomaly resolution in scientific theorizing, a process she compares with the process of redesign of human made artifacts in engineering.

system in which the mechanism functions), I have developed the notion of levels of organization and discussed the respects in which mechanistic explanation is reductionistic. Because organization of biological systems is what distinguishes them from non-biological ones, I also discussed the kinds of organization found in biological mechanisms. I ended with the issue of the discovery and testing of models of mechanisms, a topic that takes us into the main objective of this book – the discovery of mechanisms within the cell. This was a project on which major advances were made in the period 1940– 70. As I will discuss in Chapter 4, a major reason these advances occurred when they did was the development of new research techniques – cell fractionation and electron microscopy – that made discovery of cell mechanisms possible. But the project did not emerge from nowhere. In the next chapter I focus on the background prior to 1940 that both provided critical ideas that figured in the later period and also revealed the terra incognita that could not be explored until the new tools were available.

3

The Locus of Cell Mechanisms

Terra Incognita between Cytology and Biochemistry

> Until these 'accidents' occurred, workers engaged in the exploration of
> living organisms had been forced to stop at the edge of a mysterious
> no-man's-land, bounded at the upper level of the dimension scale by the
> resolving power of the light microscope, and at the lower level by the
> applicability of chemical techniques. They knew, in a frustrating sort
> of way, that the area between these two boundaries contained some of
> the essential clues without which life would remain forever ununder-
> standable. With the technical advances mentioned, this region suddenly
> became accessible, both to visual examination right down to the level
> of macromolecules, and to chemical separation and analysis right up to
> the level of microscopic entities.
>
> (de Duve, 1963–4, pp. 49–50)

Having described in abstract terms what mechanisms are and how they figure
in scientific explanation, I turn now to setting the stage historically for the contributions of cell biology. The project of identifying cell mechanisms began
in earnest after 1940 in what was then unoccupied territory between cytology
and biochemistry. Researchers were at best dimly aware that crucial cellular
operations occurred in organelles for which no direct methods of investigation
were available. These organelles were too small to be meaningfully examined with the light microscope and much larger than the reacting molecules
in homogenates that biochemists prepared from broken cells. Investigating
these organelles as mechanisms required structural tools more powerful than
those of cytology, functional tools building on those already developed in
biochemistry, and new techniques incorporating both types of tools in order
to integrate structure and function.

64

Christian de Duve, in the quotation above, refers to the "no-man's-land" in which these mechanisms resided.[1] To discover them, researchers needed new instruments and strategies of investigation. Those will be the focus of the next chapter. Nonetheless, cell biologists utilized knowledge obtained in both cytology and biochemistry in their investigations. Thus, before turning to cell biology proper, I need to consider both what cytology and biochemistry could provide as well as their limitations when it came to the activities of cell life. I will pursue this project historically, examining, albeit in abbreviated fashion, how both cytology and biochemistry reached the state they had by 1940.

The differences between cytology and biochemistry involved not just the size of the objects of their study. Both fields were engaged in decomposing living systems, but cytology emphasized the structural decomposition of tissues, first into cells and then into organelles, while biochemistry emphasized functional decomposition of metabolic activities down to individual biochemical reactions.

1. CYTOLOGICAL CONTRIBUTIONS TO DISCOVERING CELL MECHANISMS UP TO 1940

Because cells are generally too small to be seen by the naked eye, their discovery depended on the development of what was in the seventeenth century a new tool – the microscope. The origins of the microscope are obscure but likely involved someone inserting a lens into a viewing tube, which would provide an approximately tenfold increase in magnification. Anton van Leeuwenhoek became familiar with the use of lenses to magnify cloth to count its threads when he was an apprentice in a dry goods store in Holland. He established new methods for grinding and polishing lenses and made a single-lens microscope that magnified objects up to 270 times. Using this instrument, van Leeuwenhoek identified what he termed *animalcules* in blood, sperm, and water from marshes and ponds. In the same period Robert Hooke[2] developed a compound

---

[1] de Duve (1984, p. 11) again used the metaphor of a no-man's-land and referred to the region as *terra incognita* in characterizing knowledge of cells at the beginning of the 1940s: "there remained between the smallest entity discernible in the light microscope and the largest molecular size accessible to chemistry, an unexplored no-man's-land extending over two orders of magnitude, a vast region that had to be labeled terra incognita on the map of the living cell."

[2] Other investigators of the period who reported looking at plant and, sometimes, animal tissue with microscopes, included Nehemiah Grew, Marcello Malpighi, and Jan Swammerdam. See Hughes (1959) for a more detailed account of this early history. Gall (1996) presented pictures of several early microscopes as well as investigators' drawings of what they saw.

microscope, albeit one that provided less resolution than van Leeuwenhoek's single lens. Hooke published many drawings from his microscopic observations of such biological objects as insects, sponges, bryozoans, foraminifera, and bird feathers in his 1665 book *Micrographia*. The most celebrated image in the book is a drawing of cork, in which he identified small cavities that he labeled *cells*[3] (see Figure 3.1).

> I took a good clear piece of cork, and with a pen-knife sharpened as keen as a razor, I cut off... an exceeding thin piece of it, and placing it on a black object plate, cause it was itself a white body, and casting the light on it with a *plano-convex glass*, I could exceedingly plainly perceive it to be all perforated and porous, much like a honey-comb, but that the pores of it were not regular; yet it was not unlike a honey-comb in these particulars. First, in that it had a very little solid substance, in comparison of the empty cavity that was contained between... Next, in that these pores, or cells, were not very deep, but consisted of a great many little boxes, separated out of one continued long pore... (Hooke, 1665, pp. 112–13)

Two types of artifacts produced by simple lenses seriously hampered advance in microscopic observation for two centuries. Spherical aberration resulted from the fact that light rays that leave a lens at different distances from the axis come into focus at different points. Chromatic aberration resulted from the fact that when light passes through a lens, the component wavelengths refract to different degrees. These aberrations often resulted in misleading observations, especially when their effects were compounded by use of a multi-lens system. At the beginning of the nineteenth century researchers such as Henri Milne Edwards and René Joachim Henri Dutrochet spoke of seeing *globules*, which they construed as the building blocks of organisms.[4]

---

[3] The Latin term *cella* designates a small room (and was applied, for example, to the small memorial chapels erected in cemeteries). The choice of the term reflects the fact that for Hooke, focusing on plants, it was the walls surrounding a cavity that made for cells. E. B. Wilson commented on the adoption of the name: "The term 'cell' is a biological misnomer; for whatever the living cell is, it is not, as the word implies, a hollow chamber surrounded by solid walls. The term is merely an historical survival of a word casually employed by the botanists of the seventeenth century to designate the cells of certain plant-tissues which, when viewed in section, give somewhat the appearance of a honeycomb." Wilson went on to say with respect to the conception of cell that had emerged by the end of the nineteenth century: "Nothing could be less appropriate than to call such a body a 'cell'" (Wilson, 1896, pp. 13–14).

[4] Reference to *globules* as the units in animal tissue goes back at least to Leeuwenhoek. Unlike plants, the units in animals lacked walls and thus the term *cell* was not regarded as appropriate. One of the tip-offs to the fact that these observers were seeing artifacts, not actual cells, was their insistence that the units were of uniform size. Milne-Edwards contended that all animal tissue is comprised of "a rather large number of these corpuscles which might differ in their constitution,

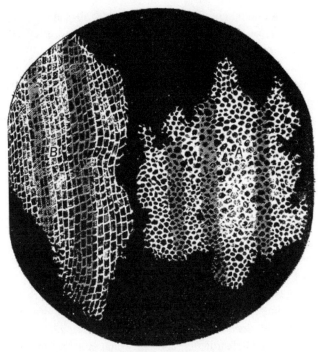

Figure 3.1. Robert Hooke's drawing of his observations of cork. He identified the small bounded areas as cells and noted in particular that the cells on the left side (labeled B) appeared to be boxes. Reproduced from R. Hooke (1665), *Micrographia: or some physiological descriptions of minute bodies made by magnifying glasses with observations and inquiries thereupon.* London: John Martin and James Allestry, Plate 11, Figure 1.

Sometimes these investigators may actually have been observing cells, but all too often what they observed were merely artifacts of their instruments.

Beginning in the late eighteenth century, a number of lens makers, including John Dollond, Joseph Lister, and Giovanni Amici, devised techniques using combinations of different glasses that greatly reduced chromatic aberration and also corrected for spherical aberrations. Microscopes using these *achromatic lenses* became available in the 1820s and 1830s.[5] (Despite their

but which vary but little in their shape and probably in their volume" (1823, translated in Harris, 1999, p. 176). He specified that they were 1/300 mm in diameter.

[5] Friedrich Gustav Jacob Henle, himself a major contributor to microscopical anatomy and histology, described the introduction of new microscopes to Müller's laboratory where he and Schwann were working: "Those were then happy days which the present generation might well envy us, when one saw the appearance of the first good microscopes from the firms of Ploessl at Vienna and from Pistor and Schlieck at Berlin, which we students bought with what money we were able to save" (Frédéricq, 1884, p. 13, as quoted in Hughes, 1959, p. 10).

name, they did not completely eliminate chromatic aberrations. In 1886 the collaboration of Ernst Abbe, Otto Schott, and Carl Zeiss resulted in the introduction of *apochromatic lenses*, which substantially reduced these residual chromatic aberrations.) The procedure for viewing something under these early microscopes was rather direct and straightforward: "With most animal tissues the usual practice was to tease out or squash fresh material into a layer of sufficient thinness, and to study this directly under the microscope" (Hughes, 1959, p. 13). With the dramatic improvements in microscopes, even such crude methods of specimen preparation were sufficient for knowledge of cells to burgeon during the nineteenth century.

Cytology in the Nineteenth Century

Numerous investigators made significant contributions to cytology as early as the first half of the nineteenth century including, for example, Johann Evangelista Purkinje and his student Gabriel Gustav Valentin, who made influential observations of neurons. However, it was Matthias Schleiden and Theodor Schwann who played the pivotal role by advancing what came to be known as the *cell doctrine* or *cell theory*. This unifying doctrine held that cells are the fundamental structural and functional units of life – adequate to comprise single-celled organisms but also serving as the building blocks of multicelled organisms. Ironically, many of the features of their accounts of cells that helped secure the initial acceptance of the cell doctrine proved false.

Schleiden's research directly built on the finding by Robert Brown (1833), in his microscopic investigations of orchids, of the ubiquitous presence of "a single circular areola, generally somewhat more opaque than the membrane of the cell" (p. 710).[6] He found one such structure in each cell and named it the *nucleus*. Schleiden, referring to it as the *cytoblast*, construed the nucleus as the most important region of the cell and that from which the rest developed. Growth from a nucleus became, for him, the characteristic feature through which he identified cells in different tissues. Schleiden was a botanist and initially limited his statement of the cell doctrine to plants.[7] The challenge

---

[6] Similar structures had been described previously in some animal tissues, e.g., in the epithelial cells by Felice Fontana in the 1780s and in the stigma of *Bletia Tankervilliae* by Franz Bauer in the 1790s.

[7] In *Beiträge zur phytogenesis*, Schleiden construed whole animals as individuals but claimed that plants are formed of "fully individualized, independent, separate beings" – the cells (Schleiden, 1838). In Schleiden (1842) he mentioned approvingly, if tentatively, Schwann's extension of the cell doctrine to animals.

facing microscopists examining animal tissues is that the cells of different tissues appear quite different. Moreover, these cells lack a cell wall, still treated by most investigators as the defining characteristic of plant cells. Schleiden's associate Schwann, however, focused on the similarities between the constituent units of different animal tissues and plant cells.[8] The most important similarity was the nucleus, which Schleiden had shown him in plant cells. Finding opaque spots in the units of different animal tissues, especially those in an embryological state, Schwann identified them as nuclei. Because all animal tissue he examined contained nuclei, at least early in development, Schwann concluded that all animal tissues were comprised of cells despite their varied appearance. Figure 3.2 presents Schwann's drawings of several different cell types with their nuclei.

At the heart of both Schleiden's and Schwann's characterization of cells was an account of their formation. Schleiden (1838) had proposed that cells formed within preexisting cells through a physical process of accretion, first of the material comprising the nucleus around the nucleolus, and then another layer, corresponding to the cytoplasm, around the nucleus.[9] Schwann (1839/1947) adopted this mechanism for animal cells, but situated the process not in existing cells but in the intercellular fluids. As a result, Schwann referred to the process in animals as *exogenous cell formation* and that in plants as *endogenous cell formation*. This account of cell development provided the foundation to the cell doctrine: "The elementary parts of all tissues are formed of cells in an analogous, though very diversified manner, so that it may be asserted *that there is one universal principle of development for the elementary parts of organisms, however different, and that this principle is the formation of cells*" (p. 165).

Given that other investigators were already describing cell division, and that Schleiden and Schwann were masters of microscopic technique, a question arises as to why they were convinced that cells formed in such a manner. Here it is important to appreciate that a central objective for Schwann in developing his cell theory was to provide a mechanistic account of the basic structures of living organisms. (Schwann was a member of the group of investigators working with Johannes Müller who pursued such a mechanistic vision

[8] Schwann was working in the laboratory of Johannes Müller, who had himself observed cells in the *chorda dorsalis* and noted the similarity between them and plants (Müller, 1835) and had drawn Schwann's attention to the similarities.

[9] One attractive feature for Schleiden of his account of cell formation was that it provided a role in development for the nucleus. Noting its regular appearance in embryos, Schleiden had made accounting for its role in development a major objective.

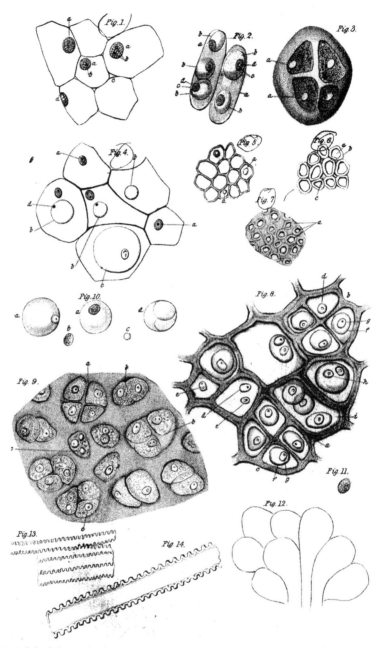

Figure 3.2. Schwann's drawings of different cell types. The first three figures are of plant cells – onion parenchymatous cellular tissue (1) and Rhipsalis salicornoides pollen (2 and 3, supplied by Schleiden), Cyprinus erythrophthalmus (a ray-finned fish) chorda dorsalis (4) and cartilage (5–7), Rana esculenta (frog) cartilage (8), Pelobates fuscus (toad) cranial cartilage (9), fetal pig crystalline lens (10–12), pike lens (13), and grass epidermis (14). Reproduced from Plate I from T. Schwann (1839/1847), *Microscopical researches into the accordance in the structure and growth of animals and plants* (H. Smith, trans.). London: Sydenham Society.

throughout physiology.)[10] Schwann drew a parallel between cell formation and crystal formation. Even though the details of crystal formation were not yet known, it was clearly a mechanical process not requiring any vital forces, and this made it a very compelling model for Schwann. He proposed that a mechanism of attraction drew a particular kind of material out of the inter-cellular fluid, with each layer (the nucleolus, the nucleus, and the cytoplasm) drawing out different substances. He discounted the importance of differences between inorganic crystals and biological cells: "If crystals were formed from the same substances as cells, they would probably, in these respects, be sub-ject to the same condition as cells" (Schwann, 1839/1947, p. 208; for further analysis, see Bechtel, 1984).

Schwann's commitment to chemical mechanisms for explaining vital phe-nomena was already manifest in his earlier research on digestion, in which he discovered pepsin, the first enzyme identified in animals (Schwann, 1836). In the eyes of some mechanists, Schwann's commitment to mechanism was com-promised when, in 1837, he presented evidence that fermentation was a pro-cess requiring living yeast cells (Schwann, 1837). Many mechanists associ-ated such a claim with vitalism. But Schwann did not see himself as embracing vitalism and in part III of *Microscopical Researches*, the book in which he introduced his cell theory, he returned to the topic of fermentation, present-ing it as an example of the kind of metabolic phenomena exhibited in cells (Schwann here coined the term *metabolism*).[11] He proposed that the ability of living cells to carry out activities that would not otherwise occur in nature was due to the distinctive chemical constitution of cells that resulted from the process of cell formation, itself a mechanical process. The significance of Schwann's contribution was to fix the cell as the locus of control (Bechtel & Richardson, 1993, Chapter 3) of basic life functions such as nutrition.[12] He

---

[10] Müller's own position regarding vitalism was more ambiguous, but among his students many, such as Emil du Bois-Reymond, Ernst von Brücke, Hermann von Helmholtz, and Carl Ludwig, were outspoken advocates of a mechanistic physiology (see Cranefield, 1957).

[11] Schwann characterized a cell as having "the faculty of producing chemical changes in its con-stituent molecules. Besides which, all the parts of the cell itself may be chemically altered during the process of its vegetation. The underlying cause of all these phenomena, which we comprise under the term metabolic phenomena, we will denominate the *metabolic power*" (Schwann, 1839/1947, p. 197).

[12] Schwann's strategy basically was to argue that whatever activities cells perform would not be performed twice, once in the cell, then again in the whole organism: "Now, as all cells grow according to the same laws, and consequently the cause of growth cannot in one case lie in the cell, and in another in the whole organism; and since it may be further proved that some cells, which do not differ from the rest in their mode of growth, are developed independently, we must ascribe to all cells an independent vitality, that is, such combinations of molecules as occur in any single cell, are capable of setting free the power by which it is enabled to take up fresh

71

also pointed the way to further developing a mechanistic account of cellular functions in terms of the component parts found in cells and their operations, but was not in position to develop a detailed account himself; that task was left for later scientists.

Although Schwann's contention that cells were the basic building blocks of all organisms was widely accepted, his and Schleiden's mechanism for cell formation became the focus of controversy. Researchers such as Barthélemy Dumortier (1832) and Hugo von Mohl (1837) reported observations of dividing cells in *Conferva aurea* even before Schleiden's publication,[13] and von Mohl as well Franz Unger and Carl von Nageli continued in subsequent years to develop ever more detailed descriptions of cell division. These included observations that the division of the nucleus preceded the division of the cell itself. Even von Mohl, though, allowed that sometimes cells formed like crystals and for a number of years the two accounts, cell division and Schleiden's and Schwann's accounts, were considered to provide two ways in which new cells were created. Rudolf Virchow (1858), a pathologist, and Robert Remak (1852; 1855) provided what turned out to be the decisive argument against Schwann's account – it amounted to spontaneous generation.[14] Virchow argued instead that each cell arises from a pre-existing cell (*omnis cellula e cellula*).[15] Virchow could offer no specific mechanism for cell division, but he was not as committed to mechanism as Schleiden and Schwann. He viewed the difference between healthy and pathological life as involving organizational properties of whole cells. Accordingly, he characterized his approach to pathology as "cellular pathology" (Virchow, 1855) and appealed

---

molecules. The cause of nutrition and growth resides not in the organism as a whole, but in the separate elementary parts – the cells" (Schwann, 1839/1947, p. 192).

[13] Schleiden actually cited von Mohl's paper, but contended that von Mohl was deceived by the smallness and transparency of the new cells that had formed in the old ones and so was led to see the final breaking free of these new (and already fully developed) cells from the mother cell as a process of cell division.

[14] Remak (1852) commented, "As for myself, the extracellular formation of animal cells struck me, from the very moment that this theory was propagated, as no less improbable than the generation aequivoca of organisms" (translated by Henry Harris (1999, p. 130). There is irony in the contention that Schwann's theory of cell formation amounted to spontaneous generation because in his paper on fermentation he had come down against spontaneous generation by demonstrating that fermentation required a microorganism that was killed by heat (Schwann, 1837).

[15] The expression was not original with Virchow. François-Vincent Raspail (1825) used it as an epigraph. When Virchow first employed the phrase, he wrote it "omnis cellula a cellula," a more ambiguous expression asserting only that all cells arise by means of cells rather than explicitly deriving from them.

to a "life force" to account for the specific activities that occur in living cells (this life force being acquired from other cells).

Although the cell wall had been the focus of investigations for plant cells from the time of Hooke through the investigations of Brown and Schleiden, Hooke had already noted the fluid contents of cells. He labeled them "*succus nutritus*, or appropriate juices of vegetables," and described cells "fill'd with juices, and by degrees sweating them out" (Hooke, 1665, p. 116). With the improved microscopes of the 1830s, featuring achromatic lenses, more investigators began commenting on the fluid contents of cells. Von Mohl, for example, described "an opaque, viscid fluid, having granules intermingled in it" as a universal feature of cells (1852, p. 37). In 1846 von Mohl had applied the term *protoplasm*[16] to the fluid. Slightly earlier, Dujardin (1835) labeled the viscid, slimy fluid found in animal tissues *sarcode*.[17] Remak (1852) pointed to the similarities between plant protoplasm and animal sarcode and employed the term *protoplasm* for both. This identification was cemented when Max Schultze (1861) characterized a cell as "a lump of protoplasm inside of which lies a nucleus" (p. 11). For Schultze, protoplasm was sufficiently nonmiscible with fluids surrounding the cell that a cell membrane was not needed; he rather viewed it as a sign of cell senility.

As I will discuss below, cytologists for the most part kept their focus on the structural components of the cell and on the processes involved in cell division. A number of chemically minded physiologists, though, made the special chemical nature of protoplasm their pursuit. Many of them also developed mechanistic accounts of cell functioning. One exemplar of this approach was T. H. Huxley's popular essay, "On the physical basis of life" (Huxley, 1869), in which he proposed a three-point unity among living things – they all exhibit the power of contractility, are composed of cells (defined à la Schultze), and are made of protoplasm. He took protoplasm to be comprised of proteins, whose chemical composition of carbon, hydrogen, nitrogen, and oxygen had been established in the 1830s. Huxley's objective was simply to provide a

---

[16] Six years earlier Purkinje introduced the same term for the embryonic material in animals, but von Mohl seems to have been unaware of this. Part of the significance of protoplasm for von Mohl was that he viewed it as providing the material for a new cell nucleus, whose formation would prompt division of the old cell.

[17] Dujardin made it clear that he was giving a name to a substance that had been observed earlier by others: "I propose to give this name to what other observers have called a living jelly, this glutinous, diaphanous substance, insoluble in water, that contracts into globular lumps, sticks to dissecting needles, and can be drawn out like mucus. It is to be found in all lower animals interposed between other structural elements" (1835, translated by Harris, 1999, p. 74).

physical grounding to biological inquiry, but other theorists were interested in the apparently special powers of protoplasm in metabolic processes. I will discuss some of the most prominent chemical accounts of protoplasm in the second part of this chapter.

Returning to cytology, the second half of the nineteenth century was a period of very active investigation. The technique of microscopy underwent significant changes in the decades after 1850. I have already noted the introduction of apochromatic lenses, which further reduced chromatic aberrations. Although chemical reagents such as chromic acid had sometimes been applied as preservatives and "hardening agents," a major step was the discovery that using osmic acid as a fixative would preserve the fine detail of cells (although it also gave rise to the question of whether it revealed existing structure or generated artifacts). Schultze (1865) pioneered this approach in his study of the luminescent organ of the glowworm. Fixation served both to kill the cells in the tissue and to stabilize the structure that would otherwise be disrupted postmortem by processes of *autolysis*. Embedding in hard materials was important because it made it possible to cut thin slices of material through a process known as *sectioning*. The contents of a given section could then be viewed without being occluded by the contents of other sections.

Equally important was the development of stains that would selectively color different components of cells. Carmine red was the first stain to be employed in this way, by Alfonso Corti and Joseph von Gerlach in the 1850s. (Produced from the crushed and dried bodies of chochineal insects in Mexico, it had been used there as a dye for centuries.) Gerlach found that the nucleus selectively absorbed carmine, resulting in a far clearer image of it than had been available previously. In the following decade Böhmer applied hematoxylin, obtained from logwood, a tropical American tree, to stain the nucleus deep blue. In the 1850s William Perkin discovered aniline dyes while he was trying to synthesize quinine, and in the 1860s–70s other investigators determined that some aniline dyes would stain parts of the cell not affected by carmine or hematoxylin. This made it possible to stain different structures in different colors.

The greatest advances in understanding cell mechanisms in the second half of the nineteenth century involved the events of cell division. As I noted previously, Brown gave the nucleus its name and Schleiden proposed a central role for it in his account of cell formation. As accounts in terms of division of existing cells supplanted Schleiden's and Schwann's accounts of crystal-like cell formation, there was considerable uncertainty about the role of the nucleus. Karl Bogislaus Reichert (1847) contended that the nucleus disappeared when cells divided and that new nuclei were created in daughter cells.

In opposition, Remak (1852) contended that the nucleus itself divided prior to cell division but noted, "Observations on the mechanisms of nuclear division are by no means so extensive as those on the behaviour of the cell membranes" (translated by Harris, 1999, p. 139). Three years later he contended that nuclear division was preceded by division of the nucleolus through a process of constriction and nuclear membrane partitioning analogous to that found in division of the cell body (Remak, 1855).

One of the first microscopists to contend that the division of the nucleus was more complex than a simple division was Wilhelm Hofmeister (1849). He observed in pollen mother and staminal hair cells of *Tradescantia*, a plant used for ground cover, that the nuclear membrane dissolved before the cell divided. If he stained the cell with iodine, however, he could observe discrete lumps (*Klumpen*, presumably chromosomes) that Hofmeister proposed were protein coagulates. He reported that a *granular mucilage* formed around these lumps and divided into two, with a membrane forming around each as the cell divided.

The availability of fixatives, stains, and finally apochromatic lenses enabled research on nuclear processes in cell division to explode in the 1870s and 1880s. Edouard van Beneden (1875) characterized the structures in the nucleus as little rods (*bâtonnets*) and observed that they moved apart in the process of nuclear division. Working with fertilized eggs of the marine mollusk *Geryonia*, Hermann Fol (1873) described the configuration of the spindle and astral rays and proposed an analogy with the lines of force found between opposite magnetic poles, suggesting thereby a dynamical perspective. Focusing on cell division in conifer embryos, Eduard Strasburger, in the first edition of his *Zellbildung und Zelltheilung*, presented drawings portraying a fibrous spindle at several stages of division.

Walther Flemming (1878; 1879) devoted considerable effort to developing new fixatives that would better reveal the details of the process of nuclear division. The product was a mixture of chromic, osmic, and glacial acetic acids that came to be known as Flemming's solution. Because the rods responded to this and other stains, Flemming referred to them as *chromatin*, contrasting them with unstained structures he labeled *achromatin*. (Waldeyer, 1888, gave them their current name, chromosomes.) Working with salamander larval epithelial cells, Flemming observed that the rods split lengthwise and proposed that one half went to each of the two daughter cells. To distinguish this far more complex sequence of events from the simple form of nuclear division described by Remak, Flemming referred to the process as "indirect nuclear division" and introduced the term "*Karyomitose*" for the changes in the chromatin. Flemming reviewed his findings and presented several detailed figures of the

75

process of mitosis in his 1882 book *Zellsubstanz, Kern und Zelltheilung*. In an obvious play on Virchow's dictum, he invoked the dictum *Omnis nucleus e nucleo* for his account.

In addition to this research on mitosis, numerous researchers observed the processes of fertilization and began to propose accounts of the continuity of chromosomes from parents to offspring. Strasburger (1884) described the chromosomes of the offspring as originating half from the father and half from the mother. Van Beneden (van Beneden & Neyt, 1887) discovered that prior to fertilization the chromosomes in both the egg and sperm nuclei were reduced in half (the process is known as *meiosis*) and that a full complement of chromosomes only reappeared after fertilization of the egg. Van Beneden was able to sustain this claim by showing that what he termed pronuclei in the nematode (*Ascaris maglocephala*) possessed only two *anses chromatiques* rather than the typical four. Flemming (1887) then described the process of reducing the number of chromosomes in sperm production.

This line of research on mitosis and meiosis represented the major nine-teenth-century success stories in describing mechanisms of cell life. Although some cytologists were simply interested in identifying stages in the overall operation of splitting cells, others were keenly interested in the potential of these operations to explain heredity. August Weisman (1885) drew attention to the fact that the procedures of cell division insured that daughter cells received one complete complement of the hereditary material by drawing half from each parent. Thus, when Carl Correns (1900) participated in the rediscovery of Mendel's work, he made explicit the connection between the steps in meiosis and Mendel's account of inheritance: Each germ cell must receive one or the other of the factors Mendel held responsible for the dominant or recessive traits. The relation between chromosomes and Mendel's factors, rechristened *genes*, was developed in much greater detail in the first half of the twentieth century by researchers of the Morgan school, who discovered such additional phenomena as the crossing over of chromosomes during meiosis.

Thus, one effect of the introduction of fixatives and stains was an initial understanding of mechanisms of nuclear division in the late nineteenth century that laid a foundation for genetic accounts of the twentieth century. Fixatives and stains also sparked observations and theorizing about structures and operations in the cytoplasm. These proved far more contentious. For example, Leydig, Carnoy, and Heidenhain claimed to identify structures they called *fibrils*, which they proposed accounted for the unusual consistency of protoplasm. Hanstein identified particles he called *microsomes* embedded in these fibrils. Critics such as Flemming, Bütschli, and Fischer charged that the fibrils were artifacts. They introduced what became a common argumentative strategy – demonstrations that they could artificially produce structures with the same appearances as fibrils by, for example, coagulating the whites of eggs or gelatin with various fixatives. I will turn to the question of how scientists appraise claims about artifacts in the next chapter.

The pattern of claim and counterclaim concerning the reality of various proposed cytoplasmic constituents continued through the early years of the twentieth century. This period also witnessed further improvement in the light microscope and the techniques for using it, as well as the introduction of variations, such as the ultraviolet microscope and phase contrast microscope. Moreover, as is clear from a 1924 textbook edited by Edmund Cowdry, *General cytology: A textbook of cellular structure and function for students of biology and medicine*, it was a period in which cytologists were increasingly interested in figuring out the function of new structures and making connections to chemistry (though often frustrated by the limitations of available tools). In the following sections I describe in more detail what was being learned about cell membranes and three organelles – mitochondria, the ergastoplasm, and the Golgi apparatus – that would later become foci of inquiry using improved tools in the early cell biology of the 1940s and 1950s.

Cell Membranes (1825–1935)

Membranes are among the most important parts of cells, serving to organize other parts and functions. Not only is there a plasma membrane surrounding the cell itself, but also membranes enclose the cell organelles, including the nucleus. Although membrane boundaries can be readily seen in electron micrographs, they were not so easily distinguished with the light microscope. Much of the evidence for them in the nineteenth century stemmed from their role in selectively admitting substances into cells. Already in the late eighteenth century William Hewson (1773) had appealed to the ability of red blood cells to shrink or swell to argue that they were envelope-bounded vesicles. Hewson's research was not well known in the nineteenth century, however, and the presence of a barrier surrounding animal cells continued to be debated. Schultze (1861), as I noted previously, characterized a cell as a lump of protoplasm which acquired a membrane only in its senility.

One of the key types of evidence pointing to the existence of membranes was the phenomenon of osmosis, the movement of water in and out of cells until the substances dissolved in it are distributed equally on the two sides. During the period when animal cells were known as *globules* and reports of their existence may have been artifacts of chromatic aberrations, Henri

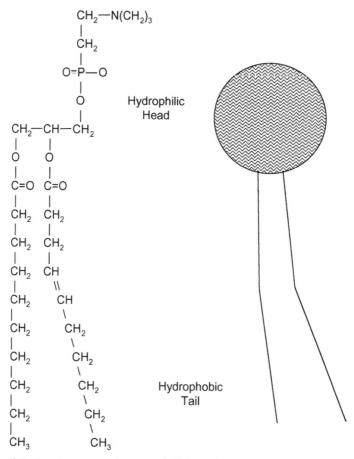

Figure 3.3. Structure of a typical phospholipid molecule with a hydrophilic head and hydrophobic tail.

Dutrochet (1826; 1828) characterized them as units of physiological exchange, selectively admitting nutrients along with the inflow of water, a process he termed *endosmosis*, and excreting waste with the diffusion out, which he termed *exosmosis*. Carl Wilhelm von Nägeli and Carl Cramer (1855) described osmosis in plant cells. Drawing on an analogy with osmometers that Moritz Traube had constructed using artificial semipermeable membranes, Wilhelm Pfeffer (1887) proposed that cells generally were surrounded by semipermeable membranes. Although early research on osmosis linked the ability of materials to cross membranes to their permeability in water, Charles Ernst Overton (1895; 1896; 1899) found in his attempts to get root hairs in plants to absorb substances that fat soluble substances passed through more

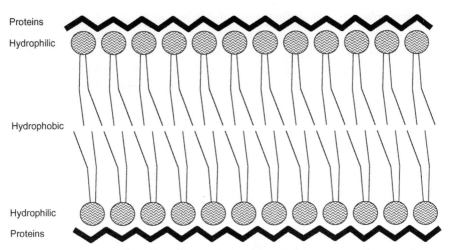

Proteins
Hydrophilic

Hydrophobic

Hydrophilic
Proteins

Figure 3.4. Danielli-Davson "sandwich" model of membrane structure. Following Gorter and Grendel (1925), phosopholipids are arranged into a bilayer in which each molecule's hydrophobic tail faces inward and hydrophilic head faces outward. Danielli and Davson (1935) added outer layers (jagged lines) composed of proteins.

easily than water soluble ones. Finding that membranes thus acted like fatty oils, Overton inferred that they were partly composed of lipids. Overton's research served to settle the debate, establishing that membranes were a universal feature of cells.

Substantial advances in the understanding of membranes continued in the first half of the twentieth century. In 1925 Evert Gorter and Grendel extracted lipid from red blood cells and concluded that there was just enough for the cell to be "covered by a layer of fatty substances that is two molecules thick" (1925, p. 443). Each molecule was a phospholipid in which the head had an affinity for water (*hydrophilic*) and the tail did not (*hydrophobic*), as illustrated both by chemical formula and abstractly in Figure 3.3. Gorter and Grendel proposed that the molecules are arranged such that the heads of the two layers face out in opposite directions while the tails face inward. This would result in the outside of the membrane being hydrophilic but the inside hydrophobic. Fat soluble substances like hydrocarbons can pass through such a membrane, whereas water soluble substances such as sodium and potassium ions and sugars cannot.

In addition to lipids, investigators found proteins and carbohydrates in membranes. In 1935 James F. Danielli, a physical chemist, and Hugh Davson, a physiologist (Danielli & Davson, 1935), advanced an extremely influential sandwich model in which, as shown in Figure 3.4, a protein layer covers

79

each surface of a lipid bilayer like that proposed by Gorter and Grendel. Specifically, each protein is attached to the outward-facing head of one of the phospholipid molecules. This model offered a ready explanation for the appearance of membranes as pairs of dark lines with a lighter region between them in electron micrographs and was generally accepted for more than thirty-five years before being replaced by a model in which proteins were embedded in the membrane.

Mitochondria (1890–1925)

Discussions of mitochondria frequently attribute their discovery to the investigations of Richard Altmann around 1890.[18] He introduced improved fixation techniques (e.g., a solution of potassium dichromate and osmium tetroxide). When used with an acid-fuchsin stain differentiated by picric acid, and with delicate heating, he was able to observe filaments in the cytoplasm of nearly all cell types. Although some of the structures Altmann saw were undoubtedly mitochondria, he positioned his discussion in a very different context than modern thinking about cell organelles. He titled his 1890 book *Die Elementarorganismen und ihre Beziehungen zu den Zellen (The Elementary Organisms and their Relations to the Cells)*. His endeavor was to revise and revive the view that living substances were comprised of elementary living granules (the filaments were, for him, strings of granules). The revisions were designed to make the granular theory compatible with cell theory, and specifically with Virchow's dictum *omnis cellula e cellula*, which he then extended to granules with the dictum *omne granulum e granulo*.

In espousing the view that granules were the basic living unit, Altmann opposed the view that protoplasm has a uniform structure. He attributes basic metabolic processes (specifically, fat metabolism and secretion) to these granules. Moreover, he claimed granules were equivalent to independently existing microorganisms and named them both *bioblasts*: Because both granules and microorganisms "represent the elementary organisms which are found wherever vital forces become active we shall name them with the joint term Bioblasts. **It seems that with the bioblast [the] morphological unit of living matter has been found**" (Altmann, 1890, from an unpublished

---

[18] This is not to say that Altmann was the first to see mitochondria. Starting around 1850 Kölliker studied granules in muscle cells. In 1888, he separated them from insect muscle and found that they swelled in water and possessed a membrane. These were probably mitochondria, but it is often very difficult to know for sure what structures (or artifacts) he and other early observers were actually observing.

translation prepared by Hanna Stoeckenius, folder 13, box 2, RU 518, Rock-efeller University Archives, RAC, p. 125).

Many investigators responded to Altmann's reports of granules with great skepticism, raising doubts in particular about his reliance on fixatives such as osmic acid. The comments of William Bate Hardy are illustrative:

> It is notorious that the various fixing reagents are coagulants of organic colloids, and that they produce precipitates which have a certain figure or structure. It can also readily be shown . . . that the figure varies, other things being equal, according to the reagent used. It is therefore cause for suspicion when one finds that particular structures which are indubitably present in preparations are only found in cells fixed with certain agents, used either alone, or in particular formulae. Altmann demonstrates his granules by the aid of an intensely acid and oxidizing mixture. (Hardy, 1899; quoted in Fruton, 1972, p. 389)

Fischer (1899) cast further doubt on the reality of the granules by showing that by applying commonly used fixing agents, especially osmium, to various homogeneous protein solutions (e.g., egg whites and gelatins, which would not contain subcellular structures, he could produce a variety of granular and filamentous structures. In part to counter charges of methodological arti-fact, Altmann also pioneered an alternative technique to chemical fixation, freeze-drying, which, while very laborious, provided an independent basis for evaluating the reality of the structures revealed by chemical fixation. This, however, did not suffice to stop the objections from the critics.

Carl Benda (1898; 1899), employing crystal violet as a stain, observed Altmann's structures and proposed the named *mitochondria*, from the Greek words for thread and granule. The name reflected the fact that in his prepa-rations they sometimes appear threadlike and at other times more granular.[19] Importantly, Benda provided evidence that these structures could be seen both in fixed and in living cells, reducing the plausibility of the objection that they were an artifact of fixation. Further, Michaelis (1899) showed that the dye Janus Green (diethylsafraninazodimethylanalin), which appears blue-green when oxidized but colorless when reduced, would turn mitochondria blue-green in living cells. This suggested that mitochondria had the capacity to oxi-dize substrates and provided the first clue as to their role in cellular respiration.

---

[19] These different appearances, we now understand, depended upon the angle at which the mitochondrion was sliced in preparing the slide. The term *mitochondrion* only gradually became accepted. Some other terms were blepharoblast, condriokonts, chondriomites, chondrio-plasts, chondriosomes, chondriosheres, fila, fuchsinophilic granules, interstitial bodies, Körner, Fädenkörner, mitogel, parabasal bodies, plasmasomes, plastochondria, plastomes, sphereo-plasts, and vermicles.

With the tools available in the early twentieth century, researchers were able to develop two types of information about mitochondria: (a) their shape and location in cells, and (b) their composition. Microscopic examination showed that the appearance of mitochondria is relatively stable across species, but varies in degree of elongation and thickness according to the cell type. Cowdry (1924) noted that in gland and nerve cells as well as embryonic cells, they were observed as filaments but that on injury, their shape changes, "providing by far the most delicate criterion of many types of cell injury at our disposal" (p. 317).

Evidence about the composition of mitochondria came primarily from reactions with various reagents. Their solubility with acetic acid, as well as with alcohol, ether, and chloroform, indicated a phospholipid constitution. The failure of mitochondria to stain with Sudan III indicated that they did not contain fat, and failure to stain with Millon's reagent indicated little if any protein.

Once dissociated from Altmann's conception of an elementary organism carrying out all basic metabolic processes, a natural question was what function mitochondria perform. Kingsbury (1913) noticed that the fixatives yielding the best visualization of mitochondria – osmic acid, potassium dichromate, and formalin – all depend on reducing substances. Picking up the thread from Michaelis, he advanced the proposal that mitochondria play a critical role in respiration.[20] Cowdry further noted that the amount of mitochondria in a cell was positively correlated with its level of activity (division, secretion, etc.) and negatively correlated with the amount of fat in it (an indication of decreased respiration):

> We have two lines of observation to harmonize: this association of abundant mitochondria with intense protoplasmic activity and a reciprocal relationship which appears to exist between the amount of mitochondria and the amount of fat. Where there are few mitochondria there is often much fat, and vice versa. Decreased oxidation favors the deposition of fat and increased oxidation hastens its elimination, which suggests at once the existence of some connection between the amount of mitochondria and the rate of oxidation; and their abundance in the more active stage of the life of the cell, when protoplasmic respiration is rapid, points to the same tentative conclusion. (Cowdry, 1924, p. 321)

---

[20] According to Bourne (1962, pp. 70–1), Kingsbury's major evidence was that "anesthetics such as ether or chloroform, which depressed cellular respiration and the respiration of the animal in general, also broke up mitochondria in the cell."

Even while reviewing this impressive evidence for the role of mitochondria in respiration, Edmund Cowdry himself was not convinced:

> Although this view, that mitochondria take part in protoplasmic respiration, has been well received by cytologists and serves as a useful and convenient working hypothesis, it is still only a theory and must be regarded as such. (Cowdry, 1924, p. 325)

Cowdry preferred Claudius Regaud's (1909) interpretation that mitochondria served to select substances out of the cytoplasm and, after bringing them inside, condensed and transformed them into different products.

In the end, Cowdry was quite pessimistic about the prospects of rapid progress in the study of mitochondria:

> it is quite obvious that the investigation of mitochondria will never achieve the usefulness which it deserves as an instrument for advance in biology and medicine until we know much more of their chemical constitution as the only accurate basis for interpretation of our findings. In other words, we must wait upon the slow development of direct, qualitative cellular chemistry. (Cowdry, 1924, p. 311)

Having made little more progress by the late 1930s, Cowdry, on the advice of Simon Flexner, Director of the Rockefeller Institute, abandoned the problem of mitochondria and moved to cancer research.

Ergastoplasm or Basophilia (1900–1930)

Investigators in the nineteenth century advanced a number of ideas about the constituency of the area of cytoplasm that largely appeared empty under the light microscope. One concept that would play a prominent role in the development of cell biology was that of *ergastoplasm*. According to Haguenau (1958), it stemmed from the thesis research of Charles Garnier, a French cytologist in Nancy who explored the effects on cells of a number of stains, including safranin, gentian violet, and toluidine blue. He identified a fibrillar material or rod-like structure in gland cells. Because he thought this structure was associated with production of secretion granules, he called it *ergastoplasm*, that is, the plasma that elaborates or transforms something (Garnier, 1897; Garnier, 1900). Because the intensity of stain varied with the stage of secretion in gland cells, Garnier concluded that ergastoplasm was not a permanent structure but emerged during the resting phase after secretion. He also proposed a link between the ergastoplasm topographically and the nucleus, claiming that at times the ergastoplasm formed around, and sometimes encircled, the nucleus. He suggested that the nuclear sap or chromatic substance from the nucleolus passed through the nuclear membrane and joined with the ergastoplasm. Garnier's proposal was further developed by Prenant (1898), the supervisor of Garnier's thesis, who termed the ergastoplasm *protoplasme supérieur* and characterized it as a zone of the cytoplasm that could differentiate into other structures.

As with the mitochondrion, investigators disagreed as to whether the ergastoplasm was a real structure. Regaud offered support for Garnier and Prenant, showing that when he added acetic acid to his fixative, the ergastoplasm appeared as Garnier reported but mitochondria were not visible, whereas if he left out the acetic acid, mitochondria appeared, but no ergastoplasm (Regaud, 1909). He also supported the idea that ergastoplasm involved cytoplasmic material which was "impregnated with chromatin or a closely related substance" (quoted in Haguenau, 1958). In opposition, Morelle (1927) denied that the ergastoplasm had a fibrillar structure and proposed that it was simply modified ground cytoplasm which, due to its chemical composition, took up basic stains. Christian Champy (1911) proposed that ergastoplasm was simply poorly fixed mitochondria.

Although, as Haguenau describes, there continued to be some publications describing the ergastoplasm, it largely faded from view in characterizations of cell structures in the first half of the twentieth century. It is not discussed, for example, either in Cowdry's *General cytology* (1924) or in Bourne's *Cytology and cell physiology* (1942).

The Golgi Apparatus (1900–1940)

What came to be known as the *Golgi apparatus* was first systematically observed in Purkinje and ventral horn cells of the barn owl and the cat by Camillo Golgi in 1898. He fixed his cells with an osmium-tetroxide-potassium dichromate mixture followed by impregnation with silver salts (Golgi, 1898). To Golgi it appeared as a fine network, which he characterized as an internal reticular apparatus (*apparato reticulare interno*; see Figure 3.5). A number of other investigators in the same period also described what was probably the same structure. Whaley provides an explanation as to why Golgi's work stands out:

> Considering the fixatives and stains being used, the amount of experimental work with them characteristic of the latter part of the 19th century, and the multiplicity of the tissue and cell types being studied, it seems reasonable to suggest that a considerable number of investigators may have seen this pleiomorphic

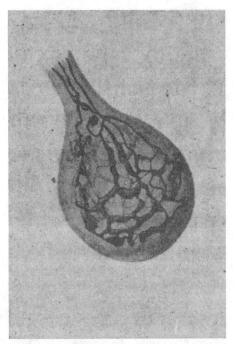

Figure 3.5. Golgi's drawing of what he identified as an internal reticular apparatus in a Purkinje cell of a barn owl. Reproduced from C. Golgi (1898), Sur la structure des cellules nerveuses, *Archives italiennes de Biologie, 30*, 60–71, Figure 2.

organelle before Golgi. It was Golgi, however, who devised a method that put the organelle in sharp contrast with other cellular components and permitted him and his students to demonstrate it as a consistent structural component in a wide variety of tissue cell types. It was also Golgi who recognized not only some of the details of its structure but also that it is variable in character and position. (1975, p. 3)

Shortly after Golgi's first publication, Emil Holmgren (1902) identified a set of clear canals within the cytoplasm that remained uncolored when the rest of the cytoplasm was stained. Holmgren construed these canals as formed by cytoplasmic processes penetrating into a cell from its neighbors and suggested that they might serve a nutritive function. He called them *trophospongium* and claimed they were the same as Golgi's reticular nets although they were located near the external membrane of the cell.

Initially Santiago Ramón y Cajal (1907; 1908) accepted the link between Golgi's reticular structure and Holmgren's canals and referred to them as the Golgi-Holmgren canals. Later he revised this assessment (Cajal, 1914) and switched to Jòzef Nusbaum's term, *Golgi apparatus*.[21] Cajal went on to examine how its appearance varied with different cellular functions and with different experimental manipulations. Without endorsing the claim that it figured in secretion, Cajal noted a correlation between the presence of the Golgi apparatus in a cell and secretory activity and reported, in intestinal goblet cells that secrete mucus over the lining of the intestine, the appearance of tiny droplets of mucus in the Golgi region.

Progress in clarifying the structure and determining the function of the Golgi apparatus was slow. In 1924, Cowdry commented,

> Even now, twenty-five years after its discovery, we can only say that the Golgi apparatus is an area of the cytoplasm frequently (especially in higher forms) of reticular shape, often as large as the nucleus, and sometimes definitely located in relation to cellular polarity. Part of the material of which it is composed is soluble in alcohol, becomes blackened after prolonged treatment with osmic acid, and, after appropriate preliminary fixation, shows a marked affinity for silver salts. In addition it may occasionally be stained with resorcin-fuchsin, iron hemotoxylin, and other dyes, but the word "apparatus" is unfortunate because it carries with it the idea of a mechanism of a rather mechanical type. (Cowdry, 1924, p. 334)

Cowdry did report that the appearance of the Golgi apparatus was relatively constant in a particular cell type but varied across cell types so that "variations in its morphology are closely related to variations in cellular organization and function" (p. 334).

Variability in the appearance of the Golgi apparatus posed difficulties for cytologists. Kirkman and Severinghaus (1938a) reported that the following descriptions of the Golgi apparatus had been offered at different times: "a fibrous reticulum, network, ring, or cylinder, a very irregular fenestrated plate, a more or less incomplete hollow sphere, vesicle, or cup, a collection of small spheres, rodlets and platelets or discs, a series of anastomising canals, a group of vacuoles, and a differentiated region of homogeneous cytoplasm crossed by irregular interfaces" (p. 419).

I noted previously that Cajal correlated the appearance of the Golgi apparatus to cell secretion. Stronger evidence for its role in secretion stemmed from the work of Dimitry Nassonov (1923; 1924), who demonstrated the consistent

---

[21] Dröscher (1998, p. 429), commented that Nusbaum "included 'Golgi' as a tribute to its discoverer, but eliminated the word 'reticulum' or 'net' because he and his collaborators at the University of Lemberg (Lwòw) found that the disposition of the apparatus, especially in invertebrate cells, was not necessarily net-like, but predominantly in the form of single dictyosomes."

association of accumulating secretory products and the Golgi apparatus. He reported that granules for secretion first appeared in the meshes of the Golgi apparatus and proposed that after reaching a particular size, parts of the apparatus broke free and collected near the boundary of the cell. In a review paper in 1929, Robert Bowen defended the claim that the Golgi apparatus figured centrally in cell secretion, building up in the organelle and separating from it in different ways in different tissues.

In the first half of the twentieth century there was a long history of claims to the effect that the Golgi apparatus was an artifact of silver or osmium staining. Thomas Strangeways and R. G. Canti (1927) made such a case on the grounds that they were unable to find any evidence of the Golgi apparatus in unstained tissue-cultured cells either by direct or dark ground illumination. Many of these claims were followed by counterclaims. Richardson (1934), for example, claimed to find such evidence in cultured cells and contended that Strangeways and Canti were viewing cells in which the Golgi apparatus was fragmented. He as well as other investigators of the time also differentiated two parts to the Golgi apparatus – an outer part that absorbs osmium and silver and an inner portion that does not.

The strongest case for the claim that the Golgi apparatus was an artifact came from demonstrations that the preparation of cells for microscopy could result in a structure with the appearance of the Golgi apparatus. Maurice Parat (1928) proposed that the artifactual structure arose when fixatives caused certain cytoplasmic vacuoles to coalesce. John Baker (1944) further elaborated this view, holding that it actually was produced by deposition of metals (osmium or silver) on the periphery of the vacuoles. Walker and Allen (1927) used chemical models of gelatin, albumen, and lecithin to obtain evidence that the appearance of the Golgi apparatus resulted from the spreading of phospholipid materials on various interfaces produced during fixation. Gicklhorn (1932) claimed that laminated, doubly refractive myelin figures resembling the Golgi apparatus could be produced by treatment of isolated tonoplasts with methods developed by de Vries and suggested that the Golgi apparatus itself was produced by a similar release of myelin and the subsequent staining with silver or osmium.

On the other hand, the strongest evidence for the reality involved showing ways in which one could manipulate it experimentally. Beams and King (1934), applying Beams' centrifuge (see Chapter 4) to the uterine gland cells of guinea pig, caused the Golgi apparatus to appear to stream through the cytoplasm. They took this as indicating a fluid or semifluid character. Bourne (1942, p. 117) accepted this as compelling evidence "that the Golgi apparatus was a definite cell organ."

During this period suggestions of a possible role of the Golgi apparatus in secretion were further developed. Researchers found that it absorbed a variety of dyes into droplets, giving rise to the idea, articulated by Kirkman and Severinghaus, that the Golgi apparatus serves as a condensation membrane:

> A great deal of work strongly suggests that the Golgi apparatus neither synthesizes secretory substances nor is transformed directly into them; but it acts as a condensation membrane for the concentration, into droplet or granules, of products elaborated elsewhere and diffused into the cytoplasm. These elaborated products may be lipoids, yolk, bile constituents, enzymes, hormones, or almost any other formed substance. (1938b, p. 85)

Others, however, argued that it was in fact the locus of synthesis – Bowen described it as "a great intracellular center of chemical synthesis or enzyme formation" (1924, p. 215). Bourne advocated the view that the surfaces of the Golgi apparatus were the key to their synthetic function.

Overall, by 1940 a consensus was developing that the Golgi apparatus was a true component of the cell, with most theorists proposing that it functioned in cell secretion.[22] These claims, however, were still open to challenge. In Chapter 4 I note a pair of 1949 papers in which Albert Claude and George Palade argued that the Golgi apparatus was indeed an artifact of staining. Eventually they accepted the existence of this structure and, as discussed in Chapter 6, Palade made major contributions to determining its function.

The State of Cytology circa 1940

By 1940, cytology had advanced about as far as it could with the resources of the light microscope. The accounts of the mechanism of cell division remained the major success story in pursuit of mechanisms responsible for cell functioning. Although several organelles had been identified in the cytoplasm, each remained controversial. The evidence was strongest for the reality of the mitochondrion and the Golgi apparatus, but dissenters remained who

---

[22] De Robertis, Nowinski, and Saez commented in their textbook, *General Cytology*, "Although the existence of a relationship between secretion and Golgi apparatus seems possible, the explanation of this relationship has not yet left the domain of hypothesis. If this is the present situation for the secretory cells, even more nebulous is the interpretation of the functional significance of the Golgi apparatus in the nonsecretory cells and particularly in the nerve cells, where it has such a considerable development. It has been thought that it may intervene in the secretion of fats, the elaboration of Nissle bodies, the metabolism of carbohydrates, and so on, but it is safer to affirm that up to the present time, there is no satisfactory theory to explain in a general form and for all cells the function of the Golgi apparatus" (1949, pp. 112–13; the same passage appears in the second edition, 1954, p. 147).

maintained that both were artifacts of the staining techniques used to see them. Even those who accepted the reality of the organelles disputed what function they played. A major reason both for the claims of artifact and for the disputes about function was that cytologists had limited resources to determine the function of organelles. The major strategy was to correlate frequency of the organelles' occurrence with activities the cell was performing. Many cytologists saw that the best prospect for moving beyond this state involved linkages with biochemistry. During the first half of the twentieth century, while cytologists were struggling, biochemists were making rapid advances.

2. BIOCHEMICAL CONTRIBUTIONS TO DISCOVERING CELL MECHANISMS UP TO 1940

Biochemistry established itself in the early decades of the twentieth century in what was then underdeveloped territory between physiology and chemistry (Kohler, 1973; Kohler, 1982). It drew upon considerable research on chemical processes in living organisms in the nineteenth century, but did not develop its own methods and conceptual framework until the beginning of the new century. I will begin by briefly reviewing some of the foundations that were laid in the nineteenth century, discuss what distinguished biochemistry as it emerged in the twentieth century, and then describe in greater detail research on glycolysis (fermentation) and aerobic respiration in the period 1900 to 1940.

Foundations for Biochemistry in the Nineteenth Century

The first attempts to build bridges between chemistry and activities of living organism preceded the chemical revolution at the end of the eighteenth century, but with the emergence of a new chemistry outlined by Antoine Lavoisier these efforts acquired a new foundation.[23] Lavoisier himself contributed significantly to charting the path that subsequent investigators would follow. The new systemization of the basic elements led to analyzing organic substances in terms of these elements. Lavoisier (1781) determined that carbon, hydrogen, and oxygen are constituents of all living organic substances. Claude Louis Berthollet (1780) identified nitrogen as another frequently occurring component. With these foundations, investigators began trying to characterize physiological processes in terms of changes in elemental composition.[24] For

---

[23] For more detailed analysis, see Holmes (1963).

[24] Since different organic substances were composed from the same elements, many chemists concluded that they differed from one another only in terms of their relative proportions. Thus,

example, Lavoisier (1789) himself characterized fermentation as involving the oxygenation of carbon in sugar to produce carbon dioxide at the expense of the deoxygenation of the remainder, yielding alcohol. Shortly thereafter Louis Jacques Thénard (1803) and subsequently Joseph Louis Gay-Lussac (1810) worked out the general formula for fermentation, represented in modern notation as

$$C_6H_{12}O_6 \rightarrow 2CO_2 + 2C_2H_5OH.$$

At the time yeast was not regarded as an organism, so fermentation, although involving organic compounds, was regarded as a strictly chemical process. In further research in collaboration with Pierre Simon LaPlace, Lavoisier measured the heat generated as animals respired and compared that with the heat of combustion of coal that generated the same amount of carbon dioxide. They concluded, "Respiration is thus a very slow combustion phenomenon, very similar to that of coal" (Lavoisier & LaPlace, 1780, p. 331 in 1886 reprint).

In the succeeding decades, a host of chemists turned their attention to the phenomena exhibited by living systems. One of the challenges they confronted is that chemical reactions occur in living organisms that do not freely occur outside of them. In some cases chemists identified and isolated substances that promoted reactions without being consumed in them. Gottlieb Sigismund Kirchhoff (1816) had shown that germinating grains of malt facilitated the conversion of starch to sugar. Payen and Persoz (1833) later extracted the chemical component that facilitated the process and named it *diastase*. Jöns Jacob Berzelius (1836) introduced the term *catalysis* for the process and, appealing to inorganic examples, proposed to apply the idea to fermentation.

> This property was not an isolated, exceptional behaviour but proved to be a more general one, exhibited by substances to varying extents. . . . We have found, for instance, that the conversion of sugar to carbon dioxide and alcohol, which occurs in fermentation through the influence of an insoluble substance known by the name of ferment . . . could not be explained by a chemical reaction between sugar and ferment resembling double decomposition. However, when compared with phenomena known in inorganic Nature, the preceding phenomenon most

Antoine François de Fourcroy concluded that substances such as oils, acids, mucilages, and fibres "differ from each other only in the number and the proportions in which the primary substances are combined in them" (Fourcroy, 1789, translation by Holmes, 1963, p. 57). One consequence of this research was the determination that nitrogen was present in far higher concentrations in animal tissue than plant tissue, inspiring the hypothesis that plant tissue had to be animalized by increasing the concentration of nitrogen to generate animal tissue.

closely resembles the decomposition of hydrogen peroxide under the influence of platinum, silver or fibrin; it was hence very natural to imagine an analogous activity in the case of the ferment. (passage translated in Friedmann, 1997, p. 74)

In the same year, Schwann (1836) isolated pepsin, a catalyst in gastric juice that breaks down proteins. Chemists also sought to synthesize substances produced by living organisms. Friedrich Wöhler (1828), for example, synthesized urea from ammonia and cyanic acid. Wöhler's enthusiasm was evident in a letter he wrote to Berzelius: "I can no longer, as it were, hold back my chemical urine; and I have to let out that I can make urea without needing a kidney, whether of man or dog" (quoted in Friedmann, 1997, p. 68).

With the tools of elementary analysis and the concept of catalysis and the example of Wöhler's success in synthesizing urea, some investigators saw the time as ripe to formulate a detailed chemical account of all processes occurring in living organisms. In many respects, the most ambitious of these attempts came from Justus Liebig. After studying in Paris with Thénard and Gay-Lussac, among others, Liebig became a professor at the University of Giessen in 1824 and established one of the most influential laboratories for the study of chemistry. One of his first accomplishments was to perfect an instrument for the elemental analysis of organic substances (Liebig, 1831). A further inspiration for Liebig's thinking was William Prout's (1827) classification of the nutrients required by animals into three classes: saccharine (carbohydrates), oleaginous (fats), and albuminous (proteins). Prout had noted that there were only minor differences in chemical composition between the nutrients animals took in from plants and the compounds that comprised the fluids and solids of their bodies. Liebig drew upon this observation to formulate a central part of the synthetic and highly speculative account of the chemical processes of animals in his *Animal Chemistry* (1842). He proposed that animals incorporated nutrients into their bodies and, as needed, broke them down to their constituents. Because animal tissues were primarily made of protein, he hypothesized they reconstituted them by simply incorporating protein from their diet. When muscle work was required, these proteins were broken down, with waste products excreted. In contrast, he thought animals burned carbohydrates and fats to generate heat. When insufficient oxygen was available for burning, animals converted them to fat and stored them. With these key ideas, Liebig articulated a general scheme, complete with detailed formulae,[25] that described the chemical operations occurring in animals.

---

[25] These detailed formulae often aroused skepticism. Even though Liebig dedicated *Animal Chemistry* to Jacob Berzelius, Berzelius derided it as "physiological chemistry . . . created at the writing table" (quoted in Fruton, 1972, p. 97).

Liebig's approach to physiological chemistry promised to reveal the chemical events in animals without requiring direct empirical investigation of internal operations. This was only plausible because Liebig had assumed that all processes in animals were catabolic – breaking down complex substances into simple ones but creating nothing new. However, when Claude Bernard set out to determine where the oxidation of carbohydrates occurred in animals, he discovered that in fact glycogen was being synthesized. This surprising discovery revealed the oversimplification in Liebig's scheme (Bernard, 1848). This was a major inspiration for Bernard's very different conception of metabolic processes. As I discussed in Chapter 2, Bernard (1878a, p. 113) proposed that each organ in an organism performs one of the operations necessary to maintain the "constancy of the internal environment."

When he coined the concept of a catalyst, Berzelius had assumed that it would be a relatively straightforward project to generate a chemical account of fermentation. The discovery by Schwann, as well as Charles Cagniard-Latour (1838) and Friedrich Traugott Kützing (1837), that yeast were living suddenly made fermentation a more challenging case for those seeking to provide chemical accounts of physiological processes.[26] Louis Pasteur followed

---

[26] Liebig, Berzelius, and Wöhler perceived the threat to the program of giving chemical accounts of physiological processes, and their response was extremely harsh. Wöhler published excerpts of a paper by Turpin (1838) following up on Cagniard-Latour's research in *Annalen der Pharmacie*, which he and Liebig edited. Following the excerpts he published a heavy-handed satire (officially anonymous, but clearly the work of Wöhler, perhaps with the collaboration of Liebig) titled "The demystified secret of alcoholic fermentation," which purported to describe detailed observations with a special microscope: "Incredible numbers of small spheres are seen which are the eggs of animals. When placed in sugar solution, they swell, burst, and animals develop from them which multiply with inconceivable speed. The shape of these animals is different from any of the hitherto described 600 species. They have the shape of a Beindorf distilling flask (without the cooling device). The tube of the bulb is some sort of a suction trunk which is covered inside with fine long bristles. Teeth and eyes are not observed. Incidentally, one can clearly distinguish a stomach, intestinal tract, the anus (as a pink point), and the organs of urine excretion. From the moment of emergence from the egg, one can see how the animals swallow the sugar of the medium and how it gets into the stomach. It is digested immediately, and this process is recognized with certainty from the elimination of excrements. In short, these infusoria eat sugar, eliminate alcohol from the intestinal tract, and CO_2 from the urinary organs. The urinary bladder in its filled state has the shape of a champagne bottle, in the empty state it is a small bud. After some practice, one observes that inside a gas bubble is formed, which increases its volume up to tenfold; by some screw-like torsion, which the animal controls by means of circular muscles around the body, the emptying of the bladder is accomplished . . . From the anus of the animal one can see the incessant emergence of a fluid that is lighter than the liquid medium, and from their enormously large genitals a stream of CO_2 is squirted at very short intervals . . . If the quantity of water is insufficient, i.e. the concentration of sugar too high, fermentation does not take place in the viscous liquid. This is because the little organisms cannot change their place in

Schwann and the others in relating fermentation to the activity of living yeast cells: "Fermentation is correlated to the vital processes of yeast" (Pasteur, 1860, p. 323). For him and for many, this pointed to the futility of trying to provide a purely chemical account of fermentation. But despite Pasteur's influence, numerous researchers (e.g., Pierre Eugène Marcellin Berthelot, Moritz Traube, Felix Hoppe-Seyler) contended that it was a chemical constituent of yeast (as pepsin is a chemical substance in the stomach) that catalyzed the reaction. Wilhelm Kühne (1877a; 1877b) introduced the term *enzyme* (from "in yeast" in Greek), which eventually came to be the term in general use for such agents.[27] Until Buchner's breakthrough in 1897 (discussed below), however, there was no compelling evidence that enzymes – that is, chemical agents – were responsible for fermentation.[28]

During this period in which convincing empirical results supporting a chemical approach to fermentation were not forthcoming, chemically minded researchers focused on the mechanisms of animal respiration. Lavoisier's research had opened a prolonged conflict about whether respiration occurred in animals' lungs (Lavoisier), blood (Liebig, Bernard), or tissues (Pflüger). (For further discussion of this conflict, see Bechtel & Richardson, 1993, Chapter 3.) Eduard Pflüger (1872; 1875) ultimately provided compelling evidence that respiration occurred in the tissues. Settling this question opened the search for the mechanism within tissues that made respiration possible. Pflüger proposed that the protoplasm that comprised the cells of tissues had a complex physical structure that resulted from the polymerization of protein. He further proposed that this structure stored energy that could be released in "explosions" in the cell: "The life process is the intramolecular heat of highly unstable albuminoid molecules of the cell substance, which dissociate largely with the formation of carbonic acid, water, and amide-like substances, and

---

the viscous liquid: they die from indigestion caused by lack of exercise" (Wöhler, 1839; passage translated in Schlenk, 1997, p. 47).

[27] For Kühne the term marked a contrast between chemical agents and *organized ferments*, living organisms that produced chemical changes. The eventual use of the term *enzyme* for chemical agents thus reversed Kühne's intention in coining the term.

[28] After Bernard's death in 1878 a manuscript was found in his country house proposing that alcoholic fermentation resulted from a soluble ferment found in ripe or rotting fruit (Bernard, 1878b). It was not clear whether Bernard was reporting actual results. This manuscript provoked Pasteur to pursue an extensive attempt to isolate such an enzyme and an extended exchange between Bertholet and Pasteur (For a review and discussion, see Friedmann, 1997; Kohler, 1971). About the same time, Marie Mikhailovna Manasseïn (1872) reported in the same journal in which Buchner was to publish that she had succeeded in producing fermentation in a cell-free extract. Her brief report, however, did not elicit much response.

which continually regenerate themselves and grow through polymerization" (Pflüger, 1875, p. 343, as quoted in Fruton, 1972, p. 284).

As Pflüger's proposal exemplifies, theorizing about metabolic mechanisms in the later part of the nineteenth century was highly speculative. In large part, this was because proteins are extremely complex and it was not possible at the time to establish that they were even molecular in character. Whereas organic chemists focused on relatively simple molecules in order to make progress in articulating the structure of organic compounds, researchers interested in the chemistry underlying physiological functions emphasized the likely importance of complex organization. One avenue toward complexity was offered by colloid chemistry, which emphasized molecules organized in arrays on surfaces. It became increasingly influential after Thomas Graham (1861) proposed that protoplasm was composed of colloidal arrays rather than less structured molecules. Investigators such as Wolfgang Ostwald (1909) emphasized the multiple phases of colloidal dispersion and the energies associated with surfaces. Some historians of biochemistry, such as Marcel Florkin (1972, pp. 279–83), have bemoaned this as the "dark age of biocolloidology." Florkin portrayed the influence of colloid chemistry on physiology as hindering the advance of a theoretically sound biochemistry and so drew a sharp contrast between colloidal chemistry and biochemistry. Yet, even as major a contributor to biochemistry as Otto Warburg began his career emphasizing the importance of colloidal surfaces for cell respiration (Warburg, 1911; Warburg, 1913b).[29]

The Emergence of Biochemistry in the Twentieth Century

The pioneering efforts to provide chemical explanations of vital processes in the nineteenth century suddenly developed into a robust research endeavor in the early twentieth century. At this time, the term *biochemistry* began to supplant older labels such as *physiological chemistry*. In the first decade of the twentieth century, new journals developed with *biochemistry* in their titles[30] and new professional societies were established.[31] In 1909 Carl Oppenheimer

---

[29] Warburg was generally dismissive of the emphasis on enzymes in mainstream biochemistry and pointedly referred to them using an older term, *ferment*. He identified one such important substance himself, which he called *Atmungsferment* (now known as cytochrome oxidase).

[30] In 1902 *Biochemisches Centralblatt, vollstandiges Sammelorgan fur die Grenzgebiet der Medi-zin und Chemie*; in 1903 *Beitruge zur chemischen Physiologie und Pathologie, Zeitschrift fur die gesamte Biochemie*; in 1905 *Journal of Biological Chemistry*; in 1906 both the *Biochemical Journal* and *Biochemische Zeitschrift*.

[31] In 1905 the American Chemical Society initiated a section of biological chemistry and in 1911 in England the Biochemical Club – later to become the Biochemical Society – was founded.

published his *Handbuch der Biochemie* in which he offered the following characterization:

> Biochemistry is the science that deals first with the constituents of living tissues, and their determination, properties, and reactions; but which also strives to draw from the chemical changes that proceed during life processes, conclusions regarding the scope of the life-process itself. (Oppenheimer, 1909, pp. v–vi)

Historians have proposed two major factors as differentiating the newly emerging biochemistry from the physiological chemistry of the nineteenth century that I have been describing. Robert Kohler (1973) emphasized the strategy of identifying different enzymes responsible for each reaction.[32] As we have just seen, investigators had already identified a host of catalysts responsible for physiological operations before Kühne coined the term, and by the 1890s physiologists generally accepted that digestion in animals was the work of enzymes. Emil Fischer's (1894) research and his lock-and-key model of enzyme action had revealed the highly specific nature of enzyme action. All of the reactions for which enzymes had been discovered to that point, however, were simple hydrolytic ones (that is, they consumed water in reactions that split apart organic substances). What Kohler emphasized is that until the 1890s no catalysts had been identified as involved in the more complex reactions of cells such as respiration or fermentation and none had been demonstrated to operate within cells (all enzymes known at that time were secreted by cells and operated outside).

Kohler presented Gabriel Bertrand (1895) as having taken the first step beyond hydrolytic enzymes when he determined that a catalyst he identified in the Chinese art of making black lacquer finish, laccase, caused the uptake of oxygen and the production of carbon dioxide. Bertrand compared the process to artificial respiration.[33] The key step in demonstrating that enzymes could operate within cells came with Eduard Buchner's (1897) report (discussed below) that in a cell-free extract made by grinding yeast cells with sand, fermentation still occurred. Appealing to enzymes rather than protoplasm to explain processes in living organisms, according to Kohler, marked the major change from physiological chemistry to biochemistry: "the very language of

---

[32] Kohler cites Franz Hofmeister (1901) as giving voice to the view that all reactions could be explained by enzymes: "we may be almost certain that sooner or later a particular specific ferment will be discovered for every vital reaction" (p. 14).

[33] Kohler identified a second major expansion of the notion of enzymes, occurring with Arthur Croft Hill's (1898) discovery that maltase, in appropriate circumstances, synthesized maltose from glucose rather than cleaving maltose to form glucose.

biochemistry changed when enzymes replaced protoplasm as the seat of vital chemistry" (1973, p. 193).

Frederic Holmes (1986; 1992) emphasized a second factor that differentiates biochemistry from its predecessors, the conception of intermediary metabolism as involving sequences of specific chemical reactions. This established the goal of discovering each intermediate reaction.[34] This had been a goal of many early chemists, but one that was difficult to realize when the units for chemical analysis remained the elements of which organic substances were composed. (Recall Liebig trying to describe the chemical operations in animals by balancing chemical formulae.) An important contribution of organic chemistry in the last part of the nineteenth century was to identify higher-level structural components of organic molecules whose addition, removal, or transfer characterized the basic reactions biochemistry would examine.

> By 1900, organic compounds relevant to metabolic processes could be characterized structurally. Characteristic groups such as hydroxyl, carboxyl, and amino groups were linked at specific sites to the carbon backbones of organic molecules. The general classes of reaction mechanisms in which each reactive group participated, and some of the influences of their relative placements on the carbon skeleton upon their reactivities, were known. This information provided strong foundations for interpreting the chemical changes linking any compounds that could be shown to take part in metabolic processes. (Holmes, 1992, p. 56)

Biochemists also needed techniques for demonstrating intermediate reactions involving such structural units. Because the intermediate products generally did not accumulate but were immediately metabolized, it was a challenge to produce empirical evidence of the intermediate steps. Holmes identified Franz Knoop as taking a major step in 1904 when he fed animals compounds similar to normal foods but ones they could not fully metabolize. Knoop found that after feeding dogs specific aromatic derivatives of fatty acids, their urine would contain an aromatic acid that was shorter by an even number of carbon compounds (unless the fatty acid chain contained only one carbon, in which case it was not broken down). He proposed a mechanism in which fatty acids

---

[34] Noting that Kohler had cited Hofmeister (1901) for emphasizing the importance of enzymes, Holmes (1992, p. 58) quoted additional passages in which Hofmeister stated that chemical reactions in cells "must take place in specific intermediate steps" which can be expressed "through chains of physical and chemical formulas."

are decomposed by the removal of two carbon atoms at a time, a process he named *β-oxidation* (Knoop, 1904).[35]

In the first four decades of the twentieth century biochemists made great strides in developing sketches of the mechanisms underlying many metabolic processes. In particular, by 1940 they had developed detailed accounts of the biochemical mechanisms involved in fermentation (glycolysis) and aerobic cellular respiration. The research on fermentation was a singular success story for biochemistry, largely because the reactions occur freely in the cytoplasm and do not depend on cellular organization. With respect to aerobic respiration, biochemists were successful in identifying the major pathways, but their inability to carry out the full process in preparations that did not include membranes suggested a linkage with cell structure that they lacked the tools to investigate. They also faced problems in explaining the generation of ATP from the component operations of respiration and it turned out this step depended critically on the linkage to cell structure.

Alcoholic and Lactic Acid Fermentation (1895–1940)

As noted previously, the orthodox view that fermentation required living yeast was finally undercut by Eduard Buchner in 1897. Buchner had added sugar to what he termed *press juice* (later called *yeast juice* or *cell-free extract*), made by adding water to ground yeast, and then filtering it under pressure. He made his discovery while collaborating with his brother, Hans Buchner, in grinding bacterial cells with the goal of isolating proteins that might be serving as the active bacterial agents. The endeavor was initially obstructed by the inability to remove remnants of cell bodies from the preparation, but in 1896 Hans Buchner's collaborator, Martin Hahn, developed a technique for filtering the cell structure debris from ground yeast. The Buchners added sugar to the resulting juice, thinking it would serve as a preservative, but the resulting formation of carbon dioxide made Eduard Buchner realize that they had achieved something of far more importance: cell-free fermentation.

Initially Buchner took fermentation to be a single chemical reaction, and coined the name *zymase* for the responsible enzyme. In further research, though, Buchner found evidence that "lactic acid plays an important role in

---

[35] Holmes noted that Knoop in this publication set for himself the "ultimate goal of a complete knowledge of the cleavage and oxidative processes of the building materials and foodstuffs of the animal body" (Knoop, 1904, p. 3, translated by Holmes, 1992, p. 59).

the cleavage of sugar and probably appears as an intermediate in alcoholic fermentation" (Buchner & Meisenheimer, 1904, pp. 420–1). This led him to propose that alcoholic fermentation was a two-step process, with zymase catalyzing the reaction from glucose to lactic acid and another enzyme, lactacidase, catalyzing the reaction from lactic acid to alcohol.

The proposal that lactic acid figures in alcoholic fermentation was particularly interesting because it was the product of a different process, that associated with the souring of milk, which Pasteur (1858; 1857) had also investigated and interpreted as analogous to alcoholic fermentation. As well, Emil du Bois-Reymond (1859) had discovered the presence of lactic acid after muscle contraction or after death of an animal. The linkage between lactic fermentation and muscle action was a continuing focus of research in succeeding decades. The linkage was not well-established, though, until Fletcher and Hopkins (1907) showed that the formation of lactic acid during anaerobic muscle contraction was followed by its removal in the aerobic phase.

Lactic acid, however, was soon discredited as an intermediary in alcoholic fermentation on the grounds that adding it to yeast failed to generate alcohol (Slator, 1906). The nature of the relation between alcoholic fermentation and lactic fermentation was not revealed until later. Research seeking intermediaries of alcoholic fermentation turned rather to several three-carbon sugars which had been identified by organic chemists who had attacked glucose with alkalis – methylglyoxal, glyceraldehyde, and dihydroxyacetone. As well, Otto Neubauer, while investigating amino acid metabolism, identified pyruvic acid as an intermediary in that process and proposed it also figured in alcoholic fermentation (Neubauer & Fromherz, 1911). Other researchers quickly corroborated this finding and determined that it was decarboxylated to yield acetaldehyde.

The emerging challenge for biochemists was both to determine which of the possible intermediates figured in the fermentation pathway and to develop a model of the pathway that related those that did occur using only known chemical operations such as oxidations, reductions, and decarboxylations. Carl Neuberg developed a comprehensive model of a sequence of reactions for generating alcohol from glucose (Neuberg & Kerb, 1914). As shown in Figure 3.6, he proposed that glucose was scissioned into two molecules each of methylglyoxal and water. The methylglyoxal then reacted with acetaldehyde (produced in a previous iteration of the process) and water, generating both pyruvic acid and alcohol. The pyruvic acid was then decarboxylated to acetaldehyde. (Neuberg also proposed an alternative route by which methylglyoxal would generate glycerol and pyruvic acid, which could then be decarboxylated to provide the initial quantity of acetaldehyde.) Neuberg's model

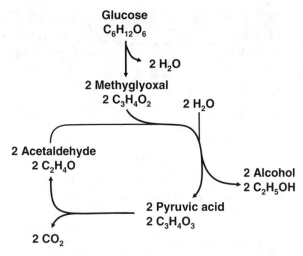

Figure 3.6. A representation of Neuberg and Kerb's proposed pathway of fermentation that brings out how it constituted a cycle.

was the focus of extensive research during the next fifteen years, much of it directed to showing that the proposed intermediates, especially methylgly-oxal, really figured in the pathway. Because intermediates typically would not accumulate, indirect evidence was required. One critical issue was whether the intermediate would react to form alcohol at least as rapidly as glucose. Methylglyoxal failed this test, leading Neuberg to propose that it was an isomer of methylglyoxal that was the true intermediate.

Neuberg's proposal also failed to explain an additional finding concerning fermentation in cell-free extracts. Fermentation in such extracts typically slowed dramatically in a short time. Arthur Harden (1903) established that adding blood serum would produce an 80% increase in fermentation. Together with William Young, Harden further demonstrated that adding phosphate would also stimulate the reaction, which would then slow down again when the phosphate was exhausted (Harden & Young, 1908). They also established that the phosphate appeared to be taken up into a hexosediphosphate ester that itself could not be further metabolized but, as illustrated in Figure 3.7, would slowly decompose through hydrolysis. Neuberg dismissed this evidence, though, on the grounds that hexosediphoshate would not ferment in living cells (Neuberg & Kobel, 1925), thereby happily invoking the same argument strategy whose conclusion he resisted in the case of methylglyoxal. Harden and Young also demonstrated the need for addition of a "dialyzable substance which is not destroyed by heat" to maintain cell-free fermentation (1906, p. 410). Because heat destroyed the enzyme itself, this was obviously an additional substance;

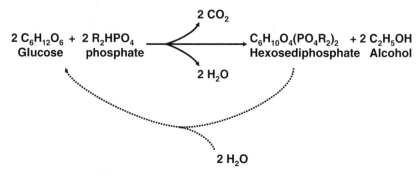

Figure 3.7. Harden and Young's conceptualization of how phosphates might figure in fermentation. Two molecules of glucose would react with two phosphate molecules to create one hexosediphosphate ester and two molecules of alcohol, with the hexosediphosphate slowly breaking back down to glucose and two phosphates, which would then be able to participate in reactions with another molecule of glucose.

they called it a "coferment." Its significance became apparent only in the 1930s (see below).

While researchers investigating alcoholic fermentation were puzzling about the roles of methylglyoxal and phosphates, Gustav Embden was investigating lactic acid fermentation in press juice from muscle. Because adding glucose failed to increase the yield of lactic acid, he proposed that the lactic acid was derived not directly from glucose but from an unknown precursor he designated *lactacidogen* (Embden, Kalberlah, & Engel, 1912). He soon found that adding hexosediphosphate resulted in a large increase in lactic acid and suggested that it was related to *lactacidogen* (Embden & Laquer, 1914; Embden & Laquer, 1921). This work by Embden, together with investigations by Otto Meyerhof (1918) demonstrating that very similar coferments were required in alcoholic fermentation and lactic acid fermentation, pointed strongly to a close connection between the two processes. Meyerhof coined the term *glycolysis* to cover both reaction pathways.

A key to the mysteries surrounding alcoholic fermentation and muscle fermentation was provided by the discovery of two new phosphorus compounds in the late 1920s. The first occurred when Philip Eggleton and Marion Grace Palmer Eggleton (1927) isolated a substance known initially as *phosphagen* and later as *phosphocreatine*. What distinguished this substance was that it rapidly hydrolyzed (i.e., broke down to creatine and phosphate with the consumption of a molecule of water), and released large quantities of energy as it did so. This indicated that it might provide the immediate source of energy for muscle contraction. Eimar Lundsgaard confirmed this hypothesis by showing that in iodoacetate-poisoned rabbits, muscle activity prior

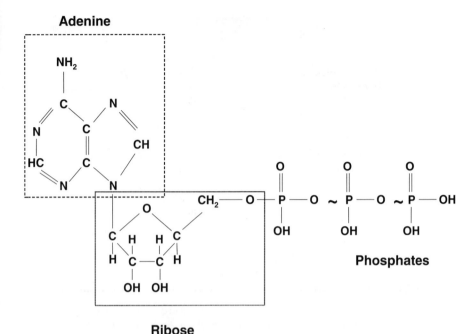

Adenine

Phosphates

Ribose

Figure 3.8. Chemical structure of adenosine triphosphate (ATP). When the high-energy bonds (~) are broken, a phosphate group is released as well as 14 kcal, about twice as much energy as released from an ordinary covalent bond (—).

to the onset of rigor mortis resulted not in accumulation of lactic acid but rather in the breakdown of phosphocreatine. Around the same time, Cyrus Fiske and Yellapragrada Subbarow (1929) and Karl Lohmann (1929) discovered another phosphorus compound initially called *adenylphyrophospate* and now known as *adenosine-triphosphate* (ATP; its chemical structure is shown in Figure 3.8). It too would undergo rapid hydrolysis (losing one phosphate group at a time) with release of considerable energy. Together with Meyerhof, Lohmann demonstrated that ATP was one of the coenzymes contributing to Harden and Young's undifferentiated coferment of yeast juice (Meyerhof, Lohmann, & Meyer, 1931). As well, Vladimir Englehardt (1932) established a linkage between ATP formation and aerobic respiration.[36]

These discoveries of phosphocreatine and ATP pointed to a crucial role for phosphorylated compounds among the products of muscle glycolysis and

---

[36] In studies on pigeon erythrocytes, Englehardt found that when he supplied cyanide, a respiratory inhibitor, ATP rapidly broke down and inorganic phosphate built up, but that ATP was resynthesized after the cells were washed. In non-respiring rabbit erythrocytes, on the other hand, the buildup occurred when he used fluoride to inhibit glycolysis.

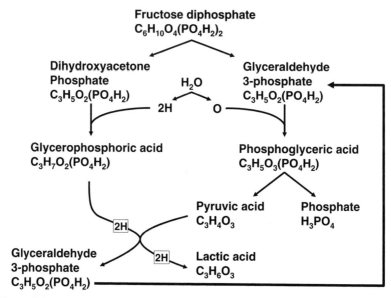

Figure 3.9. A representation of how Embden conceptualized the process of glycolysis as involving phosphorylated compounds throughout. Fructose diphosphate is first scissioned into two triosephosophates, one of which, glyceraldehyde-3-phosophate, is oxidized while the other, dihydroxyacetone phosphate, is reduced. The oxidation product, phosphoglyceric acid, then surrenders its phosphate to produce pyruvic acid. For muscle glycolysis, the product of the reduction, glycerophosphoric acid, is then oxidized at the expense of reducing pyruvic acid to lactic acid, yielding more glyceraldehyde-3-phosphate, which can reenter the reaction at an earlier step.

suggested that Harden and Young's discovery of the need for phosphates in order to maintain fermentation in a cell-free environment might reflect a more fundamental feature of glycolysis. In a paper published posthumously, Embden (Embden, Deuticke, & Kraft, 1933) produced a scheme for muscle glycolysis, illustrated in Figure 3.9, in which phosphorylated forms of several three-carbon sugars served as intermediaries. The key step in the pathway was the oxidation of one of the triosephosphate molecules to 3-phosphoglyceric acid at the expense of the reduction of the other to glycerophosphoric acid. Embden did not discuss the significance of the product of oxidation being phosphorylated, but in the next year Jacob Parnas proposed that the phosphate was not simply liberated, but transferred to phosphocreatine (Parnas, Ostern, & Mann, 1934).[37] (It was later determined that in fact it is transferred to ADP.)

---

[37] Parnas commented, "...the resynthesis of phosphocreatine and adenosine triphosphate is not linked to glycolysis as a whole, but to definite partial processes: and this leads further to the

The accomplishments of Embden and Parnas radically reshaped thinking about alcoholic fermentation and muscle glycolysis, making central the linkages between oxidation and energy by construing the process in terms of phosphorylated compounds and proposing the transfer of phosphate to ADP to form ATP. Several years later Fritz Lipmann (1941) would introduce the symbol ∼P to designate what he memorably called the "energy-rich phosphate bonds" in ATP (p. 101). Already in the 1930s, though, the connections linking oxidation, phosphates, and energy were becoming apparent. In this respect, an important step was Dorothy Needham's proposal of a second esterification of phosphate in fermentation. She noted that even when fluoride poisons blocked the step from phosphoglyceric acid to phosphopyruvic acid (which by then had been identified as an intermediate before the production of pyruvic acid), free phosphate continued to be taken up into an ester. She also observed that in normal fermentation more phosphocreatine was formed per molecule of lactic acid than the single transfer of phosphate from phosphopyruvic acid to ATP could explain. Accordingly, she proposed a second synthesis of ATP (Needham, 1937), and Negelein and Brömel (1939) demonstrated that this involved first the formation of diphospholgyceric acid (3-phosphoglyceroyl phosphate) as the immediate oxidation product of glyceraldehyde 3-phosphate.

One last change resulted in the conception of the pathway as still accepted today. In 1934 Warburg and Christian found another constituent beyond ATP in a coferment preparation made from red blood cells. In 1935 they identified it chemically as containing nicotinic acid amide (Warburg & Christian, 1935). Soon thereafter they characterized it functionally as a "hydrogen-transporting co-ferment" (Warburg, Christian, & Griese, 1935) and proposed the name *triphosphopyridine nucleotide (TPN)* (Warburg & Christian, 1936). The main purpose of this 1936 article, though, was to announce their isolation of a similar coferment with just two, rather than three, molecules of phosphoric acid per molecule of nicotinamide. They proposed the name *diphosphopyridine nucleotide (DPN)*; this key substance has also been known as *coenzyme I, nicotinamide adenine dinucleotide (NAD)*, and, in its oxidized/reduced states, *NAD+/NADH*.[38] Meyerhof then established that it was actually NAD^+, not

---

conclusion that this resynthesis does not involve a relationship that might be termed 'energetic coupling,' but more probably involves a transfer of phosphate residues from molecule to molecule" (Parnas et al., 1934, p.68).

[38] More specifically, the two coenzymes ("co-ferments") discovered by Warburg and Christian are pyridine nucleotides that have in common two phosphate-containing nucleotides (nicotinic amide mononucleotide and adenine flavin dinucleotide); however, DPN lacks an additional

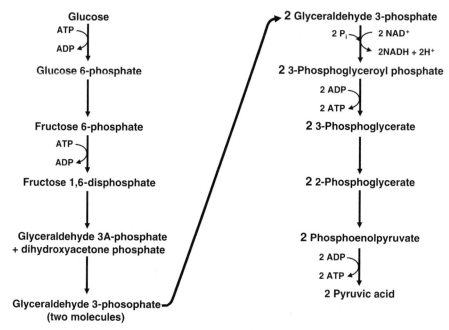

Figure 3.10. The complete Embden-Meyerhof pathway of glycolysis. The reactions in the left-hand column are often characterized as *preliminary*, transferring energy from ATP to products of glucose and splitting the molecule into two molecules of glyceraldehyde-3-phosphate. The reactions in the left-hand column involve the oxidation step and the subsequent transfer of energy to ATP.

dihydroxyacetone phosphate, that was reduced (took up two hydrogen atoms, making NADH + H$^+$) in conjunction with the oxidation of glyceraldehyde 3-phosphate (Meyerhof, Ohlmeyer, & Möhle, 1938).[39] NADH thus took over the role of glycerophosphoric acid in Embden's scheme, serving as the hydrogen donor in the reduction of pyruvic acid to lactic acid in muscle glycolysis and to alcohol in alcoholic fermentation.

With this final contribution of Meyerhof, what is known as the Embden-Meyerhof pathway (sometimes the Embden-Parnas-Meyerhof pathway) achieved its mature form as shown in Figure 3.10. The reactions from glucose

phosphate radical that is found in *triphosphopyridine nucleotide (TPN)*, also known as *coenzyme II*. The modern terms *nicotinamide adenine dinucleotide (NAD)* and *nicotinamide adenine dinucleotide phosphate (NADP)* were adopted as part of an international agreement in the 1960s (see Afzelius, 1966, p. 12, fn).

[39] Only one hydrogen atom is actually added to NAD$^+$; the second dissociates into H$^+$ and an electron, with the electron being incorporated along with the hydrogen atom into NADH.

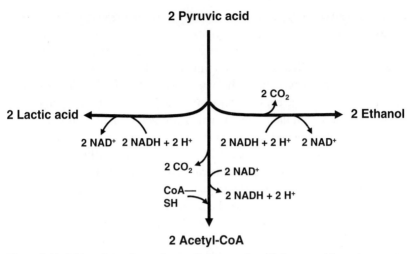

Figure 3.11. Three alternative pathways from pyruvic acid. In anaerobic environments, pyruvic acid is reduced to lactic acid (shown on the left) or to alcohol with the liberation of carbon dioxide (shown on the right). In aerobic conditions, it is decarboxylated, further oxidized, and combined with coenzyme A (CoA-SH) to produce acetyl-CoA, which then enters the citric acid cycle.

to pyruvic acid are common to both alcoholic and lactic acid fermentation and are generally referred to, following Meyerhof's proposal, as *glycolysis*. The pathways diverge in the steps after pyruvic acid (Figure 3.11). Although I have not emphasized the point, researchers along the way named enzymes and typically provided indirect evidence for their operation, so that glycolysis came to be viewed as a complex sequential pathway in which each operation in turn was catalyzed by a specific enzyme and resulted in a product that served as substrate of the next operation. As such, glycolysis became the exemplar of how to understand physiological processes in biochemical terms.

Aerobic Cellular Respiration (1910–1940)

In the nineteenth century, researchers made a sharp distinction between fermentation, which occurred under anaerobic conditions, and aerobic cellular respiration, a process that was often identified with protoplasm and was the focus of speculative accounts by researchers such as Pflüger and Nägeli. As biochemists turned their attention to the oxidation of foodstuffs to carbon dioxide and water, two models competed for attention. Both recognized that reactions between oxygen and foodstuffs would not occur under ordinary atmospheric conditions and that enzymes must be responsible. The Thunberg-Knoop-Wieland model emphasized reactions in which enzymes operated on a substrate to release hydrogen atoms (which would then combine with molecular oxygen, were it available). The other model, developed by Otto Heinrich Warburg, construed the enzyme (ferment by his terminology) as operating on oxygen, which would then combine with hydrogen from the substrate.

Heinrich Otto Wieland was an organic chemist who began his exploration of oxidation using inorganic catalysts such as palladium black. He found that when no oxygen was present, the catalyst would operate for a short time before it became saturated with hydrogen. He proposed that when oxygen was available, it served only to receive the hydrogen removed from the substrate. He substantiated this proposal by showing that methylene blue, a synthetic dye that is readily reduced to leuco-methylene blue, could substitute for oxygen in maintaining the reaction (Wieland, 1913). He proposed the following schema for oxidation of a compound RH_2 to R (Pd designates palladium, the catalyst, and Mb methylene blue, the dye):

$$RH_2 + Pd \rightarrow R + PdH_2$$

$$PdH_2 + Mb \rightarrow Pd + MbH_2.$$

When a substrate such as an aldehyde (RCHO) lacked two removable hydrogen atoms, he proposed that it was first hydrated (a water molecule was added) and then two hydrogen atoms were removed:

$$RCHO + H_2O \rightarrow RCH(OH)_2 \rightarrow RCOOH + H_2.$$

Wieland's scheme established a new conceptualization of oxidation as dehydrogenation – the removal of hydrogen rather than the addition of oxygen – and of its relation to reduction. Since the hydrogen released by the substance being oxidized had to be accepted by another substance that was thereby reduced, he proposed that these reactions were necessarily coupled as "two expressions of one process of dehydrogenation" (p. 3340). Wieland then extended this account to biological oxidation. He proposed that oxygen, when present, was the substance reduced in cells, yielding hydrogen peroxide (H_2O_2) that the enzyme catalase would quickly convert to water. Importantly, he showed that when oxygen was not present, oxidation could still occur in biological entities. Specifically, ethanol and acetaldehyde could be oxidized to acetic acid in bacteria if methylene blue was available as a hydrogen acceptor.

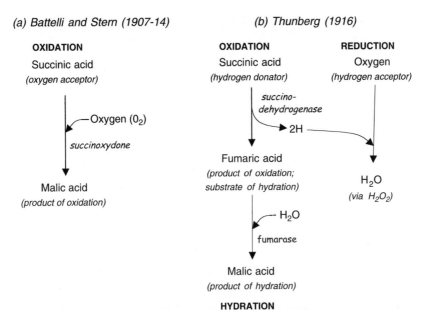

Figure 3.12. Two early accounts of the metabolism of succinic acid to malic acid. They differ in how oxidation is conceptualized and in the number of reactions involved. (a) Batteli and Stern assumed that activated oxygen is added directly to the substrate, succinic acid. (b) Thunberg identified fumaric acid as an intermediate substance between succinic and malic acids; incorporated paired reduction-oxidation reactions, as proposed by Wieland; and posited specific enzymes for which better evidence later emerged. It was also determined later that the carrier coenzyme FAD, not molecular oxygen, was the immediate hydrogen acceptor.

Wieland's proposal received little uptake from biochemists until it was further developed by Thorsten Ludvig Thunberg, who recognized that it could be applied to the experimental results that Federico Battelli and Lina Stern (1911) had obtained concerning the oxidation of succinic acid. Their approach was to add metabolic substrates to suspensions of minced animal tissue presumed to include enzymes of interest, and use manometers to supply oxygen gas and measure how much was absorbed during the resulting oxidation reactions. As illustrated in Figure 3.12(a), they interpreted their findings on succinic acid in terms of an enzyme that *activated* the dissolved oxygen so that it would combine with the substrate (succinic acid) to yield the observed product (malic acid).[40] Subsequently, however, Hans Einbeck (1914) demonstrated that fumaric acid was created in the process:

---

[40] The named the enzyme *succinoxydone*, using an alternative to the traditional ending – *ase* since this enzyme was not soluble.

$$\text{HOOC—CH}_2\text{—CH}_2\text{—COOH} \rightarrow \text{HOOC—CH=CH—COOH}$$

Succinic acid   Fumaric acid

$$\rightarrow \text{HOOC—CH}_2\text{—CH(OH)—COOH}$$

Malic acid

As illustrated in Figure 3.12(b), Thunberg recognized that succinic acid was first oxidized (dehydrogenated) by the removal of two hydrogen atoms, yielding fumaric acid, which then gained from water two hydrogen atoms plus an oxygen atom, yielding malic acid. Thus, on this view the additional oxygen found in malic acid came not from molecular oxygen but from the water molecule that was added to fumaric acid. Molecular oxygen did not figure in the oxidation reaction directly but was the recipient of the liberated hydrogen and so reduced to water. On this revised account there were three main reactions: (1) oxidation of succinic acid, (2) hydration of fumaric acid, and (3) reduction of oxygen. To show that oxidation reaction could occur in animal tissue independently of the reduction of oxygen, Thunberg adopted Wieland's strategy of using methylene blue rather than molecular oxygen as the hydrogen acceptor. To conduct these experiments, he created a special device (later known as the *Thunberg tube*) from which all oxygen was removed. When a minced muscle preparation and methylene blue were placed in the tube, nothing happened; however, when succinic acid was added, the methylene blue was reduced, as indicated by its rapidly decolorization (Thunberg, 1916). He subsequently showed that the same was true of various other organic acids (lactic acid, fumaric acid, malic acid, citric acid, etc.): They were dehydrogenated when methylene blue was supplied. Thunberg (1920) determined that reactions involving different substrates were differentially affected by heat, which he took as evidence that different enzymes were responsible for each reaction.

Warburg, in contrast, construed a reaction with oxygen as the key step in oxidation and argued that the *activation* of oxygen, which prepared it to combine with substrates, was the major catalytic event (thus employing the same model as Battelli and Stern).[41] In his earliest work Warburg focused on cell membranes, whose existence had finally been convincingly demonstrated by

---

[41] Given his focus, Warburg's challenge was to measure the amount of oxygen taken up in a reaction. In his earliest research he used titrimetric methods but during a visit to Joseph Barcroft's laboratory at Cambridge in 1910 he observed the Haldane-Barcroft blood-gas manometer, a device which kept the volume of the gas constant so that pressure would change as gas was added or lost. Warburg modified their instrument to measure rates of gas exchange and used it in most of his subsequent research. The resulting device came to be known as the Warburg manometer and has been widely used in biochemical studies of reactions involving gas exchanges, such as oxidations (Warburg, 1923; Warburg, 1925a).

Ernest Overton (1899). Overton had shown that the membrane, comprised largely of fats and lipids, served as a semipermeable osmotic barrier between a cell and its environment; he also had investigated the ability of hundreds of organic solutes to cross cell membranes. Working with sea urchin eggs, Warburg (1910) pinpointed the membrane as the site of respiration by showing that alkaline solutions increased the respiration of sea urchins without altering the alkalinity of the protoplasm, and that fatty acids and organic solvents, which affected membranes, decreased respiration. Warburg (1913b) subsequently investigated the effect of narcotics such as ethyl urethane on respiration, and from this research eventually concluded that it was not the lipids of the membranes that they affected, but solid particles in the protoplasm which turned out to be mitochondria. Even as the details of his account shifted, a constant theme at this stage in Warburg's career was his opposition to the view that soluble enzymes alone were responsible for significant biological processes – he maintained that they operated in context of structured systems (Warburg, 1913a).[42]

It was in this context of emphasizing that enzymes operated in structured systems that Warburg first introduced the term *Atmungsferment* to designate the agent responsible for biological respiration. He viewed it as operating on oxygen, activating it so that it would combine with hydrogen in the substrate undergoing oxidation. Working with Meyerhof, Warburg connected the effect of citric acid and tartaric acid in halting respiration in sea urchin eggs to their ability to chelate (form ring structures with) heavy metals. He concluded "that the oxygen respiration in the egg is an iron catalysis; that the oxygen consumed in the respiratory process is taken up initially by dissolved or absorbed ferrous ions" (Warburg, 1914, pp. 253–4, translated in Fruton, 1972, p. 302). Now Warburg proposed that *Atmungsferment* consisted of ferrous iron adsorbed onto the membranes of the cell. Warburg also developed model systems of activated charcoal or pyrolised (heat-transformed) blood in which to study the reaction.

---

[42] Warburg also criticized Buchner's characterization of zymase as a soluble enzyme. He interpreted the slowing of fermentation in cell-free extracts identified by Harden and Young as indicating the destruction of the membrane which was critical to the normal operation of the ferment. At this point, Warburg insisted that both cell structure and enzymes (for him, ferments) were required: "The question always comes to this: cell action or ferment action? Structure action or ferment action? I hope I have demonstrated to you today that there is no dichotomy here at all: both ferment chemists and biologists are right. The acceleration of energy-producing reactions in cells is a ferment action *and* a structure action; it is not that both ferments *and* structure accelerate, but that *structure accelerates ferment action*" (1913a, pp. 20–1, translated in Kohler, 1973, pp. 189–90).

At this point, military service during World War I interrupted Warburg's research. When he resumed it, he characterized *Atmungsferment* more like an enzyme and focused on how it could be inhibited, first by hydrogen cyanide (Warburg, 1925b) and later by carbon monoxide (Warburg, 1929). Although he could not isolate the enzyme, he developed techniques to fingerprint it by its absorption spectrum. Throughout this period, though, he emphasized that the critical operation involved the activation of oxygen. He fiercely opposed Wieland's contention that removal of hydrogen atoms was the crucial step in oxidation and that dehydrogenases operated on the substrate to activate and remove pairs of hydrogen atoms.

The competing theories of Wieland and Warburg each offered a simple mechanism to explain cellular respiration – one operation activated a molecule, which then reacted with the other molecule (removing it from the substrate if necessary). Their alternative accounts became the focus of a bitter controversy as each party maintained that he had identified *the* critical operation. However, other researchers began to consider the possibility that both processes were involved in cellular respiration. As proposed by Albert Szent-Györgyi (1924), "In cellular oxidation the *activated hydrogen* is burned by the *activated oxygen*. In the terminology of the hydrogen activation theory this means that molecular oxygen is not a hydrogen acceptor; the biological hydrogen acceptor is the oxygen activated in Warburg's system" (Szent-Györgyi, 1924, p. 196, translated by Fruton, 1972, p. 322). On this proposal, the claims of Wieland and Warburg did not intrinsically conflict but instead, as Figure 3.13 shows in skeletal form, figured at opposite ends of the same pathway.

As it turned out, however, the mechanism was far more complex than one obtained by simply combining the two proposals. Working from both ends of the pathway, researchers in the 1920s and 1930s identified a multitude of operations that linked them into a complex mechanism. (Bechtel and Richardson, 1993, characterized this as a move from *simple localization* to *complex localization*.) Thunberg himself took the first step by proposing the idea of a sequence of dehydrogenation reactions in which the product of a given dehydrogenation was further dehydrogenated (or otherwise operated on) in another reaction.[43] In particular, he proposed the following

---

[43] Thunberg actually had the idea of a sequences of reactions as early as 1913, before he encountered Wieland's conception of dehydrogenation: "The oxidative processes in the living cell must be thought of as forming chain reactions, a series of reactions connected to one another in such a way that, by and large, none of the links in the reaction chain can proceed more rapidly than the others" (1913, translated in Holmes, 1986, p. 68).

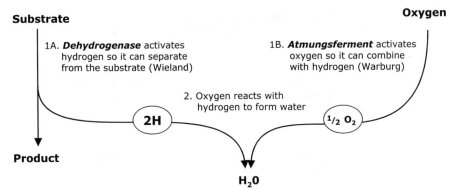

Substrate　　　　　　　　　　　　　　　　　　　　　　　**Oxygen**

1A. ***Dehydrogenase*** activates
hydrogen so it can separate
from the substrate (Wieland)

1B. ***Atmungsferment*** activates
oxygen so it can combine
with hydrogen (Warburg)

2. Oxygen reacts with
hydrogen to form water

2H　　　　　　　　　　**$1/2 \ O_2$**

Product

H_2O

Figure 3.13. Schematic account of how Wieland's account of respiration in terms of dehydrogenation could be linked to Warburg's proposal of an enzyme acting on molecular oxygen.

reaction pathway:

Succinic acid → fumaric acid → malic acid → oxaloacetic acid
→ pyruvic acid → acetic acid.

He then faced a problem in specifying what happened next – it was not possible to remove two hydrogen atoms from acetic acid. In response to this problem, Thunberg offered a bold proposal – he proposed "a reaction in which two acetate molecules are simultaneously each deprived of one hydrogen atom, with the joining of their carbon chains into one. The substance which must therefore form is succinic acid" (1920, passage translated by Holmes, 1986, p. 69). The reaction Thunberg proposed was the following:

$$2CH_3-COOH \rightarrow COOH-CH_2-CH_2-COOH + H_2.$$

In effect, Thunberg had proposed the cycle of reactions illustrated Figure 3.14.[44]

Fifteen years later Hans Krebs, together with William Johnson, incorporated the core of this idea into the citric acid cycle (also known as the *tricarboxylic acid cycle* and later as the *Krebs cycle*). Instead of two molecules of

---

[44] A related idea was advanced by Albert Szent-Györgyi (1937) on the basis of his studies adding different four-carbon dicarboxylic acids to suspensions of minced pigeon-breast muscle. He proposed a scheme in which the four-carbon acids are viewed as performing hydrogen transport, not steps in oxidation of carbohydrate:

Hydrogendonor → oxaloacetate → malate → fumarate → succinate
→ cytochrome → oxygen

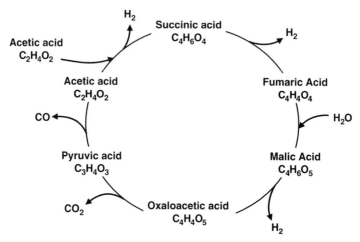

Figure 3.14. Cycle of reactions proposed by Thunberg.

citric acid combining to form succinic acid, Krebs proposed that oxaloacetic acid combined with a three-carbon substance that Krebs temporarily called *triose* to generate citric acid, with the citric acid then undergoing a sequence of reactions resulting in succinic acid (Krebs & Johnson, 1937). (See center part of Figure 3.15.) It was soon realized that pyruvic acid (pyruvate), a product at the end of glycolysis, provided the material that would react with oxaloacetic acid. Figuring out the exact linkage provided a bit of a challenge. After Fritz Lippman (1945) discovered coenzyme A, evidence began to develop that it figured in the connection. Feodor Lynen initiated research attempting to show that acetic acid figured in the pathway between pyruvic acid and citric acid, but because ordinary acetic acid would not condense with oxalacetic acid to create citric acid, he speculated that *activated acetic acid* must be involved. Lynen and Richert (1951) demonstrated that the activated acetic acid was a thio (sulfur) ester of acetylated coenzyme A, a compound now known as acetyl-CoA. With this account of the connection, the citric acid cycle was linked to the pathway of glycolysis (as well as the pathways of fatty acid metabolism and protein metabolism).

Another key advance, working from the oxygen end of the overall process, came from a very unlikely source. David Keilin was studying the respiration of the parasite horse bot-fly (*Gasterophilus intestinalis*), when he detected a disappearance of hemoglobin in later stages of metamorphosis from the pupa to adult fly stage. Spectroscopic examination of flies that died in captivity revealed a pigment with four distinct absorption bands. Keilin considered the possibility that the pigment originated from the larval hemoglobin,

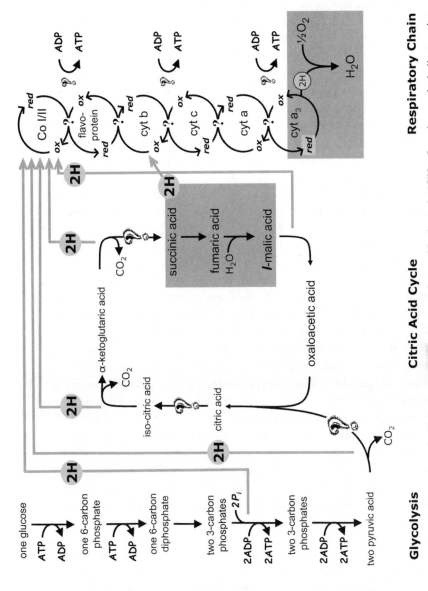

Glycolysis　　　　**Citric Acid Cycle**　　　　**Respiratory Chain**

Figure 3.15. Overview of cellular respiration as it was understood in the early 1940s. Question marks indicate points at which researchers realized there were steps they did not understand.

113

but ruled that out. He then began to look in a wide range of other species and continued to find the same four-banded absorption spectrum. While examining a yeast preparation, he observed that immediately after shaking it, he failed to find the absorption pattern. Yet, "before I had time to remove the suspension from the field of vision of the microspectroscope, the four absorption bands suddenly reappeared" (Keilin, 1966, p. 145). Searching the published literature, Keilin discovered that Charles MacMunn had made the same finding in the 1880s (MacMunn, 1884; MacMunn, 1886). Keilin (1925) went beyond MacMunn's observation in determining that these bands came from three different hemochromogens which he labeled *cytochrome a*, *b*, and *c*. Each cytochrome itself was composed of at least one protein with an iron-porphyrin prosthetic group, with the iron atoms accounting for the reversible reactions. By 1939 Keilin, collaborating with Edward Hartree, had distinguished cytochrome a_3 from a and characterized the four cytochromes as forming a catalytic chain "which, by utilizing molecular oxygen, can easily oxidize to water certain hydrogen atoms in the substrate molecules activated by specific dehydrogenase systems" (Keilin and Hartree, 1939, p. 190). They named the enzyme catalyzing this oxidation *cytochrome oxidase*. Based in part on their identical absorption spectra, Keilin and Hartree identified cytochrome oxidase with Warburg's *Atmungsferment*. They also tentatively identified it with cytochrome a_3 in particular, but commented that the b-c-a-a_3 system could be regarded as either three hydrogen carriers plus an enzyme or as a four-component chain of enzymes (Keilin & Hartree, 1939). In the 1940s, this reaction sequence was referred to as the *respiratory chain*. Later, the term *electron transport chain* became favored as biochemists discovered additional components and also wished to emphasize that it was pairs of electrons – dissociated from the protons of the hydrogen atom – that were transported down an energy gradient to molecular oxygen.

In his 1932 Nobel lecture Warburg (1932) insisted that *Atmungsferment* was *the* active agent and resisted the claim that it was the cytochromes that were oxidized by activated oxygen. Nonetheless, during this period Warburg made two critical discoveries that filled in the steps between oxidation (dehydrogenation) of substrates and the oxidation-reduction reactions of cytochromes in the electron transport chain. I noted previously that in the mid-1930s Warburg and Christian discovered two coenzymes, which are now called NAD and NADP. In addition, they established that both NAD and NADP figure in connecting the oxidative reactions in the citric acid cycle to the electron transport chain. It took a number of years, though, for investigators to work out which of these two coenzymes participated in which reactions.

In the same period Warburg and Christian (1932) had also identified a "yellow ferment" which they characterized as "oxygen-transporting." The prosthetic group (coenzyme) turned out to be flavin mononucleotide (FMN). Warburg and Christian (1938) then identified yet another yellow enzyme whose coenzyme, flavin adenine dinucleotide (FAD), consists of FMN plus adenylic acid. These coenzymes are closely related to vitamin B_2, the vitamin whose deficiency gives rise to pellagra. As with coenzymes I and II, there was considerable confusion for a number of years as to their exact functions and which was the prosthetic group of which flavoprotein. Investigators focusing on the electron transport chain in the 1940s posited two different pathways converging at cytochrome *b*. One coupled the oxidation of most citric acid cycle substrates to the reduction of NAD ($NAD^+ \rightarrow NADH$) and, in turn, its oxidation to the reduction of a flavoprotein. The other pathway, based on the fact that succinic acid was the only citric acid cycle intermediate whose oxidation did not produce NADH, coupled the oxidation of succinic acid directly to the reduction of a flavoprotein. It was thought that each of these flavoproteins was then oxidized, coupled with the reduction of cytochrome *b*. Eventually it was determined that FMN was the flavoprotein on the pathway from NADH to cyctochrome *b*, whereas FAD played a previously-unsuspected role in the second pathway: the oxidation of succinic acid was coupled to the reduction of FAD, which carried the electrons released by that oxidation into the electron transport chain. Finally, the isolation of ubiquinone, coenzyme Q, from beef heart mitochondria (Crane et al., 1957) led to the recognition that it is an additional substance undergoing reversible oxidation-reduction between both FAD and FMN and cytochrome *b* (see Chapter 6).

With the discovery of the steps involved in the citric acid cycle and the electron transport chain, the oxidation pathway from pyruvic acid to water was complete. One aspect of the process, however, was still not addressed – how the energy released in the oxidative reactions was captured in ATP. Hermann Kalckar (1939) and Fritz Lipmann (1939) both demonstrated that the oxidation of intermediaries such as succinate, malate, and pyruvic acid was accompanied by creation of ATP. Because different substrates all resulted in synthesis of ATP, this suggested that the synthesis occurred in conjunction with the oxidation reactions along the respiratory chain linking dehydrogenation of the substrate with molecular oxygen. Such ATP synthesis came to be known as *oxidative phosphorylation* to contrast it with the phosphorylation of ADP directly linked with metabolic intermediates, a process that was referred to as *substrate phosphorylation*. Both Ochoa (1940) and Belitzer and Tsibakowa (1939) showed that in oxidative phosphorylation more than one ATP molecule was formed per atom of oxygen reduced. By focusing on phosphate consumed per molecule of lactate oxidized in cat heart extracts, Ochoa argued that in fact three molecules of ATP were formed per oxygen atom reduced (Ochoa, 1943). This P:O ratio was eventually accepted after considerable conflict.

The research on cellular respiration had been remarkably fruitful in the earlier decades of the twentieth century, with major advances achieved in the 1930s. Researchers had figured out the basic schema of the overall process, but also recognized many gaps in their knowledge. Figure 3.15 shows the understanding of the component mechanisms of glycolysis, the citric acid cycle, and electron transport as they were understood in the early 1940s, with question marks indicating points at which investigators recognized important gaps that still needed to be filled in.

The State of Biochemistry circa 1940

Biochemistry not only came into its own in the first four decades of the twentieth century but made impressive advances in understanding chemical processes in the cell. The Embden-Meyerhof pathway provided a comprehensive mechanism for glycolysis, and together the citric acid cycle and electron transport chain provided a very detailed sketch of the mechanism of aerobic cellular respiration. However, two problems were emerging for filling this in. First, researchers faced a challenge in explaining the linkage between the oxidation operations and phosphorylation. Based on the example of fermentation, they assumed that intermediate phosphorylated compounds formed along the electron transport chain that would transfer a high-energy phosphate bond to ADP. As I will discuss in Chapter 6, such compounds were never found. Second, biochemists were unable to carry out the entire reaction in a cell-free extract, a critical step if they were going to satisfy the biochemical standards for successful understanding of a biochemical reaction. That required isolation of the individual components (enzymes, cofactors, etc.) and resynthesis of a functioning system from them.[45] The problem, as Keilin and Hartree (1940) discovered, was that membranes seemed to be required in any preparation that carried out oxidative phosphorylation. The recognition of the role of membranes in oxidative phosphorylation was important, because

---

[45] Twenty-five years later Efraim Racker notes the failure to satisfy this demand as a problem facing the biochemistry of oxidative phosphorylation: "The mechanism of energy production in mitochondria has long defied analysis, since a complex chemical pathway in a living organism cannot really be understood until its intermediate products have been identified and the enzymes that catalyze each step of the process have been individually resolved as soluble components" (1968, p. 32).

it pointed to the need for a bridge to a study of cell structures. Making this connection, however, required appropriate techniques to study cell structures at the appropriate level of organization, which were only in the process of being developed.

3. THE NEED TO ENTER THE TERRA INCOGNITA BETWEEN CYTOLOGY AND BIOCHEMISTRY

In this chapter, I have focused on what cytology and biochemistry were able to contribute to an understanding of how cells performed several important functions of life prior to 1940. After a promising start in the late nineteenth century, cytology stalled in the twentieth century. The chief problem was that cytologists' main tool, the light microscope, could not reveal the details of the internal structures in cell cytoplasm. Beyond the rather crude attempts to correlate structures with activities cells were performing, cytologists on their own could not address the function of these structures. Working at a lower level of organization, biochemists were making great progress in unraveling chemical mechanisms involved in cell processes such as fermentation and cellular respiration. Their main strategy for securing preparations for their investigations, homogenation, destroyed any cell structure, making contract with cytological research difficult.

By 1940, several researchers from both cytology and biochemistry recognized the potential for linking their inquiries. The first edition of Geoffrey Bourne's *Cytology and Cell Physiology* in 1942 bears witness to this desire. Nonetheless, there remained unexplored territory between reconstituted chemical pathways and what could be observed through the light microscope about the structure and behavior of cells. There had been pioneering efforts, especially in the developing fields of cytochemistry and histochemistry, that gave hope of localizing processes within cells and understanding just how structure and function related. But truly productive exploration of this territory required new tools. Two tools that were to be immensely important in exploring this territory were being applied to cells for the first time around 1940 – the ultracentrifuge and the electron microscope. Each gave rise to a problem already noted several times in this chapter: Were the results obtained with the new instruments providing real information about cells or were they merely producing artifacts? This is the topic of the next chapter.

4

Creating New Instruments and Research Techniques for Discovering Cell Mechanisms

> ... is our method of fractionation like the clumsy undertaking of a
> car mechanic who attempts to use his crude tools to analyze a watch?
> I believe that it is almost as bad as that. Nevertheless, we have no
> alternative and must hope that our tools will become refined as we
> proceed in the analysis. Meanwhile we have to look out for the signs that
> guide us in the right direction; we must try to correlate the experimental
> findings obtained with cell-free systems with the complex physiology
> of the cell; we must keep in view the metabolic 'Gestalt' of the cell; and
> finally, we must 'seek simplicity and then distrust it.'
>
> (Racker, 1965, p. 89)

"Seeing is believing." So we are often told. You do not, however, have to go far to find instances in which seeing is misleading. For example, look at Figure 4.1 and ask yourself whether the shaded surfaces of the two figures are identical in shape. If you are like most people, you will judge that the shapes differ – one long and narrow, the other closer to square. However, you can convince yourself that they actually have identical dimensions if you rotate one of them 90 degrees and measure the corresponding sides. Your visual system misled you in this case about what actually exists in the world, creating what scientists refer to as an *artifact* (or *artefact*, especially in older publications). Although the term *artifact* is used in ordinary parlance for anything made by humans, in science it denotes evidence produced by instruments and experiments that does not properly reflect the phenomenon under investigation.

Although we usually regard seeing as giving us direct access to what is nearby in the world, it may be better to think of our visual system as an instrument for guiding action and producing judgments about what we are seeing. Research over the past two centuries has revealed some of the complex brain operations that enable our visual system to perform these functions

118

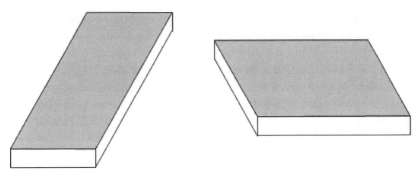

Figure 4.1. Are the shaded surfaces identical in shape? Adapted from Roger N. Shepard (1990), *Mind Sights*. New York: Freeman, p. 48.

(Bechtel, 2001) but most of us are blissfully unaware of these. Fortunately, our visual system is very well adapted to the environment in which we live; accordingly, illusions as dramatic as the one generated by Figure 4.1 are rare.

Like our visual system, the instruments used by scientists procure information by performing complex operations on phenomena of interest. They are thus prone to generating artifacts. Because, unlike the visual system, these instruments have not been honed by natural selection over a long phylogenetic history, the risk of artifact is acute. As a result, evidence advanced using new instruments and new techniques for using existing instruments[1] often is vigorously contested. Competing scientists question whether the evidence really reflects the phenomena of nature. The controversies surrounding new instruments and techniques are often the most bitter in science. Eventually they dissipate. The instruments and techniques for using them are refined so that the community of scientists agrees they are producing reliable information about the structure of the phenomena. Scientists publish manuals of standardized procedures and the community of investigators employs common tools for procuring evidence.[2] The instruments and techniques for using them become, in the language of Latour (1987), black boxes.

[1] We typically are unaware not only that our senses are instruments but that there are procedures for using them. There are exceptional contexts in which we must train ourselves in the procedures for seeing particular objects – for example, to see an afterimage by first staring at a colored patch and then shifting to looking at a white wall. But with scientific instruments, the techniques for using them are often as important as the instrument itself in determining what evidence is produced. To keep this in focus, I usually will distinguish between instruments and the techniques for their use.

[2] In biochemistry, for example, volumes in the series *Methods in Enzymology* began to be published in 1955 under the editorship of Sidney Colowick and Nathan Kaplan. In 1964 an annual publication, *Methods in Cell Physiology*, began to appear under the editorship of David M. Prescott. In 1973 the title changed to *Methods in Cell Biology* and has subsequently expanded to include multiple volumes per year.

How is such agreement reached? Naively, one might think such agreement results from a detailed understanding of how the instruments and techniques operate. But, just as most of us do not understand the operation of our visual system, scientists frequently do not understand how their instruments and techniques work, at least at the level of detail that is required to counter concerns about artifacts. Golgi staining is an extreme but illustrative example. Camillo Golgi introduced the silver nitrate stain in the 1880s and, in the hands of Ramon y Cajal, it provided much of the evidence that established that neurons are discrete cells. What made it effective is that it stains only some of the neurons in a preparation, thereby making clear that the dendrites and axons of one neuron are not continuous with those of another. Yet, more than a century later scientists still do not understand why it stains only a few neurons in a preparation. Even though in most cases scientists do figure out how their instruments and techniques operate, much of that knowledge may not be obtained until long after the instruments and techniques have been put into common use. Moreover, as Ian Hacking pointed out, changes in the understanding of how the evidence is procured has little effect on its acceptance:

> Visual displays are curiously robust under changes of theory. You produce a display, and have a theory about why a tiny specimen looks like that. Later you reverse the theory of your microscope, and you still believe the representation. Can theory really be the source of our confidence that what we are seeing is the way things are? (1983. p. 199)

These reflections pose an important question: how do scientists assess their instruments and techniques to determine whether they are providing information about the phenomena of interest or merely artifacts? This is a question about what I will call *the epistemology of evidence*. I will be exploring that question in this chapter in the context of discussing the instruments and research techniques that proved critical to the ability of cell biology to enter into the no-man's-land between cytology and biochemistry in the 1940s and beyond.

As the previous chapter demonstrated, before 1940 cytology and biochemistry had developed largely in isolation from each other. Many researchers in both fields recognized the potential for fruitful interaction, but made little headway due to the lack of research tools that could yield detailed images of cytoplasmic structures and help relate them to biochemical functions. By 1940, though, researchers were employing new instruments and developing techniques for using them toward these ends. As often happens, the instruments with the greatest impact had been developed in other fields of science. Biologists therefore had to develop specialized techniques for turning these instruments to their own use. Two techniques were especially successful in opening up the territory between biochemistry and cytology to exploration: (a) obtaining micrographs showing the fine structure of cells using the electron microscope; and (b) chemical analysis of cell fractions separated by means of the ultracentrifuge. This chapter will focus on development of these techniques in the 1940s and 1950s, focusing on the epistemic issues raised by their introduction (beyond those that had already been resolved in the fields in which the instruments had been developed). The surge of discoveries and new understandings of cell mechanisms made possible by these new instruments and techniques will then be the focus of Chapters 5 and 6.[3]

1. THE EPISTEMOLOGY OF EVIDENCE: JUDGING ARTIFACTS

I used a compelling visual illusion to demonstrate that the concern with artifacts arises even before we turn to the use of instruments to procure evidence in science. The problem is much accentuated when instruments and the techniques for using them require manipulating and altering the phenomenon under investigation. Sometimes, as amply illustrated in cell fractionation, the manipulations and alterations are severe. A major goal of cell fractionation is to pinpoint the particular organelle in which each enzyme is found and, therefore, determine the biochemical reaction associated with the organelle. Using cell fractionation to pursue this goal requires radically disrupting the cell. Force is applied to break its tough plasma membrane, and the contents are dispersed into a medium different from that in living cells. The contents are then subjected to forces several thousand times that of gravity in the centrifuge. The underlying assumption is that different cell constituents will settle in order of their mass, with the components of greatest mass settling first. Typically, the process involves several iterations in which the contents may be resolubilized in yet different media and subjected again to centrifugation. (Figure 4.3 in Section 2 illustrates a common version of the procedure.) Each iteration generates what is called a *fraction* of the original material that is then assayed for the reactions it supports and hence what enzymes it contains.

---

[3] Additional instruments and techniques that will be important at various points in the historical analysis that follows include techniques of histochemistry and cytochemistry, which used various stains to detect the location and sometimes quantify the amounts of DNA, RNA, and certain enzymes; the use of radioactive tracers to follow the migration of substances through the cell; various forms of spectroscopy that enabled the detection of reaction products; and several new types of microscopes, including phase contrast and fluorescent microscopes. A number of investigators also explored techniques for micromanipulation or microsurgery, which allowed them to dissect parts of cells, remove parts, and inject substances into them.

Other evidence is used to associate the fraction with a particular organelle. The overall procedure supports an inference as to what reactions and enzymes are located in which organelles. However, given the rather violent treatment the material undergoes, one must consider the possibility that it is the violent treatment itself that partly or wholly determines where material ends up. If so, the results are at least in part artifactual and do not provide clear evidence as to where the enzymes originated.

Because centrifugation is inherently a violent and disruptive process, one might wonder whether cell fractionation is exceptional in raising worries about artifacts. In the next section, I will show that many of the concerns target the procedures needed to prepare cells for centrifugation rather than the centrifugation procedure itself. For now it will suffice to indicate that many of the same kinds of issues arose even with electron microscopy, which superficially may seem to be a benign observational technique. The direct act of looking at a micrograph conceals what was done to generate the micrograph in the first place. In their handbook on electron microscopy, Mercer and Birbeck emphasized that in making a micrograph, a researcher is creating something artificial. The challenge is to make something artificial that can be traced sufficiently to the characteristics of the original specimen to support conclusions about that specimen:

> The final micrograph contains elements contributed both by the original object and by the preparative technique applied to it. An important part of the interpretation of the micrograph thus turns on a consideration of these latter factors. There are three distinct successive steps to be considered in the preparation of biological materials for microtomy: fixation, dehydration and embedding. To these may be added a fourth, that of staining, which may however be carried out after sections have been cut. To obviate sterile discussions concerning 'artefacts', the electron microscopists should recognize at once the nature of these operations. Their object *is* the preparation of an artefact of a type which can be examined in the electron microscope, and which bears a sufficiently well-understood relation to some structure originally present in an organism to enable this structure to be deduced from an electron micrograph of the artefact. The interpretation of the micrograph thus rests on an analysis of these steps and requires an understanding of the physical and chemical effects of each. (Mercer & Birbeck, 1972, p. 4–5)[4]

---

[4] The concern with artifact is not limited to electron micrographs but, as I had occasion to note several times in the previous chapter, was widespread for light microscopy as well. Robert Bensley identified both the reasons for suspecting artifacts and instances in which critics voiced the concern, sometimes in ways that turned out to be correct, sometimes not, in the history of microscopy: "The methods used by cytologists in the preparation of cells for microscopic

In this passage, Mercer and Birbeck tried to shift the discussion of artifacts to the more common sense in which an artifact is anything made by humans. The concern, however, as they make clear, is whether the micrograph can be interpreted so as to obtain good information about cell structure. That is, is the micrograph a reliable source of evidence or a *mere* artifact from which it is impossible to recover the structure of the cell with which the investigator began?

The concern with artifacts was widespread when cell fractionation and electron microscopy were first being introduced in the 1940s and 1950s. Writing at the close of the 1950s, at which point the techniques had acquired general acceptance, Alex Novikoff referred to a number of concerns about artifacts that had been prevalent:

> We can all recall the categorical assertion that hope for significant information concerning the *in vivo* function of subcellular particles was lost the moment the cell was disrupted or homogenized. Or, that the use of aqueous media like sucrose could yield only misleading particles, especially worthless nuclei. Or, that oxidative phosphorylation could never be retained once the organized structure of the complete mitochondrion was broken. Or, that electron microscopy was one huge blunder, based as it was on osmium artifacts. Or, that quantitative microspectrophotometry of stained tissue sections was deprived of meaning by the marked structural heterogeneity of the subcellular particles. Or, that enzyme destruction by fixative, and diffusion of reaction product during incubation, made staining methods worthless for demonstrating the intracellular *in situ* localization of enzymes, particularly important ones. (1959, p. 1)

By the time he was writing in 1959, though, Novikoff was putting all these worries in the past tense, referring to them as the "essentially destructive comments" by "gloomy critics" (p. 1). How in a relatively short time were these

---

examination are complicated and often brutal. They involve the treatment of the tissue with reagents which either violently precipitate proteins or slowly render them insoluble, followed by a series of solvents both aqueous and organic. The possibility of change is implicit in every step of this process, and it is the prime responsibility of the cytologist to inquire whether the structures which he displays in his preparation actually existed in the living cell, and if not, by what chemical process they have been produced. It is not surprising that during the history of this science, the results of investigation have been frequently challenged, sometimes on theoretical grounds based on biochemical experience, but often as the result of experiment. The list of structures which have been so challenged is a long one and includes most of the visible structures of the cell. To this list belong, among others, mitochondria, Nissl bodies of nerve cells, myofibrils of smooth and striated muscle, and the chromatin granules of the resting nuclei. All of these have been shown, by subsequent research, to be pre-existent in the living cell. On the other hand, artefacts have been produced by fixation. To this category belong the network of fibres described by the adherents of the reticular theory of the structure of the protoplasm, and the long filaments produced by Flemming in support of his filar theory" (1951, pp. 1–2).

concerns laid to rest? Despite the comments of Mercer and Birbeck cited earlier, it was not by developing a detailed understanding of how the techniques worked. An understanding of how, for example, the fixative osmium reacted with tissue preparations, would not be available until much later. If researchers lack a theoretical understanding of the procedures by which putative evidence is generated, how can they establish its reliability? The alternative is to focus on the results themselves.

At first, this suggestion to examine the results of a technique in order to vindicate it would seem to lead nowhere. If skeptics are questioning the evidential status of results, how can further consideration of them resolve the concerns? In fact, this strategy is not as circular as it might appear. To see this, assume for a moment that you are pioneering a procedure for procuring evidence. What might you demand of the results of the procedure before accepting them as evidence of something in the world?

The first thing you would likely do is examine whether there was a distinguishable structure or pattern in the results. If the results seemed randomly distributed, you would conclude that you had generated noise rather than picking up a signal from an underlying phenomenon. Likewise, if cell fractionation studies had shown that enzymes were scattered helter-skelter through the cell, scientists would not have been impressed. As it turned out, there was a pattern in the results, with particular enzymes appearing in high concentrations in one fraction and low in others, and other enzymes having higher concentrations in one of the other fractions. Such a pattern suggested that the distribution of enzymes in the fractions resulted in some way from their distribution in the original organelles within the cell. Likewise, a critical indicator that electron micrographs were revealing of cell structure was that visual patterns suggestive of structure appeared in them.

Of course, the structure could have been generated by the technique itself. That is just what the charge of artifact asserts. However, often in practice the burden of proof is on the person alleging that the results are an artifact. If a plausible account of how the technique itself could generate the structure is offered, that, as we shall see, is taken as a reason to suspect artifactual findings. But scientists are impressed by structure in their results, and when they find it, they are inclined to interpret it as reflecting the phenomenon itself and not an artifact of how the evidence was produced.[5]

---

[5] I asked Keith Porter (interview, 1987, University of Maryland, Baltimore County) whether, when he examined his first micrograph of a cell (see Figure 5.1), he had any doubt about whether it was revealing something about the cell. He responded no – the structure in the micrograph was just too impressive.

If your new procedure generates structured results, you will want to repeat the procedure before announcing them publicly. No matter how clear a pattern you obtained the first time, if attempts at replication fail, it is highly likely that the structure was merely a result of some unnoticed feature of the way you performed the procedure, not the underlying phenomenon. Likewise, if the pattern disappears with seemingly trivial variations in the procedure, that again points to the results being due to the procedure, not the underlying phenomenon. Moreover, if other researchers are not able to replicate your results, they will be suspicious that something unnoticed or unreported in your procedure was producing the result. (Suspicions do not always lead to a negative outcome. Researchers know it can be tricky to replicate results on a phenomenon that, though genuine and perhaps even important, is fragile or dependent on very particular circumstances. At the other extreme, thankfully rare, a result that is not replicated may turn out to have been fraudulent. Questions of the repeatability of results are not taken lightly in science and can take considerable time and effort to resolve.) Bearing in mind that adjudication is not always easy, we can summarize this first criterion as holding that results should exhibit a determinate pattern and be repeatable.

If your results survive this first test, your second step would likely involve comparing them to results obtained through older, already established techniques. This is akin to comparing a crime witness's account to that of other witnesses and other sources of evidence. When at least some aspects of the results produced by the new technique can be related to what was already known by other techniques, that does help establish the credibility of the new technique. This is a straightforward application of what William Whewell called the *consilience of induction*: "The consilience of Induction takes place when an Induction, obtained from one class of facts, coincides with an Induction obtained from another different class. This Consilience is a test of the truth of the theory in which it occurs" (1840, p. xxxix, Aphorism xiv).[6]

There are limitations on how far consistency can take investigators toward evaluating a new instrument. One limitation stems from the fact that the purpose of developing new instruments and techniques is to secure data about phenomena for which existing techniques are inadequate. This means that, for at least some cases in which you apply the new technique, there will be

---

[6] Cytologist John Baker issued a particularly strong version of this requirement, insisting that all results with fixation of cells be compared to results from unfixed cells: "the only evidence of what is good or bad is comparison with the living cell. This point cannot be too strongly stressed: subjective ideas of what final result is good and what bad should never be allowed to form in the mind except on the solid ground of comparison with what is visible in the cell while still unaffected by reagents" (1942, p. 4. The passage is repeated in the second edition, 1950).

no existing technique providing comparable results. You may nonetheless find certain contexts in which you can apply both techniques. That would reduce the concern about error but would not irrefutably establish that the technique is reliable with respect to the all new phenomena for which data is sought.[7]

There is, however, a more serious limitation to this approach to vindicating new techniques. The critical assumption in appealing to the consilience of different inductions is that the different inductions are independent of each other. When they are completely independent, though, they may not naturally align with each other. It often takes a great deal of manipulation to get a new technique to produce evidence corresponding to what has been procured previously with other techniques. This often requires taking the older technique as a reference point and altering the new technique so as to generate corresponding results. In this way, the older technique is used to calibrate the new one (Bechtel, 2002b). Although the techniques are not independent in this procedure, successful calibration provides evidence that the new technique is capturing the same underlying phenomenon as the older technique and thereby indicates that there is an objective phenomenon on which both techniques are aligning.

Suppose that your new procedure reliably produces results that are consistent, or can be made consistent, with older techniques in areas of overlap. That is, you have to some extent satisfied the second criterion. Now let us focus instead on results that extend beyond what other techniques provide, because these are where the new technique might advance inquiry. Imagine that these new results do not make sense. They do not suggest or fit into any plausible theoretical account (mechanistic model) of the phenomenon under investigation. Moreover, you are not able to generate such an account. In this situation, you are likely to be suspicious of the new results and seek an

---

[7] A failure of correspondence, moreover, may also be due to artifacts created by the older technique. Cosslett emphasized this point in a chapter on electron microscopy as a technique: "It is also necessary to plead for open-mindedness in comparing electron micrographs with the results of the established optical methods. Long familiarity with them unconsciously instills the feeling that they show exactly 'what the specimen really looks like.' In the case of stained specimens obviously, but to some extent also in non-stained preparations, this cannot be true, since there is always the possibility of artifact even in optical preparations. The greater danger of artifact formation in electron microscopy should not blind one to the fact that it gives pictures that are valid within their own limitations. It is only necessary to realize that these are different limitations from those of optical methods. In short, each technique provides evidence of value, each gives a partial view of the constitution and structure of the specimen under examination, and it is the task of the research worker to correlate critically all the clues and deduce from them a coherent answer to the puzzle set by Nature" (1955, p. 524).

explanation that shows them to be an artifact. On the other hand, if the results are what you would expect given an existing model of the phenomena, or you are able to advance such a model in which they make sense, that increases the likelihood that you will accept the results as evidence.

This yields the third criterion: coherence with plausible accounts of the phenomena. Relying on it seems paradoxical if you see evidence as foundational, such that models or other accounts are assessed in terms of the evidence offered for them. For what I have just proposed is that evidence is assessed by its fit to an account; specifically, it is more likely to be accepted when it fits with a plausible account than when it does not. As paradoxical as it seems, however, in practice scientists often rely on this criterion. Some may worry that if this is actually how scientists appraise evidence, they are guilty of circular reasoning. However, such reasoning would be viciously circular only if the sole basis for evaluating the account were the results of experimental techniques, which were themselves being evaluated by whether they supported the account. But that is not the only way in which an account can be evaluated; another is whether it coheres with other accounts. As we shall see, the new research instruments and techniques that were crucial for the development of cell biology often suggested the existence of new cell organelles. But did these organelles really exist? When the new organelles were identified as having a role in the emerging mechanistic model of cell function, then researchers were more likely to accept not only the existence of the organelles but also the instruments and techniques that pointed to their existence. When no role for the putative organelle was apparent, the existence of the organelle and the techniques providing evidence for them were subject to question.

I have proposed three criteria by which one might assess a new research instrument or technique merely by the results it generates: the degree to which

- the results exhibit a determinate structure or pattern and are repeatable;
- the results agree with results generated by other techniques or can be calibrated against them; and
- the putative evidence coheres with theoretical accounts that are taken to be plausible.

In practice, each of these three criteria plays an important role in scientists' evaluation of evidence (Bechtel, 1995; Bechtel, 2000; see also Creath, 1988). Having offered this rather abstract sketch of the epistemic problems encountered by scientists when introducing new instruments and techniques and answering the inevitable charges of artifact, I turn now to concrete exemplars from the early years of modern cell biology. Most prominent were two instruments – the ultracentrifuge and electron microscope – and the associated techniques that enabled scientists to gain unprecedented access to the mechanisms of the cell. I discuss the ultracentrifuge in Section 2 and the electron microscope in Section 3.

2. THE ULTRACENTRIFUGE AND CELL FRACTIONATION

As discussed in Chapter 2, to understand a mechanism scientists try to decompose it into its component parts (structural decomposition) and operations (functional decomposition). Perhaps the most obvious strategy is to isolate certain parts and see what each does. In cells, for example, simply learning what chemical compounds are typically found in each component part may be informative as to what role each part plays in the cell's functioning. If a specific enzyme appears in a particular organelle, for example, then it is likely that the reaction catalyzed by that enzyme is localized there. The critical first step is to isolate the component parts and operations and the challenge is to do so in a way that divides components as they are in the natural system. This is extremely challenging. The living cell is a highly structured milieu in which components are integrated structurally and functionally.

How then to decompose cells into their component parts? Two candidates seemed to be available – mechanical forces and chemical processes. Centrifugal force as generated in a centrifuge, which can separate particles by size and weight, was a promising mechanical candidate. However, it required some means of breaking cell membranes so as to access the contents and combine the contents of multiple cells into one container. One plausible way of breaking cell membranes was to apply chemical agents that would dissolve them. Accordingly, in 1869 the Swiss biologist, Johann Friedrich Miescher, employed first warm alcohol and then the enzyme pepsin to strip away both the cell membrane and the cytoplasm of the cell in an attempt to isolate cell nuclei. He subjected the remaining material to centrifugation to isolate the nucleus from the debris, and then subjected the nucleus itself to chemical analysis. In this manner, Miescher (1871) identified a new group of cellular substances which he named *nucleins*; when his student Richard Altmann (1889) identified them as acids, he renamed them *nucleic acids*.

Although various researchers in the early twentieth century, including Otto Warburg, employed centrifugation in attempts to isolate cell structures, the technique did not come into wide use at the time. There are a couple of reasons for this. First, many researchers were skeptical of techniques that disrupted the internal structure of the cell, which they assumed was critical to its functioning.[8] Such skepticism is often justified: if a component does manage to function when isolated in this manner, it may function differently than in the intact cell. Second, the centrifugal forces generated by the available centrifuges were too weak to separate most cellular constituents.

Beginning in the 1920s, several researchers set about improving the capacity of the centrifuge. Swedish physical chemist Theodor ("The") Svedberg turned to centrifugation in the course of his work on colloids. Svedberg and Herman Rinde (1924) built an electrically driven centrifuge from the components of a cream separator. In addition to generating higher rotational speeds and hence greater centrifugal forces, there was a window through which the process of sedimentation during centrifugation could be observed.[9] It was this potential for direct observation that led Svedberg to refer to his centrifuge as an *ultracentrifuge*. However, the meaning of the term shifted and it came to be used for any centrifuge running at high speeds in a vacuum or near vacuum.

Independently, Elime Henriot, working in Belgium, achieved high rotational speeds with an air turbine centrifuge. Jesse Beams, together with his graduate student Edward Pickels at the University of Virginia, modified Henriot's design by introducing a larger rotor enclosed in a vacuum chamber and suspended by a steel wire (see Beams, 1938). Pickels continued to innovate with centrifuges at the International Health Division of the Rockefeller Foundation, where he collaborated with Johannes Bauer in the development of a centrifuge that could be used to separate filterable viruses. Subsequently, Pickels (1942) went on to design an electrically driven ultracentrifuge which investigators found far easier to use.

By the early 1930s the centrifuge as an instrument had been sufficiently developed that it was available for use by biologists such as Martin Behrens (1932) and Robert Bensley and Normand L. Hoerr to attempt to isolate cell structures. The results of these studies were impressive. In 1934 Bensley and Hoerr could reliably produce a fraction which they claimed, based primarily on information about the size of the component particles, was mitochondrial in nature. In the late 1930s and early 1940s, Claude could produce four distinct fractions that were clearly different in both appearance and chemical

---

[8] Schneider and Kuff (1964, pp. 19–20) commented, "The attitude of most cytologists was in large part responsible for the lack of consideration of isolation methods. They argued, on the basis of their observations of living cells under the microscope, that disruption of the cell wall produced immediate and irreversible changes in the cellular components."

[9] Svedberg's initial success was in sedimenting haemoglobin and demonstrating that it was a homogeneous molecule, not a heterogeneous colloid (Svedberg & Fåhraeus, 1926). This and subsequent studies helped demonstrate that proteins were in fact macromolecules.

constitution. This was sufficient to attract attention to the technique of cell fractionation. In the next chapter, I will discuss these and other results in detail and explore how they contributed to the body of knowledge that came to characterize cell biology. It is important to note that already the first criterion for adequacy of a new technique was satisfied – the technique was generating a determinate and replicable pattern of results.

Yet, the developers of techniques for cell fractionation, and especially their critics, were very concerned about whether centrifugation procedures created artifacts. As I noted, the most commonly cited criterion in evaluating results from a new technique is correspondence with evidence from other procedures. In the case of cell fractionation researchers sought to relate the results to those obtained in microscopy by comparing the appearance of the fractionation products with the appearance of organelles in whole cells. Early fractionation studies failed this test – the isolated, centrifuged organelles did not look at all like they did under the light microscope. This was particularly true of mitochondria. Hogeboom, Schneider, & Palade (1948) explored alternative media in large part to try to make isolated mitochondria retain their typical elongated shape and to stain with Janus Green B, a traditional mitochondrial stain. Thus, the initial failures did not lead researchers to abandon the technique but to revise it so that it yielded evidence comparable to older techniques. This is what I referred to previously as seeking consilience of results in order to calibrate the new technique.

What exactly was the *technique* of cell fractionation? For Latour, a technique could be characterized as a black box when the steps involved are taken for granted as part of established practice. Cell fractionation had achieved this status by the late 1950s and 1960s, and recipes for performing it were widely available in manuals of experimental procedure. While the techniques retained a close resemblance to those initially developed by Bensley and Hoerr and by Claude, they also incorporated some substantive changes that put their mark on the results obtained. These changes were evident both in the methods for preparing cell materials and in the centrifugation regimes employed to fractionate them.

There were two major approaches to preparing materials for fractionation. (Details varied, depending on the material selected – such as bacteria, algae, or liver tissue – and which cell constituents were of interest.) In the aqueous approach, the material was put in an aqueous medium (e.g., 0.88 M sucrose solution when mitochondria were targeted) and shearing forces were used to break the cell membranes. The resulting material is called a *homogenate* (because the cells have been broken and their contents blended) or *suspension* (because many particles – intact organelles or fragments of organelles – are suspended in the solution). In the nonaqueous approach, water was removed from the material by means of freeze-drying, and then milling and grinding were used to break the cell membranes. Although the nonaqueous method was the first to be employed (by Bensley and Hoerr, for example), and manifested considerable advantages, especially for studying the nucleus, its disadvantages made it less suitable for studying cytoplasmic structures. Perhaps most critical was that the drying process utilized chemical agents that could inactivate many of the enzymes researchers were targeting. Since, following Claude, the aqueous techniques came to be preferred in studies of cytoplasmic organelles, I will focus on them and especially on two key elements: the means of breaking cell membranes and the media in which the contents were maintained.

Breaking Cell Membranes

Membranes – the cell membrane (plasma membrane) as well as various internal membranes – are relatively tough and difficult to break. Moreover, it is important to break the cell membrane without destroying the membrane's surrounding internal organelles, because breaking these could cause a redistribution of the enzymes residing in the organelles. Although its existence was not even anticipated as the earliest work on cell fractionation was being pursued, the lysosome provides a particularly vivid example of the hazards of breaking internal membranes. The lysosome contains hydrolytic enzymes, such as acid phosphatase, which are destructive of other cell organelles. Breaching the lysosome membrane would result in the rapid destruction of the other cell organelles under investigation. The desire not to destroy internal membranes largely ruled out use of the Waring blender, which was a standard tool for preparing homogenates in biochemistry. (In the dominant paradigm in biochemistry, the cell was regarded as essentially a sac of enzymes. To bring enzymes into solution for study, it was desirable to break any membranes that kept them isolated within organelles.) In addition to the objective of breaking all and only cell membranes, a further concern was the speed of operation; any delay between the initial breaking of cells and centrifugation could result in changes in the cell contents. So the goal was relatively straightforward – quickly break all cell membranes but not the membranes of any internal organelles.

Early researchers explored several different means of breaking the cell membrane. Emphasizing gentleness, in his early studies Claude gently rubbed cells against each other with mortar and pestle. However, this procedure left many whole cells in the aqueous medium, which then could not be separated from cell nuclei. This fraction was therefore essentially useless and had to be discarded, posing a serious difficulty when Claude wanted to characterize the distribution of a given enzyme by stating the amount in later, usable fractions relative to the amount in the whole cell. Two biochemists at Wisconsin, Conrad Elvehjem and Van Potter, developed a coaxial homogenizer in which finely cut or minced tissue was placed in a tube with some buffered salt solution and a closely-fitting pestle rotated either by hand or with a motor while being worked up and down in the tube. Although this may seem like a gentle procedure, it is not. It breaks the membrane through the shearing force resulting from the faster moving material near the pestle rubbing against the slower moving material near the wall of the tube. As a result, the Elvehjem-Potter homogenizer was quite effective at breaking cell membranes, but it also broke some internal membranes (Potter & Elvehjem, 1936). By the estimate of de Duve (1971), it damaged 15% of lysosomes and peroxisomes and 10% of mitochondria. It is interesting to consider how de Duve could reach such a judgment. As we will see, he made a fundamental assumption – that a given enzyme originated in a single organelle. Thus, he inferred that the fraction with the greatest concentration was the locus of the enzyme in living cells, and that any of the enzyme found in other fractions represented contamination. This was plausible insofar as the amount of an enzyme found in one fraction generally greatly exceeded that in any other. In this instance, it was the determinateness of the results that supported the judgment of what was artifact and what was evidence of an underlying phenomenon.

Several factors affect the degree of disintegration of cells achieved with the Potter-Elvehjem homogenizer, including the clearance between tube and pestle, the speed of rotation, and the number of times the pestle is moved up and down. This led to a striking number of variations in the basic design of the homogenizer, some of which are illustrated in Figure 4.2. Campbell and Epstein (1966, p. 19) commented, "it is probably true that almost every worker has his own individual preference." A variety of other means of breaking cell membranes were explored in the 1940s and 1950s, including colloid mills (employed by Mirsky to isolate chromosomes, as in Mirsky & Ris, 1951), ultrasonic or sonic vibrations, and osmotic shock. Most investigators, though, adopted the Elvehjem-Potter homogenizer or variants on it. This was largely because it provided a successful compromise between the destructiveness of the Waring blender and the insufficient disruption of membranes from simple rubbing. It produced impressive, interpretable results (clear differentiation of

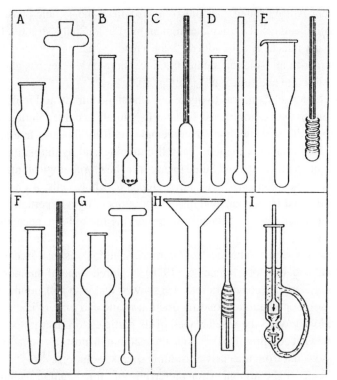

Figure 4.2. Examples of different homogenizers used to break cells prior to cell fractionation. In each, a pestle (right) is inserted and moved up and down in a tube (left). Reprinted from Vincent Allfrey (1959), The isolation of subcellular components, in J. Brachet and A. E. Mirsky (eds.), *The Cell: Biochemistry, Physiology, and Morphology*, Vol. 1. New York: Academic Press, Figure 1 on p. 214.

enzymes between different fractions). Thus, de Duve and Berthet concluded as early as 1954:

> Absolute preference should be given to homogenizers of the type described by Potter and Elvehjem (1936). Simple rubbing in a mortar, as recommended by Claude (1946), disrupts only a fraction of the cells, and mechanical choppers such as the Waring Blender cause excessive damage to the particulate components of the cells. The Potter-Elvehjem instrument is, of course, not entirely free of these drawbacks and should be used with discrimination. (p. 232)

Choice of Media

Once the cell membranes are broken, the properties of the aqueous medium into which the cell contents disperse become critical. The desired medium, as Vincent Allfrey stated, "should so approximate the soluble phase of the cytoplasm that the cell particulates remain morphologically, structurally, and functionally intact." But, he added, "Such a medium has not yet been devised, and in practice all isolation media introduce more or less serious alterations in both structure and function" (1959, p. 202).

The challenge, then, was to figure out which medium would most closely correspond to the environment in living cells. Initially a saline solution seemed physiologically realistic, and Claude employed it in his early fractionation studies. However, this medium caused clumping and agglutination of the cytoplasmic particulates and failed to preserve the morphological integrity of organelles when compared with micrographs of whole cells. As noted previously, this led Hogeboom and his collaborators to try a hypertonic 0.88 M sucrose medium, which succeeded in preserving the rod-like appearance of mitochondria (Hogeboom et al., 1948). However, the crucial function of ATP synthesis (suspected to be localized in mitochondria) was lost. Accordingly, one of the collaborators, Schneider (1948), explored some of the more commonly used sucrose concentrations. He showed that with an isotonic 0.25 M sucrose solution, the mitochondrial fraction would carry out oxidative phosphorylation (though the resemblance of its particles to intact mitochondria was somewhat compromised).[10] Soon the medium in most fractionation studies was selected from the class of isotonic sucrose solutions.[11]

The logic of the argument for preferring this fractionation technique is noteworthy. The claim that the mitochondrion was the locus of oxidative phosphorylation was based primarily on fractionation studies, but the justification for doing fractionation in sucrose and especially in isotonic sucrose was that it produced a fraction that possessed the enzymes for oxidative phosphorylation and could carry out the reactions that realized that function. The

---

[10] A similar dependence of functional activity on sucrose concentration was found in studies of isolated thymus nuclei – nuclei isolated in 0.25 M sucrose synthesized protein and RNA but those prepared in 0.4 M sucrose did not (Allfrey, 1959, p. 206). Allfrey also reported a number of other factors that affect the appearance and function of isolated nuclei, including pH and ionic strength, and provided another example of a form/function tradeoff: "It is a curious fact that when calf thymocyte nuclei are isolated in 0.25 M sucrose-0.003 M CaCl$_2$, they are granular in appearance, but active in many synthetic systems; if they are isolated in a hypertonic medium which preserves optical homogeneity they lose their synthetic capacity" (1959, p. 208).

[11] Witter, Watson, and Cottone (1955) compared electron micrographs so as to examine the fine structure of mitochondrial fractions isolated using various sucrose media and concluded that 0.44 M sucrose with pH regulated to 6.2 with citrate produced the best resolution. Based on similar studies, Novikoff (1956a) favored 0.25 M sucrose with 7.3% polyvinylpyrrollidone at a pH of 7.6 to 7.8. In general, the strategy was to tweak the method until it generated the clearest or most useful results for the current purpose.

procedure for isolating mitochondria was being calibrated to make its results correspond to those obtained by related techniques. Moreover, the fit with the plausible theory that oxidative phosphorylation was localized in the mitochondrion was a central consideration in evaluating the technique. The procedure for developing the evidence certainly was not independent of the theoretical claim it was supposed to support. Clearly, the technique was being calibrated by existing techniques and evaluated by its ability to produce evidence that fit a plausible theory.

Centrifugation Regimes

Once material is prepared, it is ready to be placed into the centrifuge and spun. Many questions of procedure remained, however. How fast should the contents be spun, and for how long? For Bensley and Hoerr, who started with the objective of isolating mitochondria, the strategy was pretty straightforward: centrifuge long enough to separate what seemed to be a reasonably pure mitochondrial preparation. However, what came to be the dominant approach involved the separation of four different fractions. This involved successive centrifugation runs in which the sediment at each stage was removed and the remaining unsedimented material (called the *supernatant*) was subjected to centrifugation at a yet higher speed (see Figure 4.3).[12] Why this procedure? Through it, as we will see in the next chapter, Claude isolated what appeared to be chemically distinct components that could be linked to four different components of living cells: the nucleus (in the sediment separated out by the first centrifugation run on the original homogenate), mitochondria (in the sediment of the second and third runs), microsomes (in the sediment of the fourth run), and cell sap or cytosol (soluble protoplasmic material remaining in the supernatant after the fourth run).

---

[12] The following is Claude's description of his procedure: "The suspension was immediately centrifuged for one minute at 2000 r.p.m. in a horizontal centrifuge. This step was found to remove most of the liver fragments, the cells which had remained intact, the free nuclei and the red corpuscles. The supernatant fluid, or extract proper, which contained practically all the organic components equal to, or smaller than, three μ diameter, was spun at 18,000 r.p.m. in the high speed centrifuge, for exactly five minutes. At that speed, a five minute run was sufficient to bring down practically all the large secretory granules. The sediment was saved for further purification in the centrifuge. The small particles, which had remained in the supernate, were sedimented by a long run purified in the higher speed centrifuge, the 'long run' in this case being five minutes centrifugation at 18,000 × gravity. The procedure consisted in suspending the material in water and sedimenting it again at high speed, four times in succession" (1941, p. 267).

Technical problems in centrifugation, as well as variations in materials and goals of the research, led researchers to vary such details as frequency of resolubilizing, length of runs, and speed of centrifugation. For example, due to differences in their initial distribution and other chance factors, some lighter particles sediment along with the heavy particles. In addition, during centrifugation particles will encounter the walls of the centrifugation tube where they may adhere or agglutinate and set up convection currents as they travel down the walls. Small particles may become entrapped and carried down with the larger particles. Empirical explorations revealed that these difficulties could, in part, be overcome by repeated resolubilization and recentrifugation of the sediment. Note again, though, that general expectations regarding the results guided the refinement of the technique.

Are there only four distinct constituents of cells that could be separated by fractionation? There were reasons to suspect more. One of the most powerful reasons was that cytologists already knew of organelles that were not distinguishable in the four basic fractions. For example, de Duve commented, "It may be recalled that the fate in differential centrifugation of such formations as secretory granules, centrosomes, and the Golgi body is entirely unsettled at the present time" (de Duve & Berthet, 1954, p. 250). Shortly after Claude's early work, other researchers developed regimes to isolate additional fractions. Hubert Chantrenne, for example, used the Henroit-Huguenard centrifuge to obtain five fractions whose chemical composition blended into each other. He concluded, "it seems that one can partition the granules in as many groups as one wishes, with no experiments or observations indicating that there exist precise demarcation lines between the different group of particles" (Chantrenne, 1947, p. 445, as quoted in Rheinberger, 1997, p. 64). Novikoff et al. (1953) separated ten fractions and studied distributions of DNA, RNA, nitrogen, and several enzymes in them. As we will see in Chapter 6, procedures for separating additional fractions played a critical role in the 1950s in the discovery of additional cell organelles.

Nonetheless, the four-fraction approach remained dominant. Perhaps the best explanation for this is that these fractions were identified with cell structures for which functions were already suspected. In particular, although it is noteworthy that the scheme provided no place for the Golgi apparatus, some researchers still suspected this structure to be an artifact. Claude in fact collaborated with Palade in a pair of papers in 1949 arguing that it was indeed an artifact. (These papers are discussed in the last section of this chapter.) Until clear evidence as to the function of the Golgi apparatus was developed in the 1960s, it remained possible to doubt its existence and to accept fractionation procedures which did not give it a place.

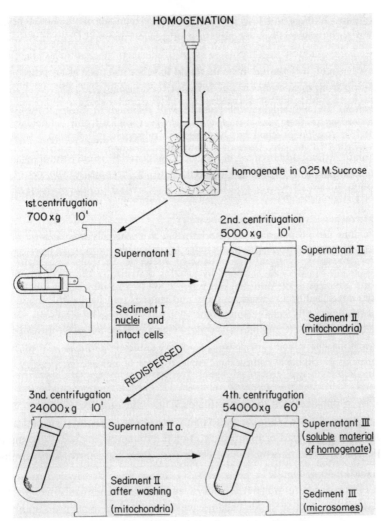

Figure 4.3. A schematic representation of a typical procedure for creating fractions by multiple centrifugations at increasing speeds. Reprinted from E. D. P. De Robertis, W. W. Nowinski, and F. Saez (1960), *General Cytology* (Third ed.). Philadelphia: W. B. Saunders Company, Figure 4.1 on p. 79, with permission from Elsevier.

Interpreting Fractionation Results

The immediate goal of cell fractionation is the determination of the chemical, especially enzymatic, composition of different cell organelles, which could then in turn support claims about the functions performed by those organelles. Typically after fractionation, however, investigators found a particular enzyme distributed across more than one fraction. One interpretation of such results is that the same enzyme occurs in multiple organelles. Claude, however, viewed the fraction in which it was in highest concentration as the likely true locus and the smaller amounts found in other fractions as contamination resulting from inexactness in the process of fractionation. de Duve construed this as a major insight on Claude's part and expressed it in the postulate "*a given enzyme belongs to a single intracellular component in the living cell*" (de Duve & Berthet, 1954, p. 239). To this de Duve himself added a second postulate: "all members of a given subcellular population have the same enzymatic composition," which he labeled the *postulate of biological homogeneity* (1963–4, p. 52).[13] Commenting on the value of these assumptions, de Duve wrote,

> granting these two postulates, especially that of biochemical homogeneity, we can now use the enzymes as markers for their host particles and conduct tissue fractionation experiments very much like any other type of chemical fractionation. We may, at least in the initial analytical phase of the work, forget all about morphological features and treat suspensions of ground cells or tissues as mixtures of different populations of physical entities to be identified, characterized, resolved, and purified, with as sole guides the enzymes. If we can reach the final preparative phase and achieve sufficient purification, then the test of our working hypotheses will come, for morphological examination will show whether our deductions were in fact valid or not. (1963–4, p. 53)

Not everyone accepted these postulates. They constitute major assumptions about how cells are organized, and many investigators found it suspicious to invoke them in the very interpretation of fractionation studies. Why should cells follow a design principle in which each chemical constituent is limited to a particular organelle?

Cell fractionation was not the only way to localize chemicals in cell structures. When selective stains became available, histochemists used uptake of the stain to determine the locus of substances in slices or even whole mounts of tissues. David Glick, a pioneer in developing quantitative histochemical methods, called for use of such methods to vindicate the results of fractionation studies: "it would be best to seek proof of the specific localization by an independent method, i.e., one not dependent on centrifugal separations" (1953, p. 451). Glick can here be seen to be requesting consilience with other techniques before accepting cell fractionation results. In the case of some

[13] Claude was characterized by de Duve and Beaufay as altering the question investigators were asking from "what is in . . . ?" to "where is . . . ?"

cell constituents, such as DNA, cytochemical techniques such as the Feulgen stain were sufficient to demonstrate a unique locus. de Duve agreed that cytochemical staining could provide a valuable test of the results of fractionation, but was unwilling to restrict fractionation studies to results that could be vindicated by histochemistry.

By the 1950s cell fractionation was generally regarded as providing reliable evidence of the enzyme composition of cell organelles and thus of their function. Yet skeptics remained. James Danielli, for example, doubted that the results of fractionation were reflective of the activities in living cells: "so far there has been an almost complete lack of proof that the bodies isolated are in the same condition as in the intact cells" (1953, p. 7–8). F. K. Sanders offered a similarly negative assessment:

> except in certain well-defined cases, little evidence has been offered that such fractions are in fact identical with known cellular structures. Moreover, it cannot be assumed that because a certain enzyme is found in isolated 'nuclei' or 'mitochondria' it is, in fact, present in this location in the living cell. Cells are highly complex colloidal systems, in which the distribution of substances between the different parts of the system is likely to be altered by procedures far milder than those [in cell fractionation]. (1951, p. 24)[14]

In the next chapter we will see that even while grounds for skepticism remained, and before a theoretical understanding of the process of fractionation had developed, researchers relied on cell fractionation in developing models of cell mechanisms. That this technique was producing (a) determinate results that (b) to some extent corresponded to results from other techniques and (c) fit into a developing mechanistic model of the cell was

---

[14] Even as enthusiastic an advocate of cell fractionation as de Duve expressed caution: "As a bridge between the fields of cytology and biochemistry, it [differential centrifugation] offers tremendous possibilities which even the most carefully worked out techniques of cytochemistry could never have been expected to fulfill. It must be remembered, however, that the application of differential centrifugation is fraught with many technical difficulties and open to a large number of errors. The methods that have been worked out today represent significant improvements over the earlier ones, but much remains to be done to augment their accuracy and selectivity. For this purpose it is important to have in mind the theoretical basis of the technique as well as the various factors of practical nature which have been found to affect the results. . . . The limitations of differential centrifugation become particularly severe when the technique is applied to the study of tissue enzymes. It is now quite clear that the observed partitions provide only the roughest sort of information concerning the true intracellular distributions of enzymes. They can only be considered as clues which have to be followed by many additional experiments in order to arrive at their real significance" (de Duve & Berthet, 1954).

reason enough for many investigators to employ it as a basic tool in their attempts to understand cells.

3. THE ELECTRON MICROSCOPE AND ELECTRON MICROSCOPY

Whereas cell fractionation enabled scientists to decompose cells into component parts in a way that permitted investigation of their function, the goal of microscopy was to visualize the component parts as they existed within the cell. (A secondary use of microscopy was to examine the fractionated parts of the cells to determine whether they corresponded to parts as they occurred in the intact cell.) Despite the advances that had been made using light microscopy in the nineteenth century, the wavelength of light presented a principled limit to the resolution that could be attained and rendered this instrument inadequate for studying the internal structure of cells. A number of variations on light microscopy developed in the 1940s, such as phase contrast[15] and ultraviolet microscopy, extended its usefulness (for reviews, see Baker, 1951; Wyckoff, 1959), but the newly invented electron microscope offered the greatest opportunity. The wavelength of an electron depends upon its voltage, but even the relatively low-powered 50 kV microscopes developed in the 1930s and 1940s theoretically permitted resolution down to 5–10 Å (50–100 mµ), in contrast to the limits of approximately 2,500 Å (.25 µ) with a light microscope using white light and 1000 Å with ultraviolet microscopes (Cosslett, 1955). Figure 4.4 shows schematically the difference in resolution between the light and electron microscope.

The electron microscope relies on the same general optical principles as the light microscope. Each requires a source of illumination, a condenser lens for controlling it, an objective lens, and one or more additional lenses for providing high magnification. The main differences between light and electron microscopes stem from the use of an electron beam, rather than a beam of light, for the illumination. One important consequence is that researchers had to adjust to not looking through the microscope, but instead viewing an image on a fluorescent screen or in a photograph. Another is that the image results not from the absorption of light of various wavelengths, but from the scattering of electrons. Scattering depends not on the chemical constitution of the object imaged, but on the amount of matter present – the

---

[15] In phase contrast microscopy, phase differences in the light refracted from various structures in the cell are converted into differences in amplitude, which can be seen as differences in image contrast. This technique is particularly useful for in vivo studies.

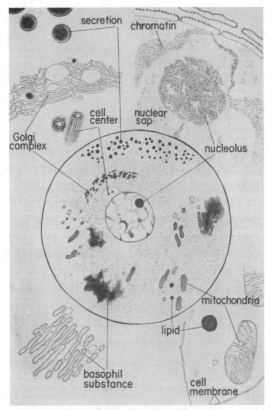

Figure 4.4. A representation of cell structure detail available in light versus electron micrographs. The center shows what can typically be seen with a light microscope, whereas the outer region reveals details of various organelles that could be seen in electron micrographs of the 1950s. Reprinted from E. D. P. De Robertis, W. W. Nowinski, and F. Saez (1960), *General Cytology* (Third ed.). Philadelphia: W. B. Saunders Company, Figure 3.9 on p. 71, with permission from Elsevier.

product of the number of atoms and their physical density or atomic number. In 1924, Louis de Broglie discovered the key theoretical concept on which electron microscopy depended, namely, that electrons have wave properties with wavelength inversely proportional to electron velocity. Two years later Hans Busch showed how to focus the electron beam by means of magnetic lenses. From these foundations, Max Knoll, Ernst Ruska, and Bodo von Borries developed transmission electron microscopes in the 1930s. In 1933 Ruska introduced the first true compound electron microscope, greatly increasing its resolving power. He then joined the firm of Siemens and Halske and, with von Borries, designed the first commercially available electron microscope, which went into production in 1939.

The onset of World War II cut off access of the U.S. and its allies to the Siemens microscope, but Vladimir K. Zworykin, head of electronics research at RCA, had himself already begun work on an electron microscope.[16] He brought Ladislaus Marton, a Hungarian physicist working in Belgium who had just fled the Nazis, to the Camden, NJ, laboratory in 1938. While at the Free University of Brussels, Marton had built two microscopes of his own in the early 1930s,[17] and with Zworykin he designed the RCA EM-A microscope. The EM-A had serious limitations in maintaining a vacuum, making it a poor candidate for commercial development. In early 1940, Zworykin hired James Hillier to replace Marton (who, with support from the Rockefeller Foundation, moved to Stanford to pursue technological improvements and chemical and biological applications of the electron microscope). As a graduate student of Eli F. Burton at the University of Toronto, Hillier was developing his own design for an electron microscope. At RCA, he designed what became the commercially viable RCA EM-B. Its availability was initially highly restricted during the war and only those with AA-1 priority were able to get on the waiting list to purchase one.[18]

Even during the war, opportunities emerged for biologists to use the new microscope and develop techniques appropriate for the study of cell structures. Pioneering these techniques was not for the faint-hearted. The most straightforward challenge was the need to master the numerous manual adjustments that were required to align the lenses to produce focused images. Two other challenges required considerably more effort and innovation. First, a way of preparing specimens sufficiently thin to be penetrated by the electron beam was needed. Second, specimens had to be transformed so as to survive conditions within the electron microscope and produce useful images.

---

[16] For details on the development of the electron microscope, see Rasmussen (1997). Rasmussen hypothesized that RCA's interest in developing an electron microscope resulted from the fact that during the war it was blocked from taking advantage of the protocols it had developed for television: "it is possible that the prestige to be won during the war for RCA's cathode-ray technique via the electron microscope may have been one reason that RCA embarked on what must at first have seemed an unremunerative program in scientific patronage" (p. 31).

[17] Marton himself pioneered in imaging biological material – in 1934 he produced and published in *Nature* the first electron micrograph of a tissue section from a plant leaf fixed with OsO_4 (Marton, 1934). Although his specimen was too thick to produce a very detailed image, Marton showed that biological material would not be destroyed by the electron beam. Nonetheless, various investigators continued to worry about that possibility in succeeding decades.

[18] The RCA EM-B and a subsequent model, the RCA EMU, which became available in 1944, were the major tools for biological electron microscopy for the subsequent ten years until the Siemens Elmiskip-1 provided an alternative.

Obtaining Sufficiently Thin Specimens

The need for thin specimens stems from the fact that electrons had to pass through the specimen to create an image. If an electron encounters an obstacle, it is deflected (or loses energy). The relatively low-powered electron beam employed in the microscopes available in the 1940s and 50s could penetrate only approximately 0.1 mm of biological tissue, a distance less than the thickness of a typical cell. This meant that the tissue had to be altered in some way before an image could be generated. Three strategies were pursued to produce sufficiently thin specimens: (a) make micrographs of structures that were already sufficiently thin; (b) section the specimens; or (c) induce specimens to spread thinly.[19] All three approaches were pursued by early electron microscopists trying to image biological material.

(A) MICROGRAPHS OF THIN BIOLOGICAL MATERIALS. Many of the earliest electron microscopists restricted themselves to thin biological materials. Two groups of researchers were noteworthy for pursuing this strategy. In order to interest biologists in using the RCA electron microscope, Zworykin had secured a National Research Council postdoctoral fellowship. It was awarded to Thomas Anderson, a recent physical chemistry Ph.D. from the California Institute of Technology. During his fellowship, Anderson worked with Stuart Mudd, head of microbiology at the University of Pennsylvania, and Wendell Stanley, from the department of plant and animal pathology at the Rockefeller Institute's Princeton Laboratory. In his first research with the electron microscope, Anderson studied bacterial and viral specimens provided by Mudd and Stanley, which were sufficiently thin for electron microscopy (Stanley & Anderson, 1941).[20] After completing his fellowship, Anderson remained at the University of Pennsylvania, and in the 1940s focused much of his work on bacteriophages.

Francis O. Schmitt's group at MIT, which included Cecil Hall, Marie Jakus, and Richard Bear, also employed thin materials. In the early 1940s they were the only biologists with their own electron microscope (brought with a $70,000 grant from the Rockefeller Foundation), which they used to

---

[19] Shadow casting, whereby a heavy metal such as gold or chromium is deposited on the surface of a specimen at an oblique angle, which generates a shadow in the micrograph whose length corresponds to the height of the material on the specimen (Williams & Wyckoff, 1946), was another method sometimes employed with success to image viruses and bacteria.

[20] At approximately the same time, Helmut Ruska, Ernst Ruska's brother, was appointed director of a laboratory for visiting scientists at Siemens and published his own micrographs of bacteriophages (Ruska, 1941).

investigate virus and protein structure. They prepared their material by fragmenting cells and drying the contents. Schmitt's group produced micrograph images of a number of cell fibers including collagen (Schmitt, Hall, & Jakus, 1942), tropomyosin (Schmitt, Hall, & Jakus, 1943), sperm tails, cilia, and flagella (Schmitt, 1944–5). The micrographs revealed a high degree of unusual and unexpected periodic and asymmetric order in the construction of these fibers.

In general, working with thin material did not create any additional risk of artifact in electron microscopy (although risks loomed from the need to work in vacuum and to fix specimens). As noted, the electron micrographs often produced unexpected results since the resolution of electron micrographs was far beyond that of other imaging techniques. Here the definitiveness of the patterns found and their reliable production seems to have been sufficient to convince the investigators that they were not artifacts.

(B) CUTTING THIN SECTIONS. The traditional means of preparing thin material for light microscopy was to cut thin slices from a tissue block with a microtome, such as the Cambridge Rocking Microtome, designed in 1885 by Charles Darwin's son Horace. The problem was that the microtomes designed for light microscopy could at best cut sections 1–2 mm thick, nearly an order of magnitude too thick for electron microscopy. During the 1940s numerous laboratories tried to create a microtome that could cut sufficiently thin sections, initially with little success. The problem is familiar to anyone who has tried to thinly slice a food like bread: it rips and tears.[21]

One approach,[22] suggested by theoretical calculations, was to use high speeds to reduce the stress on the knife edge and thereby produce cleaner slices (O'Brien & McKinley, 1943). Fullam, together with Albert Gessler, the director of research at Interchemical, developed a microtome that worked at 57,000 rpm and produced a cutting speed of 1,100 feet per second (Fullam

---

[21] Anderson (1956, p. 217) offered a more technical description of the problem: "The processes occurring at the cutting edge of the knife, for example, are poorly understood, but a knife edge with a large included angle must introduce large shearing stresses both in the section and in the face of the block as it rakes across its surface. . . . as it starts to float off onto the water in the trough, every section is visibly compressed in the direction of the movement of the knife. This compression is partly reversed as the section spreads out on the surface of the aqueous acetone or dioxane used to support it, but the uniformity of the spreading, especially on a micro scale is not known."

[22] Other approaches included cutting wedge-shape sections which would on one edge be sufficiently thin for electron microscopy (von Ardenne, 1939) and reducing the transmission ratio on standard microtomes (see Bretschneider, 1952, for a review of various microtomes introduced in the period).

& Gessler, 1946), which they used to cut thin sections from rubber, acrylic resins, and nylon.[23] Could this be made to work for biological specimens? Claude, together with Fullam, began to explore this in 1944–5, reporting that

> if the cutting blade is mounted on a disc and rotated at 50,000 r.p.m. then sections can be made less than 1 μ in thickness. It has been necessary to modify the usual techniques for preparation of the tissues. The only satisfactory fixative so far found is a dilute solution of osmic acid brought rapidly into contact with the cell by perfusion. A new type of embedding material has been used which will sublime at reduced pressure leaving the section ready for electron microscopy free from complicating foreign matter. (Annual Report, 1944–5, p. 76)

These slices, however, were still too thick for electron microscopy. In the Annual Report of his laboratory the following year, Claude indicated partial progress in creating still thinner slices: "During the year an experimental microtome, designed to give sections of 0.1 μ has been built which gives promise of achieving the objective. An entirely satisfactory blade has not yet been found and further investigation of methods of fixation and embedding of tissues to render them suitable for thin sections is in progress" (p. 86).[24] Although Claude and Fullam (1946) published some of the micrographs from their research, the pictures revealed distortions that were due to an insufficiently hard embedding agent and an insufficiently sharp blade.

Subsequently Claude collaborated with Joseph Blum, an engineer with the Rockefeller Institute, to develop a low-speed microtome in which the specimen passed the knife only once as it revolved on a disc or wheel, thereby reducing the potential for tearing the specimen. The microtome also included a liquid-filled trough mounted next to the knife into which the newly cut sections could float (see Palade, 1971). But, as Claude reported in his Harvey Lecture (1948), that did not solve the problems with fixation and embedding or of needing a sufficiently sharp blade for cutting. The embedding problems were largely resolved by Sanford Newman, Emil Borysko, and Max

---

[23] Rasmussen reported, "This ultramicrotome was first offered commercially, with an advertisement suggesting that purchasers were making a safe investment because the device could easily be converted into an ultracentrifuge. The section quality was not remarkably better than that delivered by slower microtomes, and the clouds of flying downlike sections it produced had to be collected with a fine butterfly net" (1997, p. 112).

[24] Two other groups, Daniel Pease and Richard Baker at the University of Southern California, and L. H. Bretschneider in the Netherlands, were working without knowledge of Claude and Fullam's work due to a two-year delay in the appearance of his Harvey Lecture. Pease and Baker adapted a standard Spencer 820 microtome while Bretschneider adapted the Cambridge Rocking Microtome. Both groups ultimately faced the same problem as Claude – the lack of good embedding and adequate knives. (See Pease & Porter, 1981.)

Swerdlow (1949a; 1949b), researchers at the National Bureau of Standards who introduced methacrylic resins as embedding agents as well as heat as the tool for advancing the tissue block toward the knife. Harrison Latta and Frank Hartmann (1950) solved the sharpness problem by introducing glass knifes.[25] In the wake of these developments, Keith Porter, together with Blum, took up Claude's effort to develop an adequate microtome (Porter & Blum, 1953).[26] They were hardly alone in pursuit of a suitable microtome. In 1954 the New York Academy of Sciences held a workshop at which designers from throughout the East Coast met, and about a dozen examples were on display. But Porter and Blum succeeded in developing a microtome that not only cut reliably thin sections but also was easy to use.[27] Its dominance was established when Ivan Sorvall, Inc., of Norwalk, Connecticut, began to produce it commercially a year later. The chief competitor was a European microtome developed by Fritiof Sjöstrand (1953a) and manufactured by L. K. B. Produkter AB, Stockholm. As a result of these advances, by the early 1950s

---

[25] Robertson (1987) reminisced on Latta's introduction of glass knifes: "I remember clearly one Monday morning that Harrison came into the lab with a milk bottle from home and stated that he was going to make a glass knife. We all laughed at this strange idea, but I followed him up to the machine shop where he got a hammer and smashed the milk bottle and chose a small piece to use as a knife. He mounted a fragment on a dummy steel knife using a black glue, and he and Frank Hartmann proceeded to cut very thin sections that one could use without taking out the plastic. I believe this was the first time anybody had succeeded in getting high-quality sections of biological material routinely thin enough to use for direct study without removing the plastic" (p. 139). Pease added more flavor to the early investigation of knives: "it is amusing to recall the mystique that soon developed in defining and finding the 'perfect' stain-free glass that would make ideal fracture edges. The idea developed that very old glass was apt to be better than new glass (I had a prized piece of broken, heavy, plate glass salvaged from a pre-prohibition bar widow, still with some old gold lettering on it)" (1987, p. 51).

[26] The microtome they produced was similar in some ways to the Cambridge Rocking Microtome, although Porter contended that the similarity was "simply accidental [since we] never had one of the Cambridge instruments in the laboratory at Rockefeller, and I have never seen one" (Porter, 1987). Instead, Porter and Blum credited Stanley Bennett: "In some respects also the microtomes reported on here are similar to an instrument, devoid of moveable bearings, which was observed in experimental stages of construction in H. S. Bennett's laboratory (University of Washington) in 1951. In it, the specimen was supported on the end of a bar and brought past the cutting edge by simply flexing the bar. It has no provision for avoiding the knife on the return stroke and for this and other reasons did not prove satisfactory"(Porter & Blum, 1953, pp. 687–8). Porter (1987) credited Joseph Blum with adding the gimbal for universal motion of the bar and a mechanical advance.

[27] Pease and Porter quoted the comments of Irene Manton twenty-five years later: "It was my privilege, soon after arrival in New York, to attend a meeting of the New York Academy of Sciences at which an array of devices for thin sectioning were displayed, some crude, others almost comically complex, but only the Porter-Blum behaved perfectly, cutting a clean ribbon of serial sections of the right thickness to order from a methacrylate block" (Pease & Porter, 1981, p. 290s).

thin sections could be routinely produced and use of electron microscopy for studying cells exploded. Bretschneider (1952) reviewed thirty-seven articles published since 1950, and a year later Dalton identified twenty-five additional ones, providing a sense of how quickly the field took off once thin sections could be prepared.

Because the use of microtomes to prepare specimens for light microscopy was an established practice, and looking at slices of tissues was not itself problematic, once adequate microtomes were available, they generated few epistemic concerns. Distortions due, for example, to tearing by the knife edge could usually be readily recognized in the resulting micrograph. There were, though, concerns about how to interpret the structures seen in thin-section micrographs. These issues arose especially in the context of the endoplasmic reticulum, which I will discuss in Chapter 6. Before the breakthrough in developing thin sections for electron microscopy, though, important results had already been achieved with the electron microscope using the third technique, tissue culturing of cells.

(C) GROWING TISSUES IN CULTURE. The third approach to preparing thin specimens drew upon a technique developed for a very different purpose – the growing of cells in culture. Ross Granville Harrison developed and Alexis Carrel further refined this technique to allow observers to track cell development. Tissue culture involves taking small pieces of an embryo and placing it in a medium of plasma and embryonic juice. To prepare the cultured specimen for microscopy it is deposited on a coverglass, inverted onto a slide that allows for excavation, and sealed with paraffin. In this environment, cells can grow and their development can be observed. Porter, working in Claude's laboratory, was teaching himself tissue culture techniques to provide material into which Claude could insert particles he was isolating by fractionation to test whether they would affect cell development. Porter recognized that tissue culturing provided specimens sufficiently thin for electron microscopy. He commented, "Although not a peer among the microscopists at the time, I was experienced enough to perceive that such diaphanous cells might be suitable for electron microscopy, at least in their thinner margins" (Porter, 1987).

The technique Porter developed for preparing tissue-cultured cells for electron microscopy was demanding. Cells are selected under a light microscope and

> an area of the film surrounding these, a little larger than the mesh disc, is marked out. This area of film is then cut from the surrounding film with a fine sharp instrument or a pair of watchmaker's forceps. Thus freed, the bit of film with

cultured cells is gently peeled away from the glass until only a small corner remains attached. Kept under water in this way the thin sheet of plastic retains its smooth extended form so that adhering cells are not distorted. The small wire mesh disc, immersed beforehand in the washing bath, is now slipped under the film and the two are so manipulated that the film is spread over the screen's surface. They are then lifted from the bath, drained of water, and placed to dry over phosphorous pentoxide. (Porter, Claude, & Fullam, 1945, p. 236)

Despite the authors' claim to have developed "relatively simple means" to make micrographs of cultured cells, the procedure was extremely delicate and not widely adopted.[28] It yielded, though, a dramatic result: an image of the cytoplasm eight years before comparable images were available with thin slices. As I will discuss in the next chapter, it generated a line of research on a new structure, first identified in these micrographs as a lace-like reticulum. When thin-sectioning techniques were developed, Palade and Porter relied on their experience with micrographs of tissue-cultured cells to interpret the thin-section differently than most of their peers.

There were, though, reasons to be dubious of the micrographs of tissue-cultured cells. Tissue-cultured cells are grown in abnormal conditions and the very spreading which made them suitable for electron microscopy could also engender artifacts.[29] This was especially true of the lace-like reticulum that was first observed in these preparations. Porter, however, appeared to have no doubts as to the reality of this structure. Because it did not correspond to anything seen in light microscopy, his confidence could not be based on correspondence with other techniques. Rather, what convinced him that this was not an artifact seems to have been the power of the image itself together

---

[28] Porter commented, "The preparation of adequate specimens was, at first, a discouraging process, but not totally so. I found that most cells able to grow in vitro would grow on Formvar-coated coverslips and that the Formvar film could be peeled from a glass surface and transferred under water to the EM grids then in use. When the grid, held between the points of watchmaker's forceps, was removed from the water and drained on filter paper, the Formvar film stretched over and adhered to the grid surface. The technique required a steady hand as well as determination and endurance. Approximately 50% of the specimens were satisfactory" (Porter, 1987, pp. 59–60). He continued, "When the time came to put the specimen into the electron microscope, not much of worth was expected. To our everlasting delight, however, the first specimen was surprisingly good and served to introduce the observers to more structural information than had been expected or could be interpreted" (p. 60).

[29] Novikoff commented on tissue culturing as a potential source of artifacts: "it reveals the *capacities* of the cultured cells, but perhaps not the *actualities* of those cells in the organized structure of the multicellular organism. The specialized milieu in which they are grown is quite different from that encountered naturally by cells embedded in tissue mucopolysaccharide or wedged in tightly among neighboring cells, as in epithelium – cells always under the controlling neural, hormonal and neurohumoral influence of the organism" (1959, p. 2).

with the fact that he could link the structure to speculation that cells must have a cytoskeleton that determined their shape (Needham, 1942).

Altering the Specimen to Survive Microscopy and Generate an Image

Beyond preparing a sufficiently thin specimen, electron microscopy required transforming the specimen into a condition that could both withstand the conditions in the electron microscope and produce a distinct image. In the electron microscope, specimens are subjected to vacuum, or near vacuum, conditions. The reason is that the electron microscope image is created by collisions of the electron beam with molecules, and collisions with air molecules would result in a haze in the final image. Unless researchers removed the major constituent of a cell – its water – prior to placing the cell in vacuum conditions, the water would vaporize and disrupt cell structures as it escaped. The removal of water could both shrink and distort the shape of the specimen and move or remove other cell constituents, raising serious risks of generating artifacts. In addition to dehydrating the cell, it was necessary to stabilize the structure in the cell to preserve the morphology against disruption resulting from either the removal of water or metabolic processes. This was most commonly achieved through fixation, a chemical treatment that both displaces the water and creates new chemical bonds that stabilize structures in the cell. Finally, because the resulting image was due to scattering of electrons, and because cellular materials are all of roughly the same density and will therefore scatter electrons roughly equally, it was necessary to stain those structures so that they would scatter more electrons. The chemical bonds created by many fixatives, though, also enhance scattering of electrons and thus served as a stain for electron microscopy. Thus, fixation was the central process in preparing the specimen for electron microscopy.

Because fixation involves the displacement of the water[30] and the creation of new chemical bonds in the specimen, the process radically alters conditions in the cell. The charge of artifact accompanied the use of fixation since its introduction in the late nineteenth century. These charges were fueled by the investigations of Fischer (1899), who applied different fixatives to homogenous solutions of proteins such as albumose, gelatin, egg albumin, and peptone, and generated structures that were filamentous, reticular, or granular and resembled structures observed in fixed cells. The charge of artifact was further encouraged by the fact that the same item had a different

---

[30] A major concern here was the fact that the fixative had to spread through the cell, creating currents which could displace soluble material or extract it from the cell altogether upon washing.

appearance when prepared with different fixatives, or even when prepared with the same fixative using different techniques to apply it. Further compounding the problem is that the process by which fixatives worked – for example, what substances in the cell they created bonds with – was generally unknown.[31] This made it difficult to determine whether what was seen in a fixed cell was a structure that was originally in the cell or an artifact.

Although some scientists rejected all use of fixatives, most were convinced that they could in fact reveal real structure in living tissue. Lacking a theoretical understanding of how various fixatives worked, however, these investigators typically had to assess fixatives by their results.[32] But what should the results look like? In light microscopy, investigators could sometimes compare the image of the fixed cell with that of living cells. Observation of the same structures in the unfixed cell provided evidence that the fixative was producing reliable results. Yet, the interest in fixation and staining was to observe structures that could not be seen in living cells. In such a case, researchers had to judge whether the image was likely to correspond to preexisting structures without independent access to that structure. Often, they relied on the plausibility of the image produced. Porter and Kallman (1953), for example, set out some criteria for judging from a micrograph whether the cell had been well fixed:

> Probably the optimum requirements to be made are that the cell should not detectably change its shape under the action of the fixative and that its cytoplasm should not, under high resolution, show discontinuities and lacunae of irregular size and angular form. Such structures with associated surfaces, and density differences, if present in the ground substance of the living cell,

---

[31] Porter commented on the lack of understanding of how osmium tetroxide functions despite previous studies: "Such studies as those of Hardy (1899), Monckeberg and Bethe (1899), Fischer (1899), Mann (1902), Hofmann (1912), Berg (1927), and Baker (1945) have provided only round indications of the reactions of various fixatives, including osmium tetroxide with proteins and fats" (Porter & Kallman, 1953, p. 127). Six years later Isidore Gersh commented, "It is an impressive commentary on our ignorance of the mechanism of fixation even now, that we do not know why osmium solutions should result in a homogeneous-appearing nucleus, while other fixatives like formol-Zenker result in nuclei with crisp chromatin clumps, or why Bouin's fixative preserves mitochondria only rarely, while potassium dichromate seldom fails to preserve them" (1959, p. 34). Gersh provided the following example of what was known about the operation of osmium: "Osmium oxidizes aliphatic and aromatic double bonds and sulfhydryl groups, alcoholic hydroxyl groups, and some amines. It also has an affinity for certain nitrogenous groups. Sites of oxidation of adjacent hydroxyl or ethylenic groups of adjacent molecules are thought to be bridged by reduced osmium" (1959, p. 38).

[32] Baker comments, "Nowhere in cytological technique has empiricism run riot so freely as in the invention of fixative mixtures" (1942, p. 4).

would most probably scatter light in dark field microscopy, whereas under such examination the ground substance is normally quite clear and nonrefractile. (pp. 127–8)

Early electron microscopists drew upon the armory of fixatives already employed in light microscopy. In his early studies of electron microscopy with separated fractions, Claude explored the potential of a variety of fixatives (formaldehyde, potassium dichromate, osmium tetroxide). Likewise, he and Porter, in their first study with tissue-cultured cells also tried several fixatives (chromic acid, acid formaldehyde, Flemming's mixture) and concluded that osmium tetroxide (OsO_4) vapors or solutions generated the clearest and most detailed images. Formaldehyde, they found, failed to show cytoplasmic structures, while chromic acid caused the cytoplasm to shrink and revealed only very small granules in it, and Flemming's solution yielded granules and mitochondria with a very coarse appearance. They claimed that alcohol, acetic acid, and freeze-drying (see below) were even worse in producing images resembling living cells. The standard, thus, was whether the resulting micrographs looked the way researchers expected them to look.

From the outset of electron microscopy, osmium tetroxide was by far the most widely used fixative. It had a long history of use in light microscopy, having been first introduced by Max Shultze in 1865 in his studies of the marine protozoan, *Noctiluca*. Strangeways and Canti (1927) had compared it with several other classical fixatives and concluded that it produced the least detectable change, the most evident being an increase in light scattering by the nucleus, which they attributed to the development of a "fine precipitate" (p. 9).

One way of responding to the worries about chemical fixation relied on the use of an alternative preparation technique, freeze-drying, which we have already briefly encountered in the work of Altmann and also of Bensley. The technique had been pioneered by Altmann (1890) who froze small pieces of tissue and kept them over sulfuric acid in vacuo at a temperature of −20° C for some days. The water dissipates directly into a gaseous state so that there is no intervening liquid phase that might distort the cell, as there is in chemical fixation. The result is a progressive dehydration of the tissue. Bensley and Gersh revived and improved Altmann's technique, in large part by freezing to even colder temperatures.[33] Bensley and Gersh (1933a, p. 212) argued that

---

[33] Gersh (1932) described the principles underlying the technique: "In the freeze-drying method . . . the system is cooled very rapidly to such a low temperature [−40°C] that the cohesive effects existing between ambient molecules at the instant of cooling predominate over all others; ideally all molecules are literally frozen in their tracts; they have no mobility; only occasionally does the translational energy of a surface molecule become great enough for it to escape; then it

this approach was unlikely to produce any serious distortion of the location of substances in the cell:

> It may be assumed that the redistribution of substances in the cells frozen rapidly at a temperature of liquid air is only such as is determined by the formation of ice crystals. Apart from this pushing aside of the protoplasmic constituents by ice crystals, the distribution of protoplasm itself will remain for the most part unaltered.

In addition, freeze-drying would not remove the water soluble constituents of the cell. As they noted, the technique did run the risk of forming ice crystals that could dislocate other cell structures, but such artifacts were usually sufficiently gross that they were easily recognized. The disadvantages of freeze-drying were that it was more difficult to perform and produced specimens that were extremely fragile.

Comparison of micrographs of freeze-dried cells with chemically fixed ones indicated that freeze-drying did not produce significantly different results from ordinary chemical drying.[34] Fernández-Morán (1952), for example, explored whether freeze drying yielded superior images of nerve fibers. Previous studies using chemical fixation had detected thin filaments in nerve axons and had indicated the existence of concentric laminated structures formed by thin membranes on the sheath. Fernández-Morán indicated the factors that gave rise to his concerns about these micrographs: "since the preparation techniques employed involve fixation, dehydration and in many cases embedding and sectioning of the nerve fibres, the existence of these elements cannot be definitely established until the artefacts introduced by such manipulations have been adequately evaluated" (p. 282). In this case, he concluded, *"A comparative study of fresh frozen sections from the same nerve segment, which were subjected either to freezing-drying or to parallel osmium fixation, shows that the axon and sheath structures revealed in both are nearly identical,* but better preservation and contrast was achieved in the osmium fixed sections. *For routine examinations osmium fixation of the thin fresh frozen sections would therefore appear to the method of choice"* (p. 291).

---

leaves the surface for the highly evacuated vapor phase and is trapped in a liquid nitrogen trap or expelled through the pumping system. In the freeze-drying method the pumping is continued until all the water molecules have escaped in this way" (p. 206). The tissue was embedded in paraffin while still in vacuum, and then cut into section and mounted on slides.

[34] Some researchers concluded that, if anything, freeze-drying produced greater artifacts: "Electron microscope studies of tissue culture preparations have shown that freeze-drying gives an incorrect picture of the structure of protoplasm, and it is reported that the shrinkage and distortion of certain micro-organisms such as *Bacillus megatherium* are even more marked after freeze-drying than after air-drying" (Drummond, 1950, pp. 92–3).

In this case, the correspondence of results produced by two independent techniques was the basis for resolving the chief worries about chemical fixation. It is important to note that in this case it was not a preestablished technique that provided confidence in chemical fixation but another technique being developed simultaneously. Moreover, as the quotations from Fernandez-Morán make clear, more than correspondence was at stake since he concluded that osmium yields better preservation. That judgment must have been based on independent expectations of what the image should look like, not simply on the comparison of the images produced by the two techniques. The detailed images resulting from chemical fixation evidently set the standard for freeze-drying. (For a contemporaneous review of chemical and freeze-drying techniques, see Bell, 1952.)

Although the comparison with freeze-drying resolved the general worry over chemical fixation, another concern stemmed from the fact that different chemical fixatives produced markedly different results. Bretschneider, for example, tested eighteen fixatives on the radicular cells of a plant and offered the following assessment: "The highest degree of coagulation was found to be produced only by mixtures of formalin and osmium tetroxide, preferably in combination with chromic acid or potassium dichromate. When these substances are used, the diameter of the plasma and karyolymph filaments (elementary parts of the finer structure) [is] reduced to the minimum. This condition is presumed to resemble the natural condition of plasma as closely as possible. All other fixation substances, especially acid, markedly flocculent and rapidly diffusing ones, were found to be unsuitable for electron-microscopic purposes" (Bretschneider, 1952, p. 313; see also Afzelius, 1962).

Very quickly, osmium tetroxide won favor as the fixative of choice for electron microscopy. Again, we should inquire about the basis of this judgment. One clear virtue of osmium tetroxide was that the micrographs produced with it exhibited greater detail than those made with other fixatives. (See Figure 4.5, which schematically compares osmium, permanganate, and freeze-dried preparations.) Such detail could of course have been artifactual, at least in principle, but investigators found it compelling. The images were generally consistent with those produced with light microscopy and with freeze-drying. Moreover, as we shall see in the next chapter, the structures shown in the micrographs generally fit readily into developing accounts of cell mechanisms. The one clear exception to this will be the focus of the last section of this chapter.

Settling on osmium still left open important questions about fixation methodology. In the early 1950s, both Porter and Palade began to explore variations in the use of osmium tetroxide. As part of their effort to understand

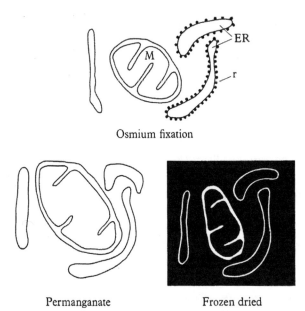

Osmium fixation

Permanganate   Frozen dried

Figure 4.5. A schematic comparison of the effects of three different preparations – staining with osmium, staining with permanganate, and freeze-drying – on mitochondria (M), endoplasmic reticulum (ER), and ribosomes (r). Reproduced with permission from G. H. Haggis, D. Michie, K. B. Roberts, and P. M. B. Walker (1964), *Introduction to molecular biology*. New York: John Wiley and Sons, Inc, Figure 5.1 on p. 114.

the operation of osmium vapors, Porter and Kallman explored the effect of extending the time of fixation from ten minutes to sixteen hours. They reported that with longer fixation "the formed bodies (the majority limited by membranes) become more sharply defined, while the diffuse and frequently fibrous components of the ground substance are removed" (Porter & Kallman, 1952). The following year Porter and Kallman (1953) proposed that this was due to the materials of the matrix being decomposed to a state that allows them to diffuse out of the cell. Finally, they used osmium tetroxide to treat various homogenous materials, both lipid and protein, that were similar to those thought to occur in cells. With albumin and fibrinogen they reported the initial formation of a gel followed by return to a liquid state and argued that this corresponded to the effect in osmium fixed cells of an initial staining of a variety of materials followed by removal of the fibrous components.

Palade (1952b) undertook a systematic study of osmium fixation at different pH values. He found, for example, that unbuffered osmium resulted in acidity upon first contact with cell tissue, but that osmium buffered with veronal caused the cytoplasm to appear more homogeneous. The resulting micrographs most closely resembled those Porter had produced with tissue-cultured cells. Palade's buffered solution of 1–2% OsO_4 in a neutral Michaelis buffer became known as "Palade's Pickle" and was widely adopted. Dalton (1955) added chromium salts to cross-link proteins and stabilize them.

Although Palade's technique for staining cells with OsO_4 was widely adopted, that did not mean that there were no controversies over its application. Sjöstrand took issue with Palade on several aspects of the technique. For example, while faulting the time lag in Palade's application of the fixative (which, he claimed, caused tissues to deteriorate), Sjöstrand also dismissed the importance Palade had placed on maintaining the pH of the specimen:

> We owe a good deal to Palade (1952) who modified the osmium fixation to give a decent preservation of the fine structure of cells. His introduction of the veronalacetate buffer as a medium for the osmium tetroxide has improved the quality of fixation for several tissues. However, the pH of the osmium tetroxide seems not to be so important as he originally claimed. Variation of pH within a wide range from pH 4, and in some cases even pH 2, to pH 8 does not substantially affect the result in many tissues, e.g., the kidney (Rhodin, 1954), the intestinal epithelium (Zetterqvist, 1956) and the sensory epithelium of the inner ear (Wersäll, 1956). Most of the great differences of the quality of fixation described by Palade (1952) in his study at a low resolution seem to represent various degrees of post-mortem changes. (1956a, p. 459)

To speed fixation, Sjöstrand injected fixative into live animals and immersed cut tissue in fixative as quickly as possible. He also claimed that the tonicity of the fixative, which Palade argued was not critical, was very important in preventing swelling:

> The osmium tetroxide solution of Palade is strongly hypotonic and in several cases ... hypotonic solutions of osmium tetroxide produce a more or less marked swelling of the cells. The importance of using isotonic solutions has therefore been stressed (Sjöstrand, 1953a). It has even been possible to obtain as good a preservation of the cell structure by using isotonic non-buffered solutions of osmium tetroxide as for instance a solution of isotonic sodium chloride. (Sjöstrand, 1956a, p. 459)

Rasmussen traced the methodological dispute between Palade and Sjöstrand in part to a difference in the use to which they put electron micrographs. For Palade, the use was primarily qualitative – to provide morphological perspective on the biochemical information generated, for example, from fractionation and functional analysis of mitochondria and mitochondrial fragments.

Sjöstrand's interest, in contrast, was quantitative – to determine with precision the size of cellular components so as to support models of the molecular architecture of membranes. This difference was manifest in their respective papers at the Third International Conference on Electron Microscopy, held in London in July, 1954. Sjöstrand asserted,

> The quantitative attitude is stressed because it is necessary for the identification and classification of the different components . . . The electron microscope is an efficient measuring device making an exact quantitative description possible, provided it is mastered to give high resolution. (1956b, p. 35)

Rasmussen argued that for this reason Sjöstrand showed only micrographs of mitochondria arrayed longitudinally. In that position the membrane would be perpendicular to the electron beam, and the micrograph would reveal the true width of the membrane. Palade, in contrast, judged the methods of electron microscopy insufficiently reliable to support quantitative conclusions:

> It seems, therefore, that the dimensions and spacings shown in fixed material by the electron microscope cannot be considered sufficiently true to nature to permit us to deduce the chemical composition and molecular architecture of a certain structure by finding out, as recently proposed (Sjöstrand, 1953a, b), which particular kind of molecule would best fit a given spacing. (1956c, p. 132)

The dispute between Sjöstrand and Palade also turned on the question of how to evaluate the accuracy of micrographs. For Palade, the critical criterion was consilience with results from other techniques, both other forms of microscopy and fractionation studies.[35] Palade claimed that Sjöstrand's criterion was the detail in the micrograph itself:

> Sjöstrand is of the opinion that fixation tends, in general, to disorganize the cytoplasm s;, in comparison with the situation *in vivo*, no structure is added, but structure may be subtracted. Accordingly the best fixative is considered to

---

[35] Rasmussen (1995) maintained that Palade's strategy undercut the ability of electron microscopy to conflict with biochemical results: "Palade removed virtually all potential for conflict between the evidence from electron microscopy and that from biochemistry. The two experimental approaches spoke to separate domains: the former to questions of arrangement; the latter to questions of molecular structure and function" (p. 400). Sjöstrand, on the other hand, by proposing to use precise measurements so as to assess the molecular structure of membranes, set up the potential for results that conflicted with biochemistry. The fact that Palade's approach supported and did not threaten biochemists, according to Rasmussen, turned them into allies and helped insure greater success for his approach over that of Sjöstrand: "I would suggest that it was the different kind of relationship that Porter and Palade designed with biochemistry that ultimately made electron microscopic cytology of the Rockefeller school the dominant configuration" (p. 419).

be that which leaves the specimen with a maximum of organization. Sjöstrand's view undoubtedly applies in numerous cases, but it supposes that there is always more order in a living than in a fixed specimen, which may not necessarily be true. (1956c, p. 133)

This controversy further illustrates how investigators evaluate techniques by their results and, in particular, by whether those results conform to expectations generated by theories of the phenomenon. For Palade, buffered osmium was the preferred fixative because it produced results that fit well with biochemistry and his own use of cell fractionation. As we will see in Chapter 6, this is particularly true of two new structures that appeared in the micrographs made with the new fixative – the cristae of the mitochondria and the ribosomes on the endoplasmic reticulum. For Sjöstrand what was critical was generating micrographs that seemed to support very precise measurements of the observed structures.

It is worth noting that even as most researchers had come to accept the reliability of micrographs produced with osmium fixation, there were still doubters. No less an investigator than Jean Brachet expressed his worries about relying on micrographs produced with osmium for definitive accounts of such cell structures as the endoplasmic reticulum:

One cannot help but admire the very beautiful photographs which have been published, during the past few years, in order to establish and demonstrate the fine structure of the ground cytoplasm; however, the interpretation of the findings is still open to question and the unpleasant problem of the possible artifacts must be raised. It is a matter of some concern that almost all of the work we have just described has been performed on buffered osmium tetroxide fixed tissue; it is hard to believe, from the daily experience of ordinary cytology, that one single fixative can give a trustworthy image of the cell. Careful studies by Frederic (1956), with the so-called 'anoptral' microscope, failed to show the existence of a reticulum in living cells; such a reticulum became conspicuous after osmium fixation only.... It is not impossible that the nice double membrane structure, which is so conspicuous in the ergastoplasm and which will be found again in other cell constituents, is nothing but a fixation artifact. Nor can we exclude the possibility that many structures, which do not fix osmium to any large extent, are invisible in present day electron microscopy. (1957, pp. 40–1)

4. A CASE STUDY OF AN ARTIFACT CHARGE

To conclude this chapter, it will be useful to look in some detail at a specific case in which investigators raised the charge of artifact. The case involves the Golgi apparatus, which, as discussed in the previous chapter, had long been the target of claims of being an artifact. Two of the central figures in developing techniques for electron microscopy, Claude and Palade, joined the legacy of those arguing that it was artifact.[36] The case is ironic, because both researchers returned to work on the Golgi apparatus several years later, and Palade played a central role in revealing the function of the Golgi (see Chapter 6). But in 1949 they charged that it was an artifact of osmium fixation. The case they made is a model argument for demonstrating an artifact. Because the conclusion was ultimately rejected even by the authors, it reveals just how epistemologically difficult the evaluation of artifacts is.

In the first of their two papers, Palade and Claude (1949a) applied ethanol to homogenates created from a wide variety of cell types from several different species. They claimed it transformed lipid inclusions that occurred in the area where the Golgi apparatus was typically found into myelin figures. Such figures were known to be very polymorphic and unstable, an interesting property given the variable appearance of the Golgi apparatus. In cells, they argued, such myelin figures are not able "to expand freely, but were forced to adapt themselves to the spaces available between the masses of precipitated cytoplasm" (p. 44). Moreover, the appearance of the myelin figures varied with the concentration of ethanol. Under some conditions, their appearance was very similar to that of the Golgi apparatus: "For a certain range of ethanol concentrations, generally comprised between 40 and 55%, the morphology and topography of the intracellular myelin figures are surprisingly similar to those assumed by the Golgi apparatus in corresponding cells" (p. 49). Palade and Claude summarized their results with ethanol:

> Myelin figures can duplicate faithfully the numerous and different forms ascribed at various times to the Golgi apparatus. Thus they can take the appearance of massive, or canalicular networks, scattered strands, canaliculi, polymorphic bodies, "poly-systems" and "mono-systems".... These facts strongly suggest that the Golgi apparatus may be a myelin figure or a complex of myelin figures artificially induced in cells by given cytological techniques. Such a hypothesis could reconcile the different aspects presented by cells of the same time when examined in the fresh condition or after the application of recognized cytological procedures. (p. 52)

In normal preparations, however, ethanol is not introduced until after staining with osmium tetroxide or silver nitrate, at which point the Golgi apparatus

---

[36] According to Porter (Interview, 1987, University of Maryland, Baltimore County), Claude had become convinced that the Golgi apparatus was an artifact and initiated the project. Palade had recently been recruited into the laboratory by Claude and was drawn into the project of establishing that the Golgi apparatus was indeed an artifact.

is already visible.[37] Accordingly, in their second paper Palade and Claude (1949b) explored the effects of the more common fixation and staining procedures employed for the Golgi apparatus. They also applied the usual Golgi techniques to lipid droplets of known chemical composition and on fractions isolated by cell fractionation. In all cases they found evidence that lipid droplets with high phospholipid content developed into myelin figures. In the homogenate preparations, the myelin figures exhibited "the same polymorphism, growth particularities, and typical intracellular distribution as the figures produced by ethanol and described in the previous paper" (p. 81). Among the morphological features exhibited was the division into two regions that a number of investigators had observed. In addition, Palade and Claude proposed to account for the success of Beams and King (1934) in segregating the Golgi apparatus through centrifugation: "The high-speed centrifuging experiments prove only that the lipid droplets can be displaced within the cell and that their specific gravity is less than that of the cytoplasm" (p. 92). Further, they attributed the evidence suggesting a role in cell secretion to the fact that "The haphazard growth of a myelin figure may occasionally bring it in contact with a bile capillary, secretion granules, or neutral fat drops. Likewise, developing myelin figures often penetrate between zymogen granules or mucus droplets which, subsequently, will seem embedded in the meshes of a Golgi apparatus" (pp. 92–3).

Palade and Claude also observed the formation process over time, and concluded that the wave of acidity moving through the preparation corresponded to the formation of the myelin figures. They also proposed a mechanism by which the acid could create the myelin figures through hydrolysis of the phospholipid molecules into more soluble substances that could then be rearranged into myelin figures. Once the acid had acted, Palade and Claude hypothesized, various electrolytes in the fixation media facilitated growth and stabilization of the myelin figures. Finally, further increase in electrolytes stopped the formation of new myelin figures and the slow-diffusing osmium tetroxide then fixed the figures.

Altogether, Palade and Claude provided as compelling a case for a structure being an artifact as one could desire. Not only were they able to generate images that resembled the Golgi apparatus and trace their development from phospholipid inclusions, they also offered a theoretical account

---

[37] Palade and Claude noted, though, "it may be recalled that Golgi himself, in a modification of his original procedure, proposed the use of an ethanol mixture (32% final ethanol concentration) as a fixative, and recommended it specifically for its better, quicker, and more constant results in the demonstration of the apparatus" (1949b, p. 72).

of how the artifact was generated. They could account for the reliable generation of definite images (as well as the variability in appearance, which was a factor pointing to it being an artifact). In this case, the limited means of demonstrating the Golgi further helped their case since, once they had shown how the preferred means of establishing its existence could be generating an artifact, there was no countervailing independent evidence. Thus, they could effectively counter the primary evidence offered for the reality of the Golgi apparatus. They also could answer the claims suggesting a functional role for the Golgi apparatus, thereby undercutting the role of the theoretical account of cell operation in justifying the microscopical evidence.

Coming near the end of the long history of contention over the reality of the Golgi apparatus, Palade and Claude's study did little to alter opinions. Yet, it was a model for establishing that something is an artifact, and at least in their laboratory had the effect of prohibiting discussion of the Golgi apparatus over the next fifteen years.[38]

5. EQUIPPED WITH NEW INSTRUMENTS AND TECHNIQUES TO ENTER TERRA INCOGNITA

According to the criteria set out earlier in this chapter, both cell fractionation and electron microscopy established themselves in the 1940s and 1950s as credible sources of evidence about component parts and operations within cells. Both provided patterns of results that were determinate and repeatable. Each secured consilience of the results it generated with those obtained in other ways and, especially, with each other. Finally, the results of each technique cohered well with emerging mechanistic models that accounted for a variety of cellular phenomena. The one major exception was when they delivered evidence on the Golgi apparatus, but this might be in part explicable as due to delays in finding a place for the Golgi apparatus in the mechanisms of the cell. In Chapter 6, I will show that once it was shown to comprise a mechanism that packaged proteins for secretion, doubts about its reality soon vanished.

---

[38] The second edition of *General Cytology* (de Robertis et al., 1954) was one of the few sources to mention Palade and Claude's work, doing so at the conclusion of a general summary of research on the Golgi apparatus in a section in fine print. Noting that Palade and Claude's work would suggest that the Golgi region was the locus of phospholipid droplets, they commented, "The confirmation of this concept would bring about the elimination of the name Golgi apparatus and perhaps its change into Golgi substance" (p. 148).

Neither classical cytology, relying primarily on light microscopy, nor biochemistry, relying on homogenates, had been able to enter the terra incognita in which resided the mechanisms responsible for cell phenomena. The new tools of cell fractionation and electron microscopy, however, provided the necessary resources. Electron microscopy made it possible to identify component parts in the cell much smaller than those identifiable by light microscopy. Cell fractionation made it possible to examine the operations of units of larger scale than could survive the homogenation process. Moreover, the techniques could be used together – fractions could be examined in the electron microscope, providing a means of relating the operational units identified by fractionation with component parts identified in electron microscopy of whole cells. The challenge for the scientists was to produce new results with these tools. In the next two chapters I turn to how scientists met this challenge.

5

Entering the Terra Incognita between Biochemistry and Cytology

Putting New Research Tools to Work in the 1940s

> In recent years, the construction of a bridge between these two levels
> of knowledge has been initiated. This has been due to the adaptation
> and employment in biology of techniques derived from physics and
> chemistry and to the breaking down of the barriers which previously
> separated these sciences. Below the structure visible to the microscope
> there exists a true organization of molecules and micelles in the different
> phases of the system which constitute protoplasm.
>
> (de Robertis et al., 1949, p. 64).

New research tools, especially cell fractionation and electron microscopy, opened for investigation the uncharted territory between biochemistry and cytology. The goal was to explain how cells carry out their basic functions. In the nineteenth and early twentieth century, the activity that had received the most attention was cell reproduction. The basic operations in cell division, including those carried out by chromosomes in the nucleus, had been described by cytologists using stains and the apochromatic lens decades before the advent of cell biology. The focus of early cell biology was rather on functions performed in the cytoplasm, especially capturing energy and synthesizing proteins. The first important steps in developing mechanistic explanations of these functions was to identify the mitochondrion and the endoplasmic reticulum as the cell organelles responsible for each, a project largely accomplished in the 1940s (although the name *endoplasmic reticulum* was not introduced until the early 1950s).

One laboratory at the Rockefeller Institute, in which Albert Claude, Keith Porter, George Palade, and others performed pioneering research, played the pivotal role in establishing these structure–function linkages. As we will see, understanding cell function was not the initial objective of research in this laboratory, and it was only at the end of the 1940s that it officially became a laboratory for cell biology. Hence, it is important to understand the transformation in this laboratory. Other laboratories, ones initially more explicitly devoted to cell physiology, also made significant contributions. At appropriate points in the chapter I will briefly profile three of them. It was the laboratory at the Rockefeller Institute, though, that established the model for what cell biology was and how it was done, and that will be the main focus of this chapter.

1. FIRST STEPS TOWARD CELL BIOLOGY AT THE ROCKEFELLER INSTITUTE: CLAUDE'S INTRODUCTION OF CELL FRACTIONATION

The pioneering investigations in cell biology at the Rockefeller Institute occurred in the cancer laboratory of James Murphy. Murphy was pursuing a line of research that traced back to Peyton Rous, whom Simon Flexner, the first director, recruited to the Rockefeller Institute in 1909. Shortly after arriving at Rockefeller, Rous was presented with a chicken with a large lump on its leg. The lump turned out to be a tumor that Rous showed could be transmitted from one animal to another by inoculation either with small portions of the tumor or with a filtered extract that strained out all cancer cells. He concluded that the tumor was carried by what were then termed "viral agents," characterized only as infectious entities that were not bacterial in nature. Although the tumor was to bear his name, Rous had been a reluctant recruit to cancer research and in 1915 turned instead to research on blood preservation (Corner, 1964).

With the change in direction of Rous' research, Murphy, who had been Rous' assistant, was promoted to associate member of the Institute and placed in charge of cancer research. (He became a full member in 1923.) For a number of years, his primary focus was on the possible role of lymphocytes in resisting cancer (an inquiry rooted in an observation he and Rous had made that embryos and brain tissue, both lacking lymphocytes, lacked resistance to transplanted tumor cells). By the late 1920s, though, Murphy turned his attention back to the agent responsible for Rous chicken sarcoma. Rous had shown not only that the suspected causal agent was found in cell-free extract but also that tumor cells could be killed with ultraviolet light without destroying the agent, which suggested to Murphy that the Rous chicken tumor agent might not be bacterial but rather have an "enzyme-like nature" (Murphy, Helmer, & Sturm, 1928). He had also determined while working under Rous that freezing and drying the tissue (a process known as lyophilization) did not kill the active agent. He now sought to purify this agent and turned that task[1] over to a new assistant, Albert Claude, who had received his M.D. at Liége in 1928 and spent a year in a tissue-culture laboratory in Berlin before Murphy recruited him in 1929.

Claude pursued a number of strategies in the attempt to purify the agent, including adsorption, precipitation, and dialysis. Precipitation of carbohydrate with gelatin followed by dialysis resulted in a twentyfold "enrichment of the tumor producing agent" (Claude, 1935). He identified a protein and a phospholipid as the principal constituents of the active residue. None of these techniques, however, produced the desired purification.

At this point Claude read the report of two British researchers, Ledingham and Gye (1935), who had tried using high-speed centrifugation to separate a tumor producing agent (also McIntosch, 1935, who employed a Henriot and Huguenard air-driven ultracentrifuge). Although the substance Ledingham and Gye isolated had less tumor producing capacity than the initial cell extract, Claude saw the promise of their approach and set out to refine it. In 1937 he first reported on using a high-speed centrifuge to isolate a more potent tumor causing agent (Claude, 1937)[2] and in 1938 he claimed even better results – a tumor-causing agent with ten to fifty times greater potency than the original extract (Claude, 1938b). In the 1938 *Annual Report* Claude began to characterize the chemical makeup of the active particle, arguing that it contained phospholipids as well as ribosenucleic acid. The chemical composition of the tumor-producing fraction was also the focus of a further paper (Claude, 1939). Viewing the fraction under a dark-field microscope, he observed it to consist of small granules approximately 70 mμ in diameter.

As a control to his work with tumor cells, Claude (1938a) centrifuged normal tissues, and discovered to his surprise a fraction with granules similar to that generated by centrifugation of tumor cells. Its chemical composition was also similar: "The chick embryo material, like the tumor fraction, is

---

[1] Murphy continued to work on tumors of fowl with Sturm and was the official head of the laboratory until his retirement in 1949. He focused on such topics as chemical induction of tumors (by dibenzanthracene, for example) and their transmissibility. He also examined the effect of such variables as season of year and genetic constitution on susceptibility to the cancers.

[2] In the 1937 *Annual Report*, Claude described introducing centrifugation to address the problem of why a related tumor, Chicken Tumor 10, could be transferred through filtrates. In 1931 Murphy and Claude had shown that alumina gel had to be added to produce an active filtrate with Rous sarcoma. They proposed that the alumina gel removed an inhibitor. Claude now argued that the problem with Chicken Tumor 10 was also due to an inhibitor by showing that the centrifuged particle when combined with water could generate the tumor. When the particle was combined with the supernatant from the centrifugation it could not do so, suggesting that the supernatant contained an inhibitor.

found to consist essentially of a phospholipid-aldehyde portion, associated with a nucleoprotein of the ribose type" (Claude, 1939, p. 214). He concluded, "These observations indicate that a phospholipid-ribose nucleoprotein complex is probably a general constituent of normal and tumor cells" (p. 215). This surprising result led Claude to shift his research away from the cancer paradigm and toward an investigation of the constituents of normal cells.[3]

In the following year Claude began to consider the hypothesis (later found to be false) that these granules might be mitochondria and advanced several bits of evidence in support: they (a) were of similar size and shape as mitochondria (based on size estimates by Cowdry, 1918), (b) had a similar composition of phospholipids and nucleoproteins, and (c) exhibited similar responses to heat and acids (Claude, 1940; Claude, 1941). In the 1940–1 *Annual Report*, he advanced an additional piece of evidence: "Janus green has a selective affinity for mitochondria under proper conditions and the isolated particles show an identical property. With methyl green the cell nucleus takes a brilliant green while the mitochondria become purple. The isolated particles take the same purple color with this dye" (p. 72).

Although Claude initially misidentified the granular particles as mitochondria, he later corrected himself and made landmark contributions to the understanding of both the granular particles and mitochondria. The misidentification does not detract from the general point that, having recognized that some of his findings with tumor cells pointed to important components of cells more generally, Claude's research program was radically transformed. This involved not simply turning in a different direction, but as Rheinberger (1995) has emphasized, arriving at a quite different conceptual framework as he reinterpreted his work: "What had been the tumor agent, 'was' now a cytoplasmic particle. The new scientific object arose and began to be delineated first as an intrusion, then as a supplement within the confines of the old cancer research system. Shortly after, it took over and transformed the system itself" (p. 61).

---

[3] The shift, however, was not total. In a talk at the American Association for Cancer Research on 19 April 1941, Claude offered a proposal as to how putative mitochondrial particles related to cancer. From the fact that what he took to be mitochondria seem to increase in number in cells when they are dividing, he inferred that they figure in the process of cell differentiation. He then proposed, "The nature of the cellular response to cancer-inducing chemicals suggests that the system which is affected is probably that part of the cell which normally influences differentiation and growth" (*New York Times*, 20 April 1941). Claude went on to note that other researchers had linked mitochondria to cell respiration, and adduced evidence that some cancer causing agents also interfere with respiratory enzymes.

2. ROBERT BENSLEY: AN ALTERNATIVE APPROACH
TO FRACTIONATION

In changing his focus to normal cells, Claude drew close to the work of a more senior investigator, Robert Bensley of the University of Chicago (apparently Claude was initially unaware of Bensley's work). Just a couple of years earlier, Bensley had arrived at – and applied – the idea of using centrifugation as a means of separating mitochondria from cells. Bensley was a traditional cytologist whose career spanned the first four decades of the twentieth century. He started to explore cell staining techniques while recovering from a hunting accident in his early teens that cost him one of his legs. Already in the first decade of the twentieth century, Bensley was conducting research on mitochondria and the Golgi apparatus. In particular, he further developed procedures for fixing cells with acetic-osmic-dichromate and staining them with anilin-acid fuchsin and methyl green and copper-chrome-hematoxylin that proved useful for targeting mitochondria. Bensley stands out from the other major cytologists active in the first decades of the twentieth century, though, in that late in his career he pioneered a means of bridging from morphology to chemistry. This was necessary if the structural decomposition of the cell offered by cytologists was ever to be linked to a functional decomposition (especially to the level of biochemical reactions). Whereas we saw in Chapter 3 that his former student Cowdry abandoned research on mitochondria in the 1930s, having become pessimistic about the prospects for finding such a bridge, Bensley pursued his idea with two of his last graduate students, Isidore Gersh and Normand Hoerr.

As discussed in the previous chapter, Gersh (1932) revived and simplified Altmann's procedure for fixing cells without chemical reagents by freeze-drying. Bensley and Gersh (1933b) used the technique to test many older claims about the effects of different solvents and heat on mitochondria. Many of the fat solvents, such as acetone, had no effect on mitochondria. On the other hand, water, 0.02% ammonia solution, and artificial gastric juice dissolved them. Bensley and Gersh interpreted this as showing that "the main mass of the mitochondria substance is protein in nature" (p. 230). Bensley together with Hoerr then took the major step of employing centrifugation to isolate mitochondria from fresh liver tissue preparations. They prepared specimens by perfusing the liver with salt solution, then grinding it in a mincing machine and pushing the result through a sieve of cheesecloth.[4] Several successive

---

[4] This paper, which reported results that did not rely on the freeze-drying method, appeared as the sixth in a series entitled "Studies on cell structure by the freezing-drying method." The connection

centrifugations at slow speed then removed nuclei, blood corpuscles, and connective tissues. They then centrifuged the remaining supernatant at higher speeds for longer periods, which caused mitochondria to sediment. Bensley and Hoerr's goal was to analyze their chemical constitution. Although they expressed doubt in their first paper about the fat composition of mitochondria (Bensley & Hoerr, 1934a), their second paper showed mitochondria to be comprised of, on average, 43.6% fat by weight, but not to contain lecithin or cephalin (Bensley & Hoerr, 1934b). They also showed that precipitation at different acidities revealed the presence of least two proteins.

In further research, Hoerr (1943) refined the techniques for isolating cell components and Bensley (1937) refined the analysis of the constitution of mitochondria. Arnold Lazarow, another of Bensley's students, continued the chemical analysis of the constitution of mitochondria and also discovered a smaller particle that appeared cherry red when separated by centrifugation. He produced quantitative analyses of both the original and smaller particles, showing that the smaller particle contained more phosphorus and fat than did the mitochondrial particles. He demonstrated that both the mitochondrial fraction and the smaller particle (which he referred to as *the submicroscopic lipoprotein component*) oxidized succinic acid, which he interpreted as showing that both contained succinic dehydrogenase, cytochrome *c*, and cytochrome oxidase. Lazarow's smaller particles were in fact the ones Claude was finding at the same time and misidentifying as mitochondria. He also investigated their enzyme constitution several years before that became a focus in Claude's laboratory at the Rockefeller Institute. With regard to the submicroscopic lipoprotein component, Bensley commented on their common timing:

> I had no suspicion at first that still smaller particulates were present in the liver cell until Lazarow, by long-continued centrifugation, obtained a glassy cherry-red pellet composed of particles so minute that they were quite invisible under the microscope, but showed in the dark-field of the cardioid condenser a shimmering field of light in which individual particles could with

is found in the previous paper, also coauthored by Bensley and Hoerr, in which emulsions of fresh liver cells prepared by three different methods – grinding in a mortar, mincing in a latapie grinding machine, or kneading through bolting silk – were used to provide a check on results from freeze-drying. Their focus in that paper was on contents that remained once the nucleus and (putative) mitochondria were removed. They argued for the existence of a protein structure they dubbed "elipsin" which "by itself maintains the cell as a unit of organic structure after the soluble globulins, mitochondria, and chromatin have been removed seriatim by solution [and] is in reality the basis of the microscopic structure and of the organic continuity of the cell body" (Bensley & Hoerr, 1934a, p. 263).

difficulty be distinguished. We were investigating this particle when Claude (1940) announced his discovery of the presence of submicroscopic particulates in clarified saline extracts of embryo chicks. (1943, p. 329)

Bensley and his graduate students pioneered many of the advances for which Claude and the Rockefeller group were to get most of the credit. In 1943, Cowdry referred to Bensley as "the acknowledged founder of the new cytology" and adds, "As the lesser figures shrink and are forgotten his stature will grow" (1943, p. 8). Cowdry's prophecy, however, turned out to be erroneous as Claude and the Rockefeller group rapidly eclipsed Bensley.[5] One reason might well be Bensley's advanced age when he was pursuing this research – he was already professor emeritus. Another, perhaps not trivial, factor is that Bensley did not think of mitochondria as permanent structures but as coacervates that appear and then are reabsorbed into the protoplasm (a view Bensley held until the end of his career, see Bensley, 1953). The most important factor that made Claude and his laboratory the crucial locus for developing cell biology is that they – not Bensley – convincingly established the role of the mitochondrion in cellular energetics, thereby opening the productive endeavor of linking function, determined biochemically, to structure, identified cytologically.

3. COMPETING INTERPRETATIONS OF FRACTIONS FROM NORMAL CELLS

Although both Bensley's group and Claude differentiated two fractions, their initial interpretations of how these related to cell structures conflicted. In appealing to Cowdry (1918) for information about mitochondrial size in arguing that his small particles were mitochondria, Claude ignored Bensley and Hoerr's more recent estimates. According to Bensley and Hoerr, mitochondria were considerably larger than Claude's small particles.

Claude (1941) discussed the differences between his fractions and Bensley's at length. He devised a way of separating two fractions through successive centrifugation runs. One consisted of smaller particles (the size of his original preparation) and the other of larger particles (the size Bensley had identified as mitochondria). Having distinguished them, Claude maintained

---

[5] The American Association of Anatomists named an award for Bensley, which is described in the following terms: "The R. R. Bensley Award recognizes 'rising stars' in cell biology, who have already advanced anatomy through the study of cell biology. The award is presented to someone who has made a distinguished contribution to the advancement of anatomy through discovery, ingenuity, and publications in the field of cell biology."

that the smaller particles were mitochondria and argued that those in what he called the "large granule fraction," which corresponded to Bensley's particles, were secretory. Claude's analysis of the chemical make up of the smaller and larger particles indicated only minor differences (for example, both seemed to contain phospholipids, although the smaller particles contained about twice the percentage, and contained ribose nucleic acids, iron, and copper). Claude tentatively adopted a proposal put forward by Noel in 1923 that the large particles have their "origin in a progressive transformation of the smaller elements, or mitochondria" (p. 269).

In the discussion following the paper, Jack Schultz raised the possible relevance of another constituent of the cell proposed in the era of light microscopy, the ergastoplasm. As we will see, Schultz had collaborated with Caspersson at the Karolinska Institute in the 1930s, where they had proposed and advanced evidence that RNA played a role in protein synthesis. In a very few years, most researchers would view the small particles, ergastoplasm, and RNA as interrelated, but Claude held instead that the ergastoplasm did not exist: "In recent years, the majority of workers have come to consider the ergastoplasm as an artefact produced, as a rule by fixatives containing strong acids. It is rare to find well preserved mitochondria and ergastoplasm simultaneously in the same preparation and the suggestion has been made that the ergastoplasm was merely poorly fixed mitochondria" (p. 270). In support of this position, he cited a review by Bowen (1929).

In the period that followed, however, Claude came to recognize that his small particles were too small to be mitochondria, which themselves were large enough to be visible in the light microscope. He also tried traditional stains on the particles in the small fraction and discovered they stained differently than mitochondria. As a result, he radically altered his interpretation, construing the small particles now as a newly discovered cell constituent:

> the evidence, so far, indicates that the mass of the small particles does not derive from the grossly visible elements of the cell but constitutes a hitherto unrecognized particulate component of protoplasm, more or less evenly distributed in the fundamental substance and which imparts to it, in well-preserved preparations, its staining properties. In order to differentiate the small particles from the other, already identified elements of the cell, it may be convenient in the future to refer to this new component under a descriptive name which would be specific. For this purpose the term *microsome* appears to be the most appropriate. (Claude, 1943b, p. 453)

Claude now needed to place mitochondria elsewhere, so he reinterpreted his "large granule" fraction as most likely an impure mixture that included mitochondria. Unlike Bensley and others, though, he continued to maintain for years that secretory particles were a major component as well:

> Secretory particles are abundant in the guinea pig liver, especially in the fasting animal, where they accumulate and seem to fill the cell completely, and it appears probable that up to the present, mitochondria have not been isolated in a pure or concentrated form, a large part of the so-called "mitochondria" fraction representing probably, to a large extent, mature secretory granules. (Claude, 1943b, p. 455)

In 1943 Claude also began to employ a different procedure, centrifuging whole liver cells from *Amphiuma tridactylum* for one hour at 18,000 *g*, causing the cell contents to separate into distinguishable layers while remaining within the cell membrane. By staining the cells with Bensley's stain (acid fuchsin-methyl green), he rendered what he took to be secretory granules and mitochondria vivid red against a purple background (Claude, 1943a). This yielded a sharp distinction between four successive layers, which Claude interpreted as (1) glycogen, (2) a combination of secretory granules and mitochondria that appeared red and also contained nuclei, (3) the "purple substance" that he took to be microsomes, and (4) the true hyloplasm (cytosol).

In general, the Bensley group welcomed Claude's entry into the field of centrifugation studies that they had pioneered. He was an invited speaker at the symposium in November 1942 in honor of Bensley's seventy-fifth birthday, and many of Bensley's colleagues referred positively to his contributions (Hoerr called his work "superb"). However, they challenged his refusal to identify his large-granule fraction as essentially mitochondrial and his insistence that many of the granules were secretory. Hoerr said, "it is obvious that he has separated the liver mitochondria" (1943), appealing to the ability of staining to differentiate between mitochondria and secretory granules when they are present in, for example, pancreas cells. He further criticized Claude's methods of separation, arguing that they failed to produce pure preparations but rather yielded mixtures. Just as Claude continued to refer to *large granules* in preference to *mitochondria*, so Bensley's group rejected Claude's term *microsomes*, instead referring to the small particles as *submicroscopic particulates*.

The following year Claude (1944) offered yet another reason for questioning whether the large granules extracted from guinea pig liver included mitochondria – they failed to stain with Janus green. Now he concluded that liver may be a poor place to look for mitochondria, and reported results of a study on leukemic cells from rats, in which he contended that the large granule fraction consisted of true mitochondria. He then generated a chemical analysis very similar to that which he had offered for the large granules that he took to be secretory in liver. Over the next several years, he persisted in referring to the *large granule* fraction, acknowledging that mitochondria were part of the fraction, but continuing to focus on the idea that it also contained secretory granules.

4. LINKING CLAUDE'S MICROSOMES TO PROTEIN SYNTHESIS

Claude's analysis of his "hitherto unrecognized particulate component of protoplasm" – what he called microsomes – indicated the presence of RNA. Although Claude was hesitant to suggest a function for these particles, two other investigators, Jean Brachet and Torbörn Caspersson, led the way. Each deployed new cytochemical tools to establish the connection between RNA and protein synthesis.

Brachet: Selective Staining of RNA and Correlation with Protein Synthesis

While still an undergraduate medical student at the Free University of Brussels in the late 1920s, Brachet tried to determine the location of what was then referred to as thymonucleic acid[6] (DNA) in growing oocytes. At the time, thymonucleic acid was thought to occur only in animals while zymonucleic acid (RNA) was found primarily in plants (although it was also recognized to be present in the pancreas in animals). The discovery of the Feulgen reaction (Feulgen & Rossenbeck, 1924), which stains DNA green, made it possible for Brachet to try to localize DNA, but it was zymonucleic acid or RNA that was to be the focus of his major breakthrough. Needham and Needham (1930) had found large amounts of nucleic acids in sea urchin eggs, which stained intensely red with pyronine. Was this RNA? Brachet found that when he treated the stained cells with ribonuclease, which specifically digests RNA, this removed the red stain, indicating that it was RNA. Brachet hypothesized, based on the fact that the amount of RNA was closely correlated with protein synthesis activity, that RNA played a role in protein synthesis: "The conclusion to which we are led is that the pentosenucleic acids [RNA] might intervene, by a mechanism as yet obscure, in the synthesis of proteins, which

---

[6] The discovery that the substance contained deoxyribose, which led to its rechristening as *deoxyribose nucleic acid* (DNA), only came with the research of Phoebus Levene and Takajiro Mori (1929) at the Rockefeller Institute.

perfectly matches the available facts" (Brachet, 1942, p. 239, as quoted in Rasmussen, p. 60).

As he was completing the ribonuclease research, Brachet and his student Hubert Chantrenne began a collaboration with biochemist Raymond Jeener in which they isolated particles similar to those Claude had identified as microsomes using a prototype of Emile Henriot's air-driven centrifuge. They called these "cytoplasmic particles of macromolecular dimension" (Brachet & Jeener, 1944) and claimed that nearly all cytoplasmic RNA was located in them, together with several hydrolytic or respiratory enzymes. They also speculated that the hydrolytic enzymes could be caused to work in reverse through the energy released by the respiratory enzymes in order to synthesize peptide bonds.

World War II had a substantial impact on Brachet and his research program. The Germans, occupying Belgium, closed Brussels University in 1942 and Brachet was arrested and imprisoned for nearly three months.[7] He briefly resumed work in Liége until Allied bombing made this impossible. During this hiatus from research, he wrote an overview of chemical processes in development entitled *Chemical Embryology*. After the war, Brachet reestablished his laboratory, although under difficult financial circumstances,[8] in small houses at the university's botanical gardens on the outskirts of Brussels. Relatively quickly, he reassembled his network of collaborators. Chantrenne was appointed professor of biochemistry and he and Brachet continued to investigate RNA metabolism during protein synthesis, the effects

---

[7] Brachet describes the period: "The interdiction, made by the Germans in 1942, to all members of the staff to enter any more the University laboratories obliged us to stop that sort of work; a difficult, but successful, task was the hiding of all the laboratory equipment, including the instruments given by the Rockefeller Foundation in 1938. A search was made for some of these American instruments by the Germans, but they were not to be found. They are of course in constant use in the laboratory now. My arrestation as an hostage and emprisonment in a fortress for $2\frac{1}{2}$ months, followed by a 3 months necessary rest after my release, meant a long interruption of my work in 1943" (Report of activity since 1940, folder 38, box 4, Series 707D, RG 1.2, Rockefeller Foundation Archives, RAC).

[8] Brachet at this time seriously considered emigrating, but concluded that the prospects for improved conditions were sufficient not to pursue that option. He reported in a summary to the Rockefeller Foundation, "As a matter of fact, Brussels University has recently been backing me as far as it can: a new technician has just been appointed in the laboratory and money (10,000 frs) has been given with the purpose of getting a larger ultracentrifuge in working order; it is likely that ultracentrifuge studies will be resumed in a few weeks. The University is also intending to build new laboratories, where I shall find my place among chemists and physicists; the Board of Trustees has also accepted my proposal of the creation of the 'groupment d'études de biologie physicochique', which will help in tightening the links between biologists, chemists and physicists" (Report of activity since 1940, folder 38, box 4, Series 707D RG 1.2, Rockefeller Foundation Archives, RAC).

of ribonuclease on living cells, and the role of the nucleus in the synthesis of RNA and proteins. Jeener became professor of animal and comparative physiology. Although space was very limited, a substantial number of foreign visitors spent periods working in the various laboratories. In the late 1950s, new buildings were finally constructed for Chantrenne, which included an electron microscope purchased with funds from the Rockefeller Foundation.

In a letter to Pomerat on 27 May 1961, describing the dedication of the new electron microscope facility, Brachet also noted that he was "fighting on other grounds (money coming from the Government) for the support of Cell Biology and the development of Molecular Biology in Belgium."[9] As Burian (1996) discussed, the research by Brachet and his colleagues in the late 1940s and early 1950s on different forms of RNA and their role in protein synthesis contributed significantly to the development of molecular biology. Despite his important contribution in linking protein synthesis to RNA and RNA to the particles Claude labeled *microsomes*, Brachet's research did not exert the impact that the Rockefeller laboratory did in developing the new field of cell biology. An important factor may have been the fact that Brachet and his collaborators focused solely on RNA and protein synthesis. Another is that electron microscopy was not central to their research. Probably even more important was that even as they employed cell fractionation, they were not committed to the differentiation of distinct fractions representing different functions. It was the breadth of Claude's emerging vision that enabled the Rockefeller laboratory to set the agenda for the new discipline of cell biology.

Caspersson: Spectrographic Analysis, RNA, and Protein Synthesis

During the same period in which Brachet was linking microsomes, RNA, and protein synthesis, Caspersson was making similar connections at the Karolinska Institute in Stockholm. Caspersson had been a student of Einar Hammarsten, professor of chemistry at the Karolinska, who had himself been conducting research on protein chemistry, including relations between nucleic acids and proteins. In 1944, Caspersson was appointed to a new professorship in cell research at the Karolinska. Another Hammarsten student, Hugo Theorell, had been appointed as the first research professor of the Medical Nobel Institute in 1938, and the Nobel Institute provided funding for a new building to house both Theorell's and Caspersson's laboratories. Substantial grants from the Wallenberg Foundation and from the Rockefeller Foundation, which had been a long-time supporter of Hammarsten's research,

[9] Folder 40, Box 5, Series 707D, RG 1.2, Rockefeller Foundation Archives, RAC.

helped establish both laboratories. Thus, well before the Rockefeller group, Caspersson had a well-funded laboratory dedicated to cell research.

Caspersson tended to have more of an interest in the development of instruments than in the biological research that they made possible. When he obtained his own laboratory space, a workshop for building instruments was a major component. Discussion of refinement of instruments typically appeared ahead of experimental results in Caspersson's annual reports to the Rockefeller Foundation.[10] While working under Hammarsten, he developed methods for ultraviolet spectrography that involved combining a spectroscope and a microscope with a quartz lens. Relying on the fact that the pyrimidines in nucleic acids strongly absorb light at 2600 Å, he was able to estimate nucleic acid content in different parts of living cells. In 1938, in awarding a further grant to Hammarsten, the Rockefeller Foundation took note of this equipment: "The equipment as it now stands is the result of accretions and modifications made as new needs and possibilities were uncovered. In a certain sense the equipment is relatively crude, although it is substantial enough to indicate the limits and possibilities of this type of analysis."[11] In applying for that support, Hammarsten described Caspersson's photoelectric apparatus for ultraviolet-microspectrography (letter to W. E. Tisdale, February 23, 1938):[12]

> Having passed the living cells the light is concentrated in the microscope with an iris and a quartz-prism over the ocular. By means of fluorescent glass and mirrors it is possible to get an orientation in the light-bundle, and by movements of

---

[10] In a report on December 9, 1953, Caspersson described the division between instrument development a ..d biological research: "the work in the institute has been carefully divided so that half the resources were devoted to developmental work on the side of the instruments and the other half to work on biological problems with the intracellular regulation of protein synthesis as key note. This arrangement has always very strictly been carried through, in spite of the fact that it has often been evident that the biological work on short sight would have benefited from a larger share of the efforts, that would undoubtedly also have made the work more easy to manage financially. The reason for this politics [sic] was that the biophysical techniques in question are the primary condition for the work, and furthermore they represent in my personal view one of the ways, which has to be gone sooner or later if we will ever get close to the basic problems of gene reproduction and gene function and thus a quite general approach from the beginning should prove the most fruitful at the end" (Folder: Karolinska Institutet Cell Research 1953, Series 800D, RG 2, Rockefeller Foundation Archives, RAC).

[11] Folder 5, Box 1, Series 800D, RG 1.1, Rockefeller Foundation Archives, RAC.

[12] Folder 1, Box 7, Series 800D, RG 1.1, Rockefeller Foundation, RAC. Plans were already in place for Caspersson to be a Rockefeller Foundation fellow with Lewis. These were initially postponed because Jack Schultz (then a postdoctoral fellow at the California Institute of Technology under T. H. Morgan) went to Stockholm to work with Caspersson. The outbreak of World War II then prevented Caspersson's travel to the United States; the Rockefeller Foundation provided Caspersson grants-in-aid throughout the war.

the prism to move the enlarged picture along one of two perpendicular co-ordinates. In this way it is possible to move the picture-magnification (by means of microscope and distance) of 15000 times enlargement over the small opening of a photocell. The readings are made directly by observing in a microscope the deviations of a filament-electrometer connected with the photocell. Very small areas – especially dependent on the opening and other properties of the photocell and on the magnification – in different structures in one cell can be defined and quantitatively determined by measurements in different focus.

Hammarsten went on to note that in some instances Caspersson employed two photocells, radio equipment, two galvanometers, and a double thermocell.

Caspersson's early application of these instruments contributed significantly to the understanding of the role of nucleic acids in the cell. He determined that under ultraviolet illumination, nucleic acids and proteins had different absorption spectra so that he could measure local quantities of both (Caspersson, 1936). One of his first findings was that it is in cell division that the amount of nucleic acid reaches its maximum. It is important to note that Caspersson did not conclude that the DNA represents genetic material. He rather indicated that the result "points with some probability towards a connection between the duplication of the genes and the presence of nucleic acid" (Caspersson, 1950, p. 96). For him, the genes were proteins. In his collaboration with Jack Schultz, he linked disturbances in nucleic acid metabolism with disturbances in reproduction (Caspersson & Schultz, 1940), but he continued to see the nucleic acids as serving only an ancillary role in the mechanism of self-replication of protein. He commented, "nucleic acids are necessary prerequisites for the reproduction of genes and . . . are probably necessary for the multiplication of self-reproducing protein molecules in general" (Caspersson, 1950, p. 98; summarizing Caspersson & Schultz, 1938).

Caspersson's main approach was to establish correlations involving the nucleic acids and protein synthesis. Some of this work focused on what he took to be the self-replication of proteins in chromosomes during metaphase, which he linked with DNA in the euchromatic region of the chromosome. He traced cytoplasmic proteins to the heterochromatic regions of the chromosome and its apparent relation to the nucleolus during the interphase stage. He interpreted the nucleolus as regulating protein synthesis that occurs in the nucleus just outside the nucleolus. In his report in September 1940, he summarized his conclusions:

Polynucleotides are a base for the protein synthesis in the cell. A central function for the cell nucleus is to be the centre for the protein production. The heterochromatin is an organon regulating the production of the proteins of the cytoplasm. This regulation works via the nucleolus.[13]

In the course of this research, Caspersson differentiated the roles of DNA (ribodesose nucleotides), which he took to be necessary for the synthesis of proteins in the genetic material, and RNA (ribose nucleotides), which figured in the synthesis of cytoplasmic proteins. He found that the "*rapidity* of the protein synthesis in the living bacteria is a simple, almost linear function of the amount of ribose nucleotides." He concluded that this "forms one of the best proofs, not only for the interplay of nucleotides and proteins at the protein synthesis but also for the general validity of this mechanism."[14] Caspersson identified such increases in RNA both in the interphase chromosome, in the nucleolus, and in the cytoplasm. Although drawing a correlation between RNA presence in these locations and protein synthesis, he remained vague about the role played by RNA and continued to construe self-reproducing proteins as the genetic material. He also localized protein synthesis in the nucleus and offered no account of the role of the RNA found in the cytoplasm at the same time as he took the newly made proteins to be diffusing into the cytoplasm. In the course of this investigation, Caspersson also investigated viruses and concluded that they took over the normal protein synthesis mechanisms of the host cells.[15]

In addition to his own research, Caspersson began working during the war with several younger researchers, including Holger Hydén, Bo Thorell,

---

[13] Letter to H. M. Miller, 1 September 1940, Folder 8, Box 1, Series 800, RG 1.1, Rockefeller Foundation Archives, RAC.

[14] Letter to Frank Blair Hanson on 8 August 1944, p. 2, Folder 9, Box 2, Series 800D, RG 1.1, Rockefeller Foundation Archives, RAC.

[15] Caspersson performed most of this research in the period 1938–45. After that, he focused even more on instrument development and offered little in the way of new findings. Prior to renewing his funding in 1956, the Rockefeller Foundation conducted a review of his work, soliciting evaluations from Francis Schmitt, Alfred Mirsky, H. Stanley Bennett, and Barry Commoner. Warren Weaver expressed surprise at the negativity of these assessments and provided Caspersson one last grant at a much smaller amount than Caspersson had requested. Mirsky's evaluation is instructive: "In 1943 it seemed as if Caspersson's work was highly imaginative and also precise. As time went by it became apparent that Caspersson's claims for accuracy were on the whole spurious. . . . About all that has remained of this work is the suggestion that ribonucleic acids are in some way correlated with protein synthesis. This fruitful idea was advanced independently by Brachet, and it is worth noting that Brachet's observations were made by a simple, qualitative staining procedure – quite a contrast to the imposing instrumentation of Caspersson. For the past fifteen years Caspersson has produced little, if anything of significance" (letter to Gerald R. Pomerat, December 4, 1956; Folder Karolinska Institutet Cell Research, 1956–7, Series 800D, RG, Rockefeller Foundation Archives, RAC). While also noting the dearth of recent biological advances by Caspersson, Bennett commented that Caspersson's institute had trained several important junior scientists.

and Arne Enström. He also regularly hosted a number of visiting scientists. With the move into his new laboratory in 1946–7, Caspersson expanded the range of research techniques employed, adding both electron microscopy and cell fractionation. Of particular note is that electron microscopist Fernández-Morán, whose research I will discuss in the next chapter, worked with the Caspersson group for a couple of years on nerve fiber ultrastructure. Despite this, and despite his becoming the first editor of a new journal, *Experimental Cell Research*, Caspersson did not set the agenda for cell biology. Like Brachet, the focus of his research was limited to nucleic acids, not the full range of cell activity. A major reason for his limited impact may have been his preoccupation with instrumentation and failure to apply the instruments in developing new biological ideas after the early 1940s.[16]

One interesting feature of the pursuits of Bensley, Brachet, and Caspersson is that in all three cases the research projects emerged from existing recognized problems in subdomains of what would become cell biology. Perhaps as a consequence, none of them created a laboratory with the breadth of research in cell biology as emerged at the Rockefeller Institute in the 1940s. Caspersson came the closest by incorporating instruments for both cell fractionation and electron microscopy. However, these techniques played a rather peripheral role as the laboratory's focus remained on cytochemistry.

5. ADDING A BIOCHEMICAL PERSPECTIVE TO THE ROCKEFELLER LABORATORY

Around the time that Claude's focus began to change from cancer to normal cells, the group working with him in Murphy's laboratory expanded. The first to join the laboratory, in 1939, was Keith Porter (I will return to Porter in Section 6). Then, in 1941, George Hogeboom, who had completed his medical degree at Johns Hopkins in 1939, brought skills in cytochemistry and biochemistry. Initially he worked most closely with Murphy on his cancer projects. For example, in his first year, he carried out chemotherapy studies on tetra-methyl-ortho-phenylene-daimine (OTM) and rotenone, substances

---

[16] An approach very similar to Caspersson's, but more fruitful, was pursued at Columbia University under the leadership of Arthur Pollister and Franz Schrader. Hewson Swift, a graduate student at Columbia during this period, did pioneering research establishing the constancy of the relation between number of chromosomes and amount of DNA measured spectrographically (Swift, 1953). After finishing his degree, Swift moved to the University of Chicago and set up a similar laboratory focused on measuring DNA content of nuclei. He subsequently employed electron microscopy and created an important center for cell biology. He served as program chair for the first meeting of the American Society of Cell Biology and became its third president.

known to inhibit oxygen consumption in lymphosarcoma. He also carried out studies of the comparative permeability and respiration of normal and malignant cells afflicted with lymphatic leukemia. In some cases, though, Hogeboom's work on cancer became a tool for understanding features of normal metabolism. For example, tyrosine oxidase had been shown to catalyze the formation of melanin in plants and insects, but not in humans. In collaboration with Mark H. Adams, Hogeboom used cell fractionation techniques to isolate two enzymes from mouse melanoma cells. They first obtained a supernatant that catalyzed reactions of both tyrosine and dihydroxyphenylalanine (dopa). They then created two different precipitates with different saturations of ammonium sulfate, one of which catalyzed the tyrosine reaction while the other catalyzed the dopa reaction. In addition to identifying some of the characteristics of the two enzymes, they also observed that when centrifugation continued for several hours, most of the tyrosinase activity appeared in the sediment. They concluded, "This finding suggests that the enzyme may be associated with a particulate component of the cell (microsomes, Claude) and offers an explanation of its insolubility after precipitation by ammonium sulfate" (*Annual Report*, 1942–3, p. 85, see also Hogeboom & Adams, 1942).

In the same period Hogeboom was carrying out this work, Claude himself was becoming increasingly aware of the need to collaborate with biochemists in order to ascertain the enzyme constitution of his fractions. In the *Annual Report* for 1940–1, he noted, "Reports from other laboratories have demonstrated the association of these cell components with cytochrome oxidase, succinic acid dehydrogenase and phosphatase activity. Development of methods to extend the study of function is in progress" (p. 74). The following year he reported on a study with Dean Burk from Cornell which showed that both the large and small granule fractions were capable of taking up oxygen and concluded,

> This fact and the presence of relatively large amounts of iron and copper may indicate that these bodies are associated with the oxydoreduction activity of the cell. The larger granules isolated undoubtedly are what have been referred to as "secretory granules." Attempts are now being made to ascertain if special functional activities of various cells are centered in the granules. (Annual Report 1941–2, p. 74)

In 1942–3, Claude began to collaborate with Rollin Hotchkiss, a biochemist who had been in Dubos' laboratory until Dubos left for Harvard in July, 1942. Claude and Hotchkiss focused on d-amino acid oxidase, showing that it was

Table 5.1. *Results of Enzyme Studies by Claude Hogeboom, Hotchkiss, &*
Hoagland from the Annual Report for 1943–1944

Type of Enzyme	Substrate	Supernate	Large Granule	Microsomes
Nuclease + Phosphatase	Nucleic Acid	0	+++	+++ [1]
Ribonuclease	Nucleic Acid	0	+++	+
Phosphatase	Nucleic Acid	0	+++	Trace
Phosphatase	Phosphate esters and ATP	0	+	Trace
Nucleopyridinase	Coenzyme I	0	+	+++
Cytochrome Oxidase	Ascorbic Acid	0	+++	Trace
Succinoxidase	Succinic Acid	0	+++	Trace
Dehydrogenase	α-glycerophosphate	0	+++	Trace
Oxidase	d-amino Acids	0	+++	0
Phosphate Transfer	ATP to Phosphate esters	+++	Trace	0
Catalase	Peroxide	+++	+	Trace
Malic Dehydrogenase	Malic Acid	+++	Trace	Trace
Coenzyme I	(as growth factor)	+	++	+

found only in what they referred to as the secretory or large granule fraction. In the *Annual Report* for that year, Claude indicated an intention to extend the investigation to other enzymes. These studies, involving Claude, Hogeboom, Hotchkiss, and Charles L. Hoagland, began in earnest in 1943–4 and Table 5.1 shows their results (as stated in the *Annual Report* for that year).

During the following year, these studies, except for investigations into ribonuclease, were suspended as Hogeboom was diverted to war-related research. When Hogeboom finally returned in September 1945 there was a critical change in the way the investigators pursued this research. In addition to simply indicating how much activity a particular fraction exhibited, the researchers compared the amount of activity quantitatively with the amount exhibited by the initial extract (the supernatant from which tissue debris, free nuclei, and red blood corpuscles had been removed in the initial centrifugation). This was done using the Warburg manometer to supply a substrate (e.g., succinic acid salt) to the fraction and determine the resulting rate of oxygen gas uptake. This rate was taken as a measure of the extent to which the relevant enzyme or enzyme system was present and active in the fraction. For example, succinic acid salt provided a measure for succinic oxidase (the enzyme system that includes succinic dehydrogenase plus enzymes in the cytochrome chain). Similarly, cytochrome c was supplied to the same fraction to obtain a measure for cytochrome oxidase. They then divided the amount of activity in a fraction with that from the initial extract. They reported that over three experiments on average 70% of the cytochrome oxidase activity and 74% of the succinoxidase activity was found in the large granule fraction, while less than 4% of the cytochrome oxidase activity and 7% of the succinoxidase activity was found in the microsome fraction (Hogeboom et al., 1946). They attributed the bit of activity in the microsome fraction to large granules or large granule fragments that corrupted it, and claimed that dark-field microscopic examination of the microsome fraction indicated the presence of a sufficient number of large granules to support this explanation. They traced most of the remaining activity of these enzymes to large granules in sediments that were discarded in the process of purifying the different fractions.

Whereas prior to the war the team was satisfied to determine what amount of the activity associated with an enzyme could be found in a given fraction, they now sought to link the activity of a given enzyme with only one fraction. This was the beginning of the one enzyme–one fraction approach to interpreting cell fractionation results discussed in Chapter 4. Adopting this approach, they claimed that "Taken together, these observations suggest that the cytochrome oxidase and succinoxidase systems . . . are entirely localized in the so called large granules" (Hogeboom et al., 1946, p. 626).

Subsequent to this research, two additional investigators joined the laboratory, Walter Schneider and George Palade. Schneider had completed his Ph.D. at one of the top biochemistry departments in the U.S., the University of Wisconsin, in 1945. Palade, a native of Romania, had studied medicine at the University of Bucharest, where he carried out physiological research on the kidneys of dolphins. During the war, he taught at the Department of Anatomy at the University of Bucharest, then emigrated in 1946 to New York City. There he first worked with Robert Chambers at New York University on cellular membranes but, upon seeing Claude's first micrographs (discussed below), started to volunteer in Claude's laboratory.

Together Hogeboom, Schneider, and Palade continued the investigations into the biochemistry of cell fractions.[17] A serious shortcoming of the earlier

---

[17] While his colleagues were refining the techniques for identifying enzymes with particular cell fractions, Claude pursued the question of whether the large granule fraction could itself be differentially fractionated. This would soon become a major endeavor of biochemists (see Chapter 6), but as early as Claude (1946), he reported separating a small particle component from mitochondria containing most of the ribose nucleic acid associated with large granules. He claimed these small particles from within mitochondria could be identified with elements approximately

studies was that it was still not possible to conclude with certainty that Claude's large particles constituted mitochondria. A major reason was that when examined microscopically, the particles in the fraction did not exhibit the rod-shaped appearance of mitochondria and did not respond to the usual mitochondrial stains. One of the main strategies of the new team of researchers was to vary the media used for cell fractionation. Although Claude had tried several variations in fractionation techniques, he had continued to use either distilled water or saline solutions as the media. Hogeboom, Schneider, and Palade (1948) explored sucrose as well as other sugar solutions. With approximately isotonic (0.25 M) sucrose, they noted that the large granules did not agglutinate and were roughly of the same size as with isotonic saline. The particles, however, did not retain the elongated shape of mitochondria. When they tried more hypertonic sucrose solutions, the particles became more elongated, with the percentage of rod-like shapes reaching a maximum with 0.8 to 1.0 M sucrose solutions. The researchers therefore adopted 0.88 M sucrose solutions for their research.

Not only did the large granules now more closely resemble mitochondria as observed in whole cells, they also stained with Janus green B, a stain selective for mitochondria. The researchers also dispelled Claude's claim that the large granule fraction was partially or even largely comprised of secretory granules. Secretory granules stain with neutral red and once mitochondria were made to retain their rod-like shape, the two could be distinguished. The secretory granules appeared to break up after the rupture of the cell membrane and to migrate centripetally; they were thus not part of the large granule fraction. Having developed a fraction that was demonstrably mitochondrial, the researchers replicated the findings of the earlier study that this fraction housed the succinoxidase system. They now said, more boldly than in 1946, "The mitochondrion can therefore be considered as a complicated functional unit possessing two of the most important respiratory enzyme systems of the cell . . ." (Hogeboom et al., 1948, p. 360).

With the improved fractionation procedures, Hogeboom and his colleagues continued the quantitative analysis of the composition of the different fractions. In addition to confirming that the mitochondria contained nearly all the succinoxidase, they established that the microsome fraction contained about 50% of the pentose nucleic acid. From his earliest studies of the small particle fraction, Claude had noted these particles were high in ribose nucleic acid (RNA). That, however, seemed to be the only clue to their function, and so

0.1 μ in diameter that he and Fullam had identified in mitochondria in their electron microscope studies (see below).

Claude as late as his Harvey Lecture (1948) generally pleaded agnosticism about their function:

> Because of their abundance and their universal distribution, it is reasonable to assume that microsomes play a fundamental role in the economy of the cell.... [S]ince the microsomes were isolated in the laboratory, some ten years ago, no sure clue has been found to reveal their function although several attempts have been made, on theoretical grounds, and especially because of their high content in nucleic acid, to have them play some important role in the cells, either as plasmagenes, or agents of protein synthesis. (p. 142)

Claude doubted the findings of both Caspersson and Brachet regarding the role of RNA in protein synthesis, because he questioned how protein synthesis could proceed once separated from the site of oxidative respiration. Instead he suggested that RNA may be involved in anaerobic respiration "either in some phase of the anaerobic mechanism, or act as intermediate in the energy transfer for various synthetic reactions" (p. 163). The basis for this speculation was the correlation of both RNA and fermentation in yeast and some bacteria.[18]

While agnostic about the function of the microsomes, Claude was far from agnostic about the function of the mitochondrion. The localization of key oxidative enzymes in the mitochondrial fraction led him to conclude that it was the locus of the key oxidative processes that provide the bulk of the energy for cell functioning. As he put it in his Harvey lecture, "mitochondria may possibly be considered as the real power plants of the cell" (Claude, 1948, p. 137).

6. ADDING ELECTRON MICROSCOPY AS A TOOL

As discussed in Chapter 4, in 1943–4 Claude took advantage of an opportunity to use the electron microscope at Interchemical Corporation to examine the

---

[18] "From these considerations one might venture the conclusions, partly facts and partly hypothesis, that, whereas most of the metabolic activity of the cell is found in the cytoplasm, the supply of energy may be segregated in various cytoplasmic entities: the aerobic respiration in the mitochondria, as already demonstrated, the anaerobic processes in the ground substance. This might explain the intense basophilia of cells in young embryos, in fast growing tissues, and in tumors, especially in areas where the circulation and the fresh supply of oxygen may be inadequate or defective. Demonstration that a relation exists between the power of anaerobic glycolysis and ribose nucleic acid distribution would permit us to consider further the possibility that the nucleus, where the cytochrome-linked respiratory system is apparently lacking, derives its energy at least in part, from loci where ribose nucleic acid is present, especially the nucleolus, and certain chromosomal regions" (Claude, 1948, p. 163).

large and small granules generated by his fractionation procedure. Claude described his first observations in the *Annual Report* for that year:

> The large particles from liver and mitochondria from leukemic cells under the electron microscope appear as perfect opaque spheres 0.5 to 1.6 μ in diameter. No definite membrane is evident in the prints. In instances where the granules are injured and some loss of substance has occurred, internal structure may be seen. Among these are small particles in the size range of microsomes. This observation taken with the fact that in the test tube disrupted large granules yield what appear to be microsomes suggests a relationship between these formations. (pp. 69–70)

He was not able to see any structure in the undamaged mitochondria because of their thickness, but if they had damage that allowed some material to escape, or were flattened somewhat during mounting, then the electron beam could penetrate them. Although Claude admitted in his published report (Claude & Fullam, 1945) that he could not see a membrane in the micrographs, he inferred its presence from the behavior of mitochondria in which some contents had escaped: "In some cases, loss of substance seems to release the tension which keeps the mitochondrial body spherical and, in such circumstances, the impression is gained that mitochondria possess a differentiated covering which may become more or less completely separated from the mitochondrial mass" (p. 57). Claude observed that mitochondria retained their shape and remained discrete if he used sufficient salt in the media employed in fractionation to insure proper tonicity. In opposition to Bensley's position, he contended that this disproved "the assumption that mitochondria may be formed by the concretion of substances preexisting in the cytoplasm or that they may disappear and reappear in living cells because of changes of equilibrium occurring in the surrounding protoplasm" (p. 58).

Claude's first use of the electron microscope provided little information about either mitochondria or the microsomes, but another approach would prove much more fruitful. The key person in this development was Keith Porter, who had joined the laboratory in 1939. Porter had completed his doctorate a year earlier at Harvard, pursuing research on the development of frog embryos with only a haploid set of chromosomes. This research required him to develop skills in micromanipulation of cells (e.g., removal of the nucleus before insemination so that the egg would develop with only the sperm's chromosomes). After he received his Ph.D., Porter was a postdoctoral fellow at Princeton, where he began to transplant the haploid nucleus of a frog of one geographically isolated race or subspecies into enucleated frog embryos of different races (Porter, 1941a; Porter, 1941b). Porter's ability to carry out these transplants suggested to Murphy that Porter might be capable of transplanting the hypothesized tumor particles from an infected cell to other cells. When he came into the laboratory, though, Porter continued the transplant studies he had been pursuing. His goal was to demonstrate cytoplasmic influences upon development by comparing embryos into whose cytoplasm he transplanted the nucleus of a different race with normal embryos of that race. The results indicated a genetic effect of the cytoplasm as well as of the nucleus. Given the focus of the laboratory on cancer, though, Porter also began to investigate the inhibition of growth from carcinogenic agents, including X-rays and such chemicals as methylcholanthrene. He examined both the effects of different dosages on tail regeneration in the newt and the character of the tissue reaction induced.

For a number of years, Porter's research seemed to be largely independent of other investigators in the laboratory. However, as Claude was exploring the use of the electron microscope, the value of having Porter in the laboratory was realized when Porter proposed using tissue-cultured cells to produce sufficiently thin specimens for electron microscopy (see Chapter 4). Working with Claude and Fullam at Interchemical Corporation, Porter produced a composite electron micrograph (from several pieces imaged separately) of a whole chick embryo cell. Figure 5.1 shows this micrograph and a comparable one from a light microscope that they included for comparison. The nucleus was generally too thick to observe anything but the nucleoli, which appeared as less dense than the rest of the nucleus. Parts of the cytoplasm, though, generated a detailed image. The following is the figure caption in which Porter et al. presented their interpretation:

Electron micrograph of a fibroblast-like cell, and nerve fibers cultured from a chick embryo tissue. Differential absorption and scattering of electrons by the cytoplasmic area has silhouetted a number of structural details among which are: filamentous mitochondria of various lengths and fairly constant width; scattered, small elements of high density especially abundant around the nucleus and presumably representing Golgi bodies; and a delicate lace-work extending throughout the cytoplasm. The nucleus is visible but multiple scattering of electrons due to excessive thickness results in considerable blurring. Three nerve fibers can be seen: one crossing the upper part of the picture and having no connection with the cell; one ending in contact with the cell surface at the right; and one at the lower part of the picture also in contact with the cell surface. This latter has the appearance of a growth cone. Details of the cell's margin and extensions are clearly defined. The arrows point to extensions mentioned in the text as "jagged points" (a) and "finger-like processes". (1945, p. 246)

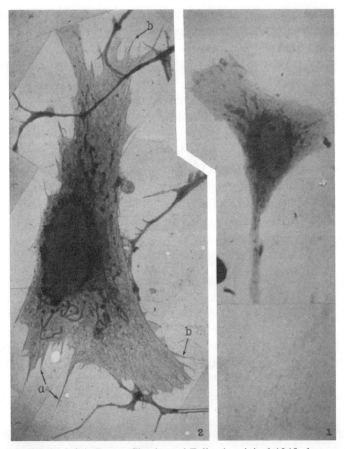

Figure 5.1. On the left is Porter, Claude, and Fullam's original 1945 electron micrograph of a fibroblast-like cell and nerve fibers cultured from chick embryo tissue. On the right is a photomicrograph of a similar cell using a light microscope. Reproduced from Porter, K. R., Claude, A., & Fullam, E. F. (1945), A study of tissue culture cells by electron microscopy, *Journal of Experimental Medicine, 81*, pp. 235–255, Figures 1 and 2 on plate 10, by copyright permission of The Rockefeller University Press.

As they noted, the mitochondria appeared as elongated rod-like structures that appeared to have areas of increased density as well as extremely small granules 10 to 20 m μ in diameter. They proposed that "these may be composed mainly of inorganic salts, or that they represent centers where reduced osmium has accumulated" (239). Due to the osmophilic character, they interpreted the dense bodies with angular outlines in the micrographs as Golgi bodies. In the *Annual Report* for 1944–5 they appealed to these observations to challenge the interpretation of the Golgi apparatus as an artifact of the way

185

the bodies cluster around the nucleus, noting that "the resolving power of the light microscope would not separate them and as a consequence they would appear as a complex" (p. 72). (Recall from Chapter 4, though, that five years later Claude would join with Palade to argue that the Golgi was indeed an artifact.)

Porter (1955–6, p. 175) characterized the goal in making these early micrographs as "to see what there might be in the optically empty parts of protoplasm."[19] Over the years several authors had proposed the existence of a cytoskeleton that was responsible for maintaining the structure of the cell (Needham, 1942; see also Bonner, 1952; Peters, 1930; Picken, 1940). With respect to this claim, the micrographs produced novel and controversial results. In the *Annual Report* Porter and Claude referred to the ground substance as *spongioplasm* and reported that in the micrographs it appeared to be comprised of particulate elements 30 to 150 m μ in diameter. In the published paper they commented on the status of these structures:

> It is not known whether the particulate elements just mentioned pre-exist in the living protoplasm, or whether they are artifacts arising from the cell body or the cell wall, as a result of fixation or drying. In this connection, however, it is of interest to recall that experiments in this laboratory have shown that the chromophilic ground substance is sedimentable and, therefore, probably particulate in nature. Touching on this problem also is the fact that small particles, or microsomes, estimated to average about 70 m μ in diameter have been previously isolated from extracts of normal chick embryos and Chicken Tumor I. (Porter et al., 1945, p. 238)

The authors then proceeded to a second observation – a lace-like reticulum running through the thinner parts of the cell.[20] They reported that "vesicle-like

---

[19] Porter (Interview, 1987, University of Maryland, Baltimore County) recalled that he had no particular hypothesis in mind in generating the first micrographs, although he was convinced from dark-field light microscopy that there had to be more structure than light microscopy could reveal. Rather, the availability of tissue-cultured cells provided "an excuse" to see what the electron microscope might reveal. The experience of seeing the first micrographs apparently created a "flash-bulb memory" for Porter. He related to me with obvious excitement, "I remember the night so distinctly I could play it back. It was in the war. New York City was blacked out. It was cold and raining and black, was it black. And we were trying to find the entrance to the building on 48th Street, West Side. We got this thing, I think there was only one cell on the grid, but it was between the bars and we took picture after picture of it, we must have taken 30 or 40 micrographs of it so that we could piece together the whole thing. And I was fascinated, I was really fascinated. I don't think I slept at all that night. We didn't leave there until 3:00 A.M."

[20] Porter (Interview, 1987, University of Maryland, Baltimore County) related that initially they had "no idea what the lace-like stuff was inside the cytoplasm." Of particular interest was the fact that it fragmented in tissue-cultured cells, which suggested that while it was incapable

186

bodies, i.e., elements presenting a center of less density, and ranging in size from 100 to 150 m μ, can be seen arranged along the strands of the reticulum just mentioned" (p. 238).

Electron microscopy gave Claude, again in collaboration with Porter, an opportunity to return to cancer research by examining tumor cells. Already in the *Annual Report* for 1944–5, they described finding in tumor cells "small granules of uniform size and characterized by a density appreciably greater than that of the surrounding cytoplasm . . . which are certainly not common in normal cells" (p. 73). The next year they conducted studies on Chicken Tumor 1 and Chicken Tumor 10 cells, identifying granules of similar structure that "are readily distinguished from normal cell constituents, especially from microsomes, by their regular granular shape, their relatively uniform size and because of a greater density in osmium preparations" (p. 91). The particles were differently distributed in the two tumor types, with those in Chicken Tumor 10 being clustered in colonies while those of Chicken Tumor 1 were distributed more widely in pairs or in rows of up to six particles. By the time of their published paper in 1947 they were willing to conclude that these particles were the causative agent of the respective tumors (Claude et al., 1947). At the end of the paper they speculated that the particles replicate by division and noted that the variation in size of the particles was compatible with their growing larger prior to such splitting and yielding smaller particles after the split.

After their joint investigation of cancer cells, Porter and Claude went in their own directions. Claude directed his attention to developing a technique for thinly sectioning cells while Porter continued to develop the technique for studying cultured cells and understanding the "lace-like reticulum" found in them. Already in the *Annual Report* for 1945–6 he had become more definite about the relation between granules and the lace-like reticulum of the endoplasm:

> Possibly the most important discovery from these studies is that in relation to the endoplasm. This cell system has been heretofore unobservable & only dark field illumination has provided any hint of its existence. As seen in osmium-fixed preparations under the electron microscope, it appears as a finely divided reticulum along the strands of which are scattered tiny bodies 60 to 100 m μ in diameter. The particulate components are believed to be the same as those isolated by high-speed centrifugation and known as microsomes. This endoplasm

of stretching as much as required in thin-spreading tissue-cultured cells, it was embedded in something that did stretch. It was not just suspended in the cytoplasm, because otherwise it would not be stretched but, rather, would remain in the middle.

187

appears from recent observations to be derived from the activities of small dense unnamed granules which in a few normal cells can be seen in what appears to be the act of division. (p. 88)

The following year (1946–7) Porter said of the endoplasm that

the evidence is convincing that it is the most active part of the living unit. It is present in one form or another in every cell so far examined. It makes up a fibrillar apparatus concentrated around the central body and sends fine extensions of its substance throughout the cell. The vesicular structure of the small units suggests a secretory activity. In as much as it is the only cytoplasmic system showing some structure and continuity, it may be assumed that it determines the mosaic character of the egg cell. (p. 80)

Porter also engaged in his own studies of cancer cells. In his first efforts he, together with visiting fellow Helen Thompson, examined cultured cells from three different rat sarcomas (the *Annual Report* for 1946–7 indicates four sarcomas). These cells, they claimed, exhibited a much greater density of endoplasmic granules located on shorter strands (Porter & Thompson, 1947). A second study involved mammary carcinoma in mice. John Bittner (1936) had discovered that this cancer was transmitted through their mother's milk. With Thompson, Porter used the electron microscope to examine mammary gland tumor cells grown in tissue culture from mice. They identified within them distinctive particles about 130 m μ in diameter with a dark, well-defined central core. Although the evidence was only circumstantial, they proposed "tentatively" that the particles were the viral agent in the milk (Porter & Thompson, 1948).

7. THE STATE OF CELL STUDIES AT THE END OF THE 1940s

As I will discuss at the beginning of the next chapter, in 1949 and 1950 the research laboratory at Rockefeller underwent a major transformation and the research efforts broadened to other laboratories. So we reach a natural transition and it is worth drawing together just what was accomplished during the 1940s in that laboratory (see Claude, 1950, for a useful recapitulation). The decade began with Claude's determination that normal cells contained particles of the same size as he had isolated in his tumor-causing fraction from Rous sarcoma cells. His research increasingly focused on the constitution of normal cells (although he would return to the attempt to identify cancer particles later in the decade). Initially cell fractionation was his primary tool, and with it, he developed standardized procedures to segregate two fractions of cytoplasmic particles plus a nuclear fraction and a super-natant. Eventually Claude revised his initial assessment of the two fractions, accepting that the larger particles were primarily comprised of mitochondria while the small particles were a new cell component he labeled *microsomes*. In the second half of the decade, he and his collaborators refined the procedures for fractionation, especially the choice of media, and began to associate particular enzymes with specific fractions. Claude also introduced another new technique, electron microscopy, initially to examine the isolated fractions and then, in collaboration with Porter, to examine whole cells grown in tissue culture. The latter approach permitted identifying both mitochondria and a lace-like reticulum. They found that the reticulum was related to the microsomes isolated by cell fractionation.

The view of cytoplasm developed through this research was that it was comprised of two primary types of structures – mitochondria and the lace-like reticulum/microsomes – plus a gel-like aqueous component – the "cell sap" or cytosol – corresponding to the supernatant. The two structures were associated with different cell activities. Both morphologists and biochemists rapidly accepted Claude's characterization of the mitochondrion as the power plant of the cell (as we will see in the next chapter). Although Claude remained agnostic about the function of microsomes and related structures in the endoplasm, others such as Brachet and Caspersson had concluded that they figured in protein synthesis. This differentiation of function provided the foundation for a functional decomposition of the cell into organelles in which were situated mechanisms that contributed differentially to cell life. The further development of this account of cellular mechanisms required two additional steps: (1) decomposing the already discovered organelles to show how their component parts contributed differentially to their functioning, and (2) finding the organelles presumed to be associated with other cell functions. A research community that rapidly increased in size energetically pursued these goals in the 1950s and 1960s.

6

New Knowledge

The Mechanisms of the Cytoplasm

> This continuous body of knowledge, which should be properly named cellular and molecular biology, could be compared to a bridge which, like its equivalents in civil engineering, has two bridgeheads: one in traditional anatomical-morphological sciences and the other in equally traditional biochemistry. The cautious and careful have stayed close to the bridgeheads because the area around them had been consolidated over centuries by the work of their predecessors. The bold and venturesome have ventured on the bridge itself from both directions, because they believed that there was where the action was going to be.... As in the old Latin proverb, fortune favored the bold: the bridge proved to be strong enough to support the intense occasionally frantic activity of whole armies of explorers.
>
> (Palade, 1987, pp. 112–13)

In the 1950s and 1960s the initial ventures into the terra incognita between classical cytology and biochemistry developed into the robust bridge Palade identified in the above quotation.[1] In large part this involved building on the localization of cellular energetics in the mitrochondria, and of protein synthesis in the microsomes, that had been established in the 1940s by decomposing these organelles and figuring out the operations associated with their parts. I will focus principally on these developments, but in the 1950s investigators identified the function of two other organelles – the Golgi apparatus and

---

[1] For Palade this was not just a perspective adopted in retrospect. Already in 1956 he commented on the integration of morphological and biochemical research: "The ample information obtained in each of these two fields has stimulated research in the other, with the result that a number of cell components have acquired a new biochemical and physiological significance. The concept of functional differentiation among cell organs has been more firmly established, and the previously sharp boundary between cell morphology and cell physiology and biochemistry has, to a large extent, faded away" (Palade, 1956a, pp. 186–7).

the lysosome – and established research programs to determine how they performed their functions. In the final part of the chapter, therefore, I will provide a brief account of this research as well.

As I will show, the research extended far beyond the Rockefeller laboratory that provided the focus for most of the research described in the previous chapter. Nonetheless, it continued to play a central role, although the shape of the laboratory changed dramatically. In 1949 Claude accepted an invitation to direct the Jules Bordet Institute at the Université Libre de Bruxelles and left Rockefeller. In 1950 Murphy reached the mandatory retirement age (he died later the same year). At the Rockefeller Institute the usual procedure was to close a laboratory after the departure of a senior laboratory director (Member). However, in this instance, presumably in recognition of the pathbreaking work of the junior researchers and the investment in an electron microscope, Gasser took the unusual action of retaining Porter and Palade and promoting Porter to Associate Member and director of the laboratory. The characterization of the laboratory that year in the *Annual Report* reads, "During the period covered by this report studies on cell fine structures and related problems have been continued." (p. 143). The newly constituted Laboratory of Cytology moved to the basement of Theobald Smith Hall, where both the RCA EMU microscope that had been bought by the Rockefeller Foundation and a new RCA EMU-2A were installed.

Initially the group working with Porter and Palade was quite small. George Pappas spent two years as an Eli Lilly postdoctoral fellow. Maria Rudzinska, a protozoologist, worked with Porter. Sanford Palay and Don Fawcett, already faculty members at Yale and Harvard, respectively, spent considerable time visiting the laboratory. In 1955 Philip Siekevitz, a biochemist, joined the laboratory. In the middle 1950s the Rockefeller Institute was transformed from an exclusively research-oriented institution into a graduate university. The laboratories, previously staffed principally by scientists and postdoctoral researchers, now served as training centers for graduate students as well. Among the first graduate students in the laboratory were Mary Bonneville, Howard Rasmussen, Aaron Shatkin, Lee Peachey, and Peter Satir.[2]

---

[2] Peter Satir (Interview, 29 November 1995, Albert Einstein School of Medicine) related the unusual nature of the application process at Rockefeller in its first years as a university. President Bronk solicited nominations for graduate students from the top institutions in the country and interviewed applicants himself before directing them to the appropriate laboratories. According to Satir, Porter and Palade had been among the less eager investigators to make the transition to a graduate institution, but were exceptional in the support they provided to graduate students once they accepted them into the laboratory.

1. THE MITOCHONDRION

Biochemists Confront Particulate Structure: Mitochondrial Enzyme Systems

During the same period in which the Rockefeller researchers were situating succinoxidase and other oxidative enzymes in the mitochondrion and Claude was identifying its role as the power plant of the cell, numerous biochemists were following up on the thread from Keilin and Hartree's (1944) demonstration that they could not eliminate cell particulates from extract preparations capable of performing oxidative phosphorylation. Keilin and Hartree (1949) themselves interpreted this as indicating that the respiratory function was connected with the physico-chemical structure of the cell.[3] For most biochemists, however, the involvement of cell particulates was, as Lehninger put it, "a nuisance" (1951).[4] Their goal was to work out the purely chemical steps in oxidative phosphorylation in a manner comparable to that already provided for glycolysis.

As discussed in Chapter 3, by the 1940s there was a good understanding of the major operations in the three metabolic mechanisms that worked in sequence to oxidize carbohydrates – glycolysis, the citric acid (Krebs) cycle, and the electron transport chain (see Figure 3.16). After the investigations of Kalckar (1939) and Lipmann (1939), it was recognized that oxidative metabolism was linked to the storage of energy via ATP formation. In a theoretical paper, Lippman (1941) characterized the phosphate bonds in ATP as *energy-rich bonds* and introduced the symbol $\sim$P for these; later they were more commonly referred to as *high-energy* bonds. The process by which energy was stored in such bonds was understood earliest (and turned out to be simplest) for the first two mechanisms, glycolysis and the citric acid cycle. As detailed in Chapter 3, in glycolysis the initial substrate (glucose) and the penultimate product (pyruvic acid) do not contain phosphates, but the various intermediates do. Phosphates are added at three different steps in the process and in later steps are transferred to ADP, yielding energy-rich ATP. This process later became known as *substrate-level phosphorylation*.

Knowledge of the process by which phosphorylation was coupled to the third oxidative mechanism in the sequence, the electron transport chain, was

---

[3] Keilin and Hartree (1949) themselves proposed that the particulate nature of their preparation facilitated respiration by assuring the mutual accessibility of the different enzymes encapsulated within each particle. Cleland and Slater (1953) determined that the Keilin and Hartree preparation included membranes from mitochondria.

[4] See also Lehninger (1964): "It was a part of the *Zeitgeist* that particles were a nuisance and stood in the way of purification of the respiratory enzymes" (p. 6).

more elusive and took much longer to achieve. One early constraint came from the work of Ochoa, who established that up to three molecules of ATP were formed per oxygen atom consumed. This indicated that the reactions forming ATP must occur at multiple steps along the electron transport chain. But what were these reactions? Lippman (1946) proposed that ATP synthesis along the electron transport chain would follow a scheme similar to that already known for glycolysis. He presented it in abstract form as a sequence of two reactions (the formulae above the line) achieving the overall effect of adding a phosphate bond to ADP to yield ATP (the summary formula below the line):

$$AH_2 + B + P_i \rightleftharpoons A \sim P + BH_2$$
$$A \sim P + ADP \rightleftharpoons A + ATP$$
$$\overline{AH_2 + B + ADP + P_i \rightleftharpoons A + BH_2 + ATP.}$$

In these formulae, A denotes a substrate which is oxidized in the first reaction coupled with the reduction of another substrate, B. (A and B may be successive cytochromes in the electron transport chain, for example.) As A is oxidized it forms a high-energy bond with phosphate. In the second reaction, ATP is synthesized via the transfer of this bond to ATP.

This scheme was elaborated when it was discovered that the first of the two ATP-producing steps in glycolysis was more complicated than originally thought. The intermediate first gained energy as it was oxidized; only thereafter did it provide the energy for adding phosphate to ADP, yielding ATP. This led E. C. (Bill) Slater (1953) to revise Lipmann's proposed scheme for ATP formation coupled to the electron transport chain. He proposed that an additional compound, C, first formed a high-energy bond with A. The energy from that bond then facilitated the uptake of phosphate into ATP:

$$AH_2 + B + C \rightleftharpoons A \sim C + BH_2$$
$$A \sim C + ADP + P_i \rightleftharpoons A + C \pm ATP$$
$$\overline{AH_2 + B + ADP + P_i \rightleftharpoons A + BH_2 + ATP.}$$

This version of the scheme set the agenda for many biochemists for the next twenty years – the race was on to identify C, the hypothesized intermediate. The search turned out to be futile, though, as no such intermediate exists.[5]

---

[5] Douglas Allchin has offered a detailed account of the quest for the nonexistent intermediate as well as lessons learned (1996; 1997). More recently he developed the idea that, although the search for a chemical intermediate could not succeed, other important results came out of the attempt. In particular, Allchin (2002) analyzed how Paul Boyer's research led to a number of discoveries, such as phosphohistidine, which figured as an intermediate not in oxidative phosphorylation, as

Phosphorylation along the electron transport chain has a very different explanation, as first outlined in the chemiosmotic hypothesis of Peter Mitchell (1961; 1966). Now called *oxidative phosphorylation*, this process turned out to depend upon the presence and structure of the inner membrane of the mitochondrion. The energy used to drive oxidative phosphorylation is stored, not in a chemical intermediate, but rather in a proton gradient across the membrane. With the eventual acceptance of this hypothesis, there no longer was any need to find the hypothesized chemical intermediate. Much more relevant were discoveries concerning the inner structure of the mitochondrion and the dependence of certain biochemical processes on that structure. These developments resulted from the interaction of biochemists and morphologists during the 1950s, which will be the focus of this section.

While Claude, Hogeboom, and their colleagues were making progress at the Rockefeller Institute, at the University of Chicago Albert Lehninger and his graduate student Eugene Kennedy were also conducting biochemical studies on particulate structures fractionated from rat liver. They, however, targeted the oxidation of fatty acids rather than carbohydrates. In ordinary liver preparations, a complex enzyme system catalyzes oxidation of the fatty acid octanoate (octanoic acid) via two possible pathways, one producing ketone bodies such as acetoacetate (acetoacetic acid) and the other proceeding through the citric acid cycle and respiratory chain, producing CO_2 and consuming oxygen to generate H_2O. Prior to Hogeboom et al.'s introduction of isotonic sucrose solution, Schneider, who was then still a graduate student at Wisconsin, had collaborated with Lehninger in testing a mitochondrial fraction of liver for evidence of fatty acid oxidation. Schneider tried supplying both water and saline suspensions of the mitochondrial fraction with octanoate and necessary supplementary substances (KCl, $MgSO_4$, a "sparking" Krebs intermediate, cytochrome c, phosphate buffer, ATP), but found no significant oxidation. Instead, he found oxidation occurring primarily in a fraction containing nuclei, erythrocytes, and some intact cells. After Hogeboom et al.'s (1948) paper, Lehninger set Kennedy to repeating this earlier work. Kennedy showed that with 0.88 M sucrose as the fractionation medium, octonoate oxidation occurred in the mitochondrial fraction (Kennedy & Lehninger, 1949).

---

previously hypothesized, but rather in substrate phosphorylation. Likewise, Boyer formulated an alternative mechanism of oxidative phosphorylation involving conformational changes that he initially advanced against both the chemical intermediate view and Mitchell's chemiosmotic view. He later recognized that his proposal was in fact compatible with Mitchell's account if he limited its scope to the actual synthesis of ATP. These efforts ultimately brought Boyer a Nobel Prize, even though his success did not lie in identifying the central mechanism of oxidative phosphorylation.

Lehninger (1951, p. 7) concluded from these investigations that "mitochondria contain a complete complement of the individual enzymes of the fatty acid oxidase system and in such amount that they could easily account for the known rates of oxidation in the intact cell."[6] Lehninger and Kennedy also found that mitochondria contain all the needed enzymes for oxidizing intermediates of the citric acid cycle, although they could not rule out the possibility that other fractions might also contain some of these enzymes. They also made an important negative finding – the reactions of glycolysis were catalyzed by the supernatant, not the mitochondrial fraction or other fractions with insoluble cell components. They concluded that glycolysis occurred in the aqueous cytosol of the cell's cytoplasm. Its product, pyruvic acid, would therefore need to be transported into the mitochondrion to join fatty acid products in the common pathway of the citric acid cycle.

In further research carried out with another graduate student, Morris Friedkin, Lehninger established, using labeled P^{32}, that isolated mitochondria synthesized ATP (Friedkin & Lehninger, 1949). In these initial studies, though, the rate of ATP formation was very slow. Lehninger (1951) discovered that when DPNH (NADH) was added to the mitochondrial preparation, it did not penetrate the mitochondrial body. He found that DPNH could enter the mitochondrion if he placed the preparation into a hypotonic KCl, sucrose, or distilled water medium for a short period before restoring isotonicity. Under these circumstances, oxidative phosphorylation occurred robustly and, in accordance with what Ochoa had found for tissue extracts, producing approximately three ATP molecules per atom of oxygen consumed.

A distinctive feature of Lehninger's theoretical outlook was that he interpreted difficulties in developing biochemical preparations for studying oxidative phosphorylation as clues to the importance of mitochondrial structure for those processes. This applied not only to the difficulty in getting DPNH to enter the mitochondria, but also to the fact that oxidative metabolism could only occur in preparations in which cell particulates remained. For him the failure to extract the responsible enzymes for oxidative metabolism when

---

[6] Noting the need to supply metabolites such as malate (malic acid), $MgSO_4$, cytochrome c, KCl, and ATP to maintain the reaction, Kennedy and Lehninger concluded, "Although the mitochondria appear to be the major site of these activities, it would appear from our examination *in vitro* that these bodies are not completely autonomous with respect to their respiratory behavior, since they must be supplemented with certain cofactors such as adenosine triphosphate and Mg++. It appears likely that in the cell there is a rapid interchange of these factors, substrates, and inorganic phosphate between the cytoplasm and the mitochondria. It also would appear that these bodies are dependent on the cytoplasm for certain preparatory metabolic activities such as glycolysis, since, as our data show, they are almost completely lacking in glycolytic activity" (1949, pp. 970–1).

liver tissue was ground in the Waring blender indicated that the mitochondrion was not merely a "sac containing a solution of soluble oxidative enzymes but . . . an organized structure with an insoluble matrix in which are embedded the individual catalytically active proteins making up the complex enzymatic machinery of oxidation and phosphorylation" (Lehninger, 1951, p. 12). Further influenced by the difficulty of accounting for the rates of oxidative phosphorylation by diffusion of intermediates, Lehninger advanced a bold proposal: "It would appear that these carrier proteins must be fixed in space so that hemes are juxtaposed, increasing the probability of fruitful collision, or that special mechanisms must exist to allow passage of electrons through the protein moieties" (p. 12).[7] Such a proposal departed radically from the traditional biochemical focus on soluble enzymes and helped set the stage for linking biochemical processes with cell structures.

David Green was another biochemist who responded to the difficulty of isolating oxidative enzymes by advancing the idea that the enzymes involved in oxidative phosphorylation constitute an organized system. Green had established his reputation through empirical work purifying and characterizing enzymes involved in cellular respiration while working with Malcolm Dixon at Cambridge in the 1930s (Green, 1936a; Green, 1936b; Green & Brosteaux, 1936; Green, Dewan, & Leloir, 1937; Green & Dixon, 1934) and by the publication of his 1940 book, *Mechanisms of Biological Oxidations*, which provided many in North America with their primary introduction to enzyme chemistry.[8] In 1948 he advanced the immediately controversial claim that the

---

[7] The next step in Lehninger's program was to determine where along the electron transport chain ATP synthesis occurs. He attempted to study the reactions starting with ascorbate (ascorbic acid, a nonenzymatic reductant of cytochrome c), and confronted the same difficulty – only when preparations were pretreated was the electron transport accompanied by phosphorylation of ADP. But with pretreatment, Lehninger was able to establish that one phosphorylating event occurred between cytochrome c and oxygen (see Lehninger, 1954). This implied that the other two phosphorylating events must occur earlier in the chain. Slater (1950) had previously found evidence for at least one phosphorylation earlier in the chain by using α-ketoglutarate (α-ketoglutaric acid) as the hydrogen donor and ferricytochrome c as the acceptor, and the question of exactly where in the electron transport chain the coupling with phosphorylation occurred remained a focus of inquiry throughout the 1950s. A major technique for approaching the problem was developed by Chance and Williams (1956). They used spectrography to determine the oxidation state of intermediates and, employing various inhibitors to impede the overall reaction and then adding ADP, they were able to identify zones in which each phosphorylation occurred.

[8] Green returned to the United States at the beginning of World War II and after a short period at Harvard, was appointed in 1941 assistant professor of biochemistry in the Department of Medicine at the Columbia College of Physicians and Surgeons in New York City. While he was in the midst of formulating the cyclophorase concept, he was approached by the University of Wisconsin, already a top institution in biochemistry, to become a founding member of the Institute for Enzyme Research at the University of Wisconsin. Green recruited an extremely

primary mechanisms of aerobic respiration (fatty acid oxidation, citric acid cycle, respiratory chain, and oxidative phosphorylation) were carried out by a single, physically structured system of enzymes that he called the *cyclophorase system* (Green, Loomis, & Auerbach, 1948). His primary evidence that these enzymes constituted a structured system was his failure to isolate the enzyme catalyzing pyruvic to acetic acid using cell fractionation techniques. Instead, his preparation from rabbit kidney, which involved homogenation with potassium chloride using alkali to neutralize the acid that formed, followed by multiple resuspensions in saline and centrifugation, metabolized pyruvic acid all the way to carbon dioxide and water. By referring to a cyclophorase *system*, he meant to contrast the enzymes involved in aerobic respiration with those involved in other biochemical processes such as glycolysis, purine synthesis, and the pentose and urea cycles. In those cases, the enzymes can be isolated and an operative system reconstituted from the isolated components. He explained the term *cyclophorase* as "literally meaning the system of enzymes carrying through the (citric acid) cycle" (Green, 1951b, p. 17). Green acknowledged that the ending "ase" is usually applied to individual enzymes, but cited precedent for his extension to a "team of enzymes":

> Keilin and his school have been referring for more than two decades to the succinic oxidase and cytochrome oxidase systems. Neither the one nor the other represents a single enzyme. They represent a considerable group of enzymes all of which are associated with the same particulate elements. (1951b, pp. 17–18)

Green conceptualized the cyclophorase system as involving a precise physical arrangement that would facilitate cooperative action between spatially proximal enzymes. He also maintained that this arrangement would enable the components to behave in ways they could not otherwise: "the chemical organization by which the many constituent enzymes are integrated confers properties on the various enzymes which they may not necessarily enjoy when separated from the complex and isolated as single enzymes" (p. 18). Although he initially claimed that the cyclophorase system represented a newly discovered constituent of the cell,[9] after he learned of Lehninger's work Green accepted

talented team of researchers to the Institute who helped identify many of the critical aspects of both fatty-acid metabolism and oxidative phosphorylation. His own research, however, became increasingly suspect (his preparations of his cyclophorase system were contaminated with many other cell components) and his theorizing less grounded in experimental evidence than other researchers thought appropriate.

[9] Both Van Potter (Interview, 6 November 1987, Madison) and Helmut Beinert (Interview, 5 November 1987, Milwaukee) noted that Green had to be convinced that his cyclophorase system was linked to the mitochondrion.

the Rockefeller group's linkage of these enzymes with the mitochondrion. He later wrote, "The mitochondrion and the cyclophorase system thus turned out to be the structural and functional sides of the same unit" (Green, 1957–8, p. 178). He nonetheless advocated using the name "cyclophorase system" "for the functional attributes of the same entity" (1951b, p. 19, n. 2). Green credited Harman (1950a), who was working with him at the Enzyme Institute, with establishing the proportionality of cyclophorase activity and the presence and number of mitochondria.

An important aspect of Green's conception of the cyclophorase system was that it not only linked together the enzymes but also bound them to the coenzymes that figured in the reactions. Washing the preparation would remove the coenzymes and, as well, most of the NAD, NADP, FAD, and ATP in the cell that was normally bound in the cyclophorase system. Green proposed further that a coenzyme was bound as a *prosthetic group* to the protein component of an enzyme, which he referred to as the *apoenzyme*, and that when the two were split, the enzyme was modified. Green suggested that such an arrangement was most efficient in that it required only one coenzyme molecule per enzyme molecule, whereas if they were dissociated and relied on random processes such as diffusion to encounter each other, many times more coenzyme molecules would be required. Green did note a serious problem posed by binding of the coenzyme to the enzyme:

> ... pyridinenucleotide must be capable not only of being reducible by the substrate of the oxidase with which the former is combined, but also in its reduced form has to interact with the flavin prosthetic group of diaphorase – the enzyme which catalyses the oxidation of dihydropyridinenucleotide by one of the cytochrome components. When the pyridinenucleotide is free as in the case of the classical, soluble systems, this sequence of reactions poses no difficulty. The coenzyme is free to shuttle back and forth ... In the cyclophorase system with bound pyridinenucleotide, the extent of shifting back and forth is severely limited. Some mechanism must be invoked to explain how a coenzyme fixed in a rigid structure would be capable of interacting with a variety of systems. (1951a, p. 429)

Most biochemists rejected Green's cyclophorase proposal as an excessively speculative response to the difficulty of rendering the enzymes of oxidative metabolism soluble and isolating them. Nonetheless, like Lehninger, those biochemists working on oxidative phosphorylation in the early 1950s came to recognize, if only as a nuisance factor, that the enzymes of oxidative phosphorylation were localized in the mitochondrion and in some intimate way connected with mitochondrial membranes. The question was how. The next

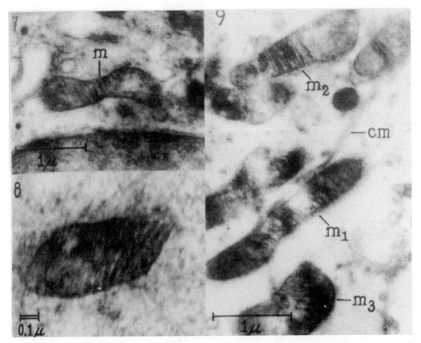

Figure 6.1. Three of Palade's 1952 micrographs revealing what he called *cristae* and construed as infoldings of the mitochondrial membrane. Reproduced from G. E. Palade (1952a), The fine structure of mitochondria, *Anatomical Record, 114*, 427–51, Plate 3, Figures 7–9, p. 449, with permission of John Wiley.

important clue stemmed from a discovery of additional structure in the mitochondrion made possible by improvements in electron microscopy.

More Structure: The Discovery of the Cristae of the Mitochondrion

As noted in Chapter 4, Palade (1952b) conducted a comparative investigation in the early 1950s that resulted in a new buffered osmium fixative. In a study using it as well as Porter's new microtome to cut thin slices, Palade announced the discovery of a "system of parallel, regularly spaced ridges that protrude from the inside surface of the membrane towards the interior" (1952a, p. 428).[10] They are visible in the micrographs in Figure 6.1. Palade

---

[10] In the same study Palade found clear evidence for the mitochondrial membrane, although he initially took it to be a single membrane. The existence of a mitochondrial membrane had been previously defended both on the basis of electron microscopy of isolated mitochondria and on the basis of biochemical findings about the soluble character of chemical compounds that are retained in the mitochondria, but it was also questioned by others (Harman, 1950b; Huennekens & Green, 1950).

noted that the ridges appeared most clearly when sections were cut longitudinally through each mitochondrion and that they were oriented "more or less perpendicular to the long axis of the mitochondrion" (p. 432). He observed that the ridges have a "trilaminar structure" with "a central layer 8 to 10 mμ thick" surrounded by two thinner and denser layers. He called these ridges or protrusions "cristae mitochondriales" (p. 433) and, although noting that their appearance varied between cell types, concluded that they were likely to be regular features of mitochondria. He even suggested that they could serve as a "criterion for the identification of mitochondria in electron microscopy where the characteristic staining reactions of these organelles are of no avail" (p. 438).

By differentiating the cristae from the fluid matrix within the interior of the mitochondrion, Palade offered a structural decomposition that raised the prospect of advancing the mechanistic account of mitochondrial function by localizing different biochemical operations in different structural components. Palade himself put forward the suggestion that the aerobic oxidation reactions were directly tied to the structure of the mitochondrion, especially the cristae:

> It is well established that isolated mitochondria are able to carry out *in vitro* complicated oxidative processes that imply the action of a considerable number of enzymes. As these oxidations are apparently well co-ordinated and, in addition are dependent on the morphological integrity of the mitochondria . . . , it has been postulated that the enzymes involved in such processes are maintained in a 'definite spatial relationship' (Schneider & Hogeboom, '51) inside the organelles. It may be assumed that they are arranged in the proper order in linear series or chains – a disposition comparable in design and efficiency to an industrial assembly line. Such enzymatic chains have to be built at least partially in the solid framework of the mitochondrion because some of the component enzymes, namely succinic acid dehydrogenase (succinoxydase) and cytochrome *c* oxydase, are known to be insoluble and structure-bound. If we integrate the present morphological information with what is known from the general behavior of the mitochondria (e.g., their flexibility, response to variations in osmotic pressure and results obtained by centrifuging disintegrated organelles), it may reasonably be assumed that the mitochondrial matrix is fluid and that the membrane and the ridges represent the solid framework. In the present state of our knowledge, the internal ridges of the mitochondria appear as the most probable location for the postulated enzyme chains. (pp. 438–9)[11]

---

[11] In a footnote, Palade commented on Green's proposal of a cyclophorase system: "The work referred to deals with some dehydrogenases of the 'cyclophorase system,' a tissue residue

In a paper the following year Palade (1953) offered a number of arguments for localizing the enzyme systems in the cristae. First, he noted that a number of experiments, such as those of Lehninger, indicated that a substrate has to penetrate inside the mitochondrion before it is acted upon. Second, he pointed out that particles comparable in size to individual cristae and containing most of the succinoxidase systems of the mitochondria had been isolated from suspensions of disintegrated mitochondria (see below). Finally, he suggested that if the enzyme systems were located in the cristae rather than the outer membrane, they would be less exposed and protected from disruption.

A Competing Perspective on Mitochondrial Morphology

Although the Rockefeller laboratory of Porter and Palade played the central role in developing the new conceptions of cell structure and function discussed so far, there were competing laboratories, especially in Europe, which challenged several of their claims. The most vocal opponent was Fritiof Stig Sjöstrand, who in the 1950s established a major electron microscopy laboratory at the Karolinska Institute in Stockholm. As a medical and doctoral student at the Karolinska during World War II, Sjöstrand had begun working with an electron microscope built by Manne Siegbahn, a physicist at the Royal Swedish Academy of Sciences. Based on his early attempts to develop thin sections he had published micrographs of muscle in *Nature* in 1943 that were suggestive but provided little detail (Sjöstrand, 1943). In September 1947, in a meeting with R. R. Struthers, he (unsuccessfully) appealed to the Rockefeller Foundation for an electron microscope, noting that the only functioning microscope in Stockholm was in the Department of Histology. Struthers noted in his diary that Sjöstrand "appears more than usually intelligent and diligent and makes an excellent impression."

Supported by a Swedish State Research Council Fellowship, Sjöstrand spent the 1947–8 academic year at MIT working with Francis Schmitt on

lately identified as mitochondria. It is known (Schneider & Hogeboom, '51) that this system is actually a mixture of cell debris, nuclei, and mitochondria, a fact that renders more difficult the interpretation of the results mentioned. What happens in a 'cyclophorase system' does not necessarily take place exclusively in mitochondria" (p. 437). For his part, Green credits the electron micrographs of Palade and Sjöstrand with providing "independent confirmation of the organization deduced from functional considerations. These microscope studies readily disposed of the then current hypothesis that all the enzymes and coenzymes of the mitochondria were present as freely diffusible molecular species without any special organization in the fluid interior of the mitochondrial 'bag,' which was surrounding by a semipermeable membrane" (Green, 1957–8, p. 178).

the ultrastructure of the retinal rods.[12] There he further developed his abilities in electron microscopy. After returning to Sweden in 1948, Sjöstrand secured an appointment as permanent Docent in the Department of Anatomy at the Karolinska Institute, a position with limited teaching obligations that permitted him to devote most of his time to research. He set out to establish a laboratory for what he called *ultrastructure research* and in 1949 secured funding from the Alice and Kurt Wallenberg Foundation for an electron microscope (RCA EMU 2C). In the context of an application to the Rockefeller Foundation for additional equipment, he identified three components of his continuing research: the "structural basis of irritability" in sensory cells (the project Sjöstrand began at MIT), thin sectioning, and fixing cells for electron microscopy. The principal specimens for this work were the rods and cones of the guinea pig, especially a thin membrane which he thought was the primary part of the cell that was stimulated by light.

In 1949, as part of a trip to the Electron Microscope Society of America meetings in Washington, Sjöstrand visited the Rockefeller laboratory for a month and had access there to Claude's new microtome. Upon his return to Sweden, he developed his own microtome, one that employed an eccentrically located tissue mount revolving sixty times per minute that was advanced by a thermally expanding column behind the eccentric wheel. With this microtome, Sjöstrand claimed to be able to cut sections as thin as 70 Å on a regular basis. According to Porter,[13] Sjöstrand returned to Rockefeller for a month in 1952, where he learned of Palade's new buffered osmium fixative.

---

[12] Schmitt's impressions of Sjöstrand based on that year, reported in a letter to Gerald Pomerat of the Rockefeller Foundation on 26 July 1950, were certainly mixed. Schmitt said,

> I think there is little doubt that he is competent in [electron microscope research]. He is rather slow and sometimes appears phlegmatic, but this is probably illusory for he acquits himself well in discussions or debates, especially when his own work is in question.
>
> I am not sure that Sjöstrand himself will make any brilliant advances, but I do hope that his laboratory will become an active center for tissue fine structure work. Sjöstrand is well grounded in the field and will doubtless make substantial contributions, but I think his leadership among younger students of anatomy may pay even greater dividends. (Folder 1947, 1949–51, Box Karolinska Institutet, Molecular Biology, Series 800D, RG 2, Rockefeller Foundation Archives, RAC.)

Two years later Schmitt was more positive in his assessment of Sjöstrand's work at MIT. In a letter of 25 November 1952 to Ture Petrén, head of the Anatomy Institute in which Sjöstrand's laboratory was housed, Schmitt said, "I found that he is a sound scientist with the patience necessary to develop the complicated techniques required for the successful application of electron microscopy to the study of cell structure" (folder 1952–6, Box Karolinska Institutet, Molecular Biology, Series 800D, RG 2, Rockefeller Foundation Archives, RAC).

[13] Interview with Keith Porter, 1987, University of Maryland, Baltimore County. According to Porter, Sjöstrand also induced Porter's technician to return to Sweden with him, although she later returned to resume work with Porter.

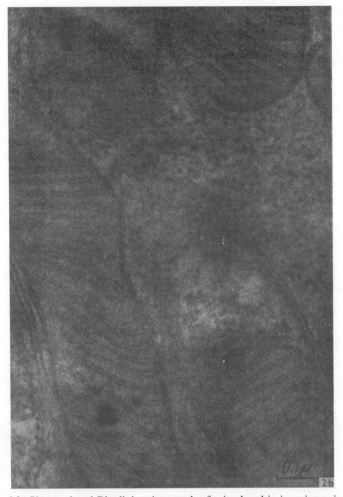

Figure 6.2. Sjöstrand and Rhodin's micrograph of mitochondria in guinea pig retina
revealing what they called *internal membranes* and construed as separate structures,
discontinuous with the outer mitochondrial membrane and with each other. Reproduced
from F. S. Sjöstrand and J. Rhodin (1953), The ultrastructure of the proximal convoluted tubules of the mouse kidney as revealed by high resolution electron microscopy,
Journal of Experimental Cell Research, 4, 426–56, Figure 2b, p. 434, with permission of
Elsevier.

That autumn he made a major splash at the Electron Microscope Society
of America meetings by presenting new, very high resolution micrographs
from guinea pig retina, later published in *Nature* (Sjöstrand, 1953b) that
revealed Palade's cristae more clearly than Palade's own micrographs of the

203

period.[14] Similar micrographs appeared in a paper that was part of the doctoral dissertation of one of his students, Johannes Rhodin (Sjöstrand & Rhodin, 1953). One of these micrographs is reproduced in Figure 6.2. Sjöstrand and Rhodin proclaimed,

> This investigation has demonstrated an internal structure within the mitochondria, which, as far as we know has not been described before. The system of transversally orientated double membranes and the clear cut demonstration of a similar outer membrane around the mitochondria indicate a high degree of organization of these cell organelles. (p. 449)

He described the structure more fully in his paper with another student, Viggo Hanzon (Sjöstrand & Hanzon, 1954):

> In the interior of the mitochondria densely packed inner membranes or plates are seen mainly oriented perpendicularly to the long axis of the mitochondrion. The inner membranes also appear double edged. One end of the membranes is in contact with the outer surface membrane and the other end in most cases is free from this membrane. (p. 406)

Although acknowledging occasional contact, Sjöstrand went on to state, "There is with few exceptions no continuity observed between the central space in the inner and outer membranes" (p. 406).

Sjöstrand's interpretation of what his micrographs revealed about mitochondria differed with Palade's views on two major points.[15] First, he maintained that the mitochondrial membrane was comprised of two layers, not one. Second, he rejected Palade's claim that the cristae were infoldings of the mitochondrial membrane: "There are no indications that the inner membranes represent folds of a single edged surface membrane. The inner membranes are

---

[14] Sjöstrand's presentations at these meetings were a major boost to his reputation. His former mentor Schmitt, who had previously been cautious in his appraisals of Sjöstrand, commented, "We have been mildly skeptical of his claims that he can section to 200–500 Å consistently. This skepticism was completely removed at the Cleveland meetings of the Electron Microscope Society of America earlier this month. At these meetings he described his results on the pancreas cells, the convoluted tubule cells of the kidney, the retinal rods and cones, and the nerve myelin sheath. Any of these papers would have been a great credit to the most experienced authorities in the field. However, the combination of the four was easily the best work reported at the meetings – a real triumph for Sjöstrand." (Letter of 25 November 1952 to Ture Petrén, Head of the Anatomy Institute in which Sjöstrand's laboratory was housed, Folder 1952–6, Box Karolinska Institutet, Molecular Biology, Series 800D, RG 2, Rockefeller Foundation Archives, RAC.)

[15] Rasmussen (1995; 1997) offered a detailed account of the dispute between Palade and Sjöstrand in which he argued that differences in their interpretive styles, including how they related electron microscopy to biochemistry, were more important than differences in technique: "The structure of nature and the structure of the sciences were at stake simultaneously in the struggle over which interpretive method should be made standard, which practices 'proper'" (p. 151).

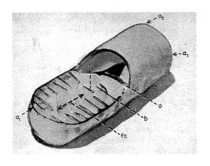

B

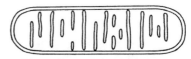

Figure 6.3. Contrasting interpretations of the internal structure of the mitochondrion as 3D models (top) and as 2D medial longitudinal sections (bottom). Palade's 3D model (left) is reproduced with permission from G. E. Palade (1953), An electron microscope study of mitochondrial structure, *Journal of Histochemistry and Cytochemistry, 1*, 188–211, p. 197. The 2D diagram is drawn by the author. Sjöstrand's 3D and 2D diagrams (right) are reproduced from F. S. Sjöstrand (1956), Electron microscopy of cells and tissues, in *Physical techniques in biological research*, G. Oster and A. W. Pollister, Eds. New York: Academic Press, pp. 241–98.

individual structures with only topographic relations to the outer membrane. Therefore the term, *cristae mitochondria*, is misleading" (p. 413).

The following April, at a symposium on the structure and chemistry of mitochondria at the Histochemical Society meetings in Chicago, Palade (1953) accepted Sjöstrand's claim that the mitochondrion was bounded by a double membrane,[16] and proposed that it was only the inner membrane which folded into the interior to form cristae. Sjöstrand remained opposed to the idea of infolding, arguing that the double-layered membranes traversing the mitochondrion were not attached to the (also double-layered) outer membrane. He also claimed that these double-layered traversing membranes went fully across the mitochondrion, whereas Palade proposed that there was an open channel extending through the interior of the mitochondrion, which he

---

[16] Sjöstrand himself referred not to two membranes but to a single double-layered membrane which he interpreted in terms of Danielli and Davson's sandwich model (see Figure 3.4). He held that the outer layers of proteins accounted for the two dark bands while the inner phospholipid layer accounted for the lighter area between them. Palade, not trying to provide a physical interpretation of membrane structure, simply interpreted each dark band as a separate membrane.

thought might serve for diffusion of substrates and products.[17] As shown in the top half of Figure 6.3, in 1953 both men produced 3D models – Palade a wax model and Sjöstrand a diagram – that illustrate the differences in how they conceived of these structures. Note especially the area in Palade's model labeled *fc* (for *central free channel*, later called the *mitochondrial matrix*). The two sketches in the bottom half of Figure 6.3 schematically illustrate a medial longitudinal section through each 3D model (discussed below).

The controversy between Sjöstrand and Palade was often personal and acrimonious.[18] Sjöstrand argued vociferously for the higher quality of his micrographs and for artifacts in Palade's micrographs. In particular, at the Third International Conference on Electron Microscopy, held in London in 1954, he proposed that the appearance of an open channel in the middle of mitochondria in Palade's micrographs was due to poor preservation as a result of delays in fixing the specimens (Sjöstrand, 1956b). These delays, he alleged, led to swelling during the postmortem interval before fixation was complete. To demonstrate the process he thought gave rise to Palade's micrographs, he prepared a series of micrographs at different intervals and claimed that the later micrographs showed increasing disruption of the internal membranes:

> Studies of post-mortem changes taking place within 15–30 to 45 minutes after death have shown that the mitochondria are changed very soon after death. They swell after the shutting off of the blood supply, the inner mitochondrial membranes appear fragmented and pulled apart leaving a more or less extensive central space free from inner membranes. It might be that such postmortem changes are responsible for the central space described by Palade. (Sjöstrand, 1956a, p. 463)

Sjöstrand maintained that a primary factor in generating high-resolution micrographs was keeping the tissue in as life-like a condition as possible. Thus, he injected fixative directly into living animals and as quickly as pos-

---

[17] In his 1952 paper Palade had commented, "In longitudinal sections that cut close to the mitochondrial membrane, the appearance of the lamellae might suggest that they are septa that traverse the mitochondrion from side to side. Oblique sections, however, indicate that the lamellae are actually ridges or folds protruding from the inside surface of the mitochondrial membrane towards the interior of the organelle without reaching the other side" (p. 432).

[18] At least in the eyes of Keith Porter, Sjöstrand's success was derivative from his and Palade's efforts. He complained to Pomerat that Sjöstrand failed to sufficiently acknowledge the assistance he had received. Yet, in his paper with Rhodin on the ultrastructure of mouse kidney tubules Sjöstrand says, "We feel very much indebted to Dr. K. Porter and to Dr. G. E. Pallade [employing the original spelling for Palade's name] for their kindness in giving the necessary information regarding the fixation technique used in the investigation. During the printing of this paper the paper by G. E. Pallade describing the fixation technique in *J. Exptl. Med.* **95**, 285 (1952) has become known to us" (Sjöstrand & Rhodin, 1953, p. 427, n. 1).

sible cut the tissue and immersed it in a phosphate-buffered saline solution maintained at salt and pH levels comparable to those found in living organisms.

In 1954 Sjöstrand presented a paper at the VIIIth Congress of Cell Biology in Leiden in which he set out to "survey the studies of the mitochondria structure that have been performed by Palade at the Rockefeller institute in New York and by our group at the Karolinska institutet in Stockholm" (1955a, p. 16). He presented only his own micrographs and began the paper by offering his interpretation of mitochondrial structure. But he then turned to the differences, noting first that "These discrepancies were more pronounced earlier but have been reduced with the improvement of the technique used by Palade" (Sjöstrand, 1955a, p. 19). Sjöstrand reiterates his contention that the inner membranes are not continuous with the membranes surrounding the mitochondrion and that there is no central space extending the length of the mitochondrion. He allowed that sometimes the membranes "do not form complete septa" so that there is communication between some of the compartments created by the inner membranes. But, he contended,

> The central space as depicted by Palade appears to us as a fixation artefact due to swelling of the mitochondria. Palade, seems not to have observed the susceptibility of the mitochondria to hypotonic media as the buffered osmium tetroxide solution originally recommended by him is strongly hypotonic. A similar swelling also occurs in a fairly rapid post mortem change and, therefore, appears deaper [sic] than 40–50μ below the surface of the tissue block. Palade's pictures show an appearance of the mitochondrion which is similar to the one we have observed as a result of post mortem changes. (p. 21)

In that paper Sjöstrand also took exception to proposals (such as Palade's, although he did not name Palade) linking the structural features of mitochondria to their biochemical function:

> What does this organization of the mitochondria mean? We may talk about the membranes as useful in realizing an orderly arrangement of the enzyme molecules to give fortunate spatial relations between enzymes taking part in chain reactions. We do not know, however, where the enzyme molecules are located and I think the speculations regarding the functional significance of these structures at this state [sic] may be reserved for very informal discussions or private contemplation. (p. 29)

Palade (1953), for his own part, explicitly credited Sjöstrand only for his determination that the mitochondrial membrane is a double membrane. Without specifically naming Sjöstrand, he offered a critical test between Sjöstrand's and his own proposals regarding the cristae. If they were true septa, he argued, "they should appear as continuous, traversing bands in all longitudinal sections, irrespective of their position in relation to the axis of the organelle" (p. 205). However, if they are ridges that project only partway into the interior, then "they should appear as continuous, traversing bands, only in longitudinal sections cutting close to the mitochondrial membrane" (p. 205). In other sections "passing lengthwise through the middle of the organelles, the lamellae should show free ends partially outlining a central channel or cavity" (p. 205). Palade contended that his micrographs revealed the second pattern. Palade also appealed to sequential serial sections to establish the claim that the cristae are really ridges, showing that different ridges disappear as one moves from slice to slice. Palade also presented evidence that the cristae are actually infoldings of the inner mitochondrial membrane and that "the light space between the two mitochondrial membranes is found to have approximately the same width as the central light layer of the cristae, with which it appears to communicate freely" (p. 207). Later this was called the *intermembrane space*.

In a 1956 paper in which he provided frequent citations to Sjöstrand's work, Palade, after indicating that most of the knowledge of fine structure came from his and Sjöstrand's groups, drew out the contrasts and ended by taking as conciliatory a stance as possible:

> The structural details described are the same, but there are, as expected, certain differences in interpretation and nomenclature. For instance, the Swedish group presents the two membranes at the periphery of the mitochondrion as a single structural unit, as a "double-edged" membrane under the name "outer double membrane." The cristae are not recognized as infoldings but described as "individual structures with only topographical relations to the outer membrane." The term "inner double membranes" is used for their designation. The points in disagreement are decreasing in number, however. For instance, the existence of two dense lines at the periphery of mitochondrial profiles, revealed by Sjöstrand's work, was subsequently confirmed by us, and the cristae ("inner double membranes"), originally described as complete septa by the Swedish group, are now recognized as incomplete partitions, at least in some cases. (1956a, pp. 194–5)

The sketches in Figure 6.3 of medial longitudinal sections through the two men's models illustrate how each inferred a different internal structure from micrographs that were not sufficiently detailed to resolve the issue to everyone's satisfaction. Palade's interpretation (bottom left) yielded two aqueous areas, each of which is in communication with a large surface area on one side or the other of the convoluted inner membrane (whose projections into the matrix were the cristae). One theme in Palade's paper was that these structural characteristics provided ample opportunity for the types of biochemical reactions identified in cell fractions. In contrast, Sjöstrand's interpretation (bottom right) yielded multiple pieces of inner membrane that sometimes touched the inner layer of the outer membrane but were structurally separate – they were discontinuous both with the outer membrane and with each other. In the type of cell Sjöstrand used for his earliest micrographs, exhibiting plate-shaped rather than puzzle-piece-shaped inner membranes in 3D, the inner membranes typically extended across the entire 2D section, touching at both ends – the "complete septa" referred to by Palade. Regardless of how completely the mitochondria were partitioned in Sjöstrand's account, the resulting topography would turn out to be inconsistent with an important biochemical account proposed in 1962 and eventually accepted (Mitchell's chemiosmotic hypothesis; see Figure 6.8).

Sjöstrand never capitulated, but the intense conflict soon abated as Palade's interpretation gained ascendancy (for details, see Rasmussen, 1997[19]). Just a few years later Bourne (1962) could write matter-of-factly, "after a little controversy, it was agreed that the inner of these two membranes was extended into the interior of the mitochondria, in some cases touching or almost touching the other side" (p. 59).

Biochemists Further Fractionate Mitochondria

The cristae offered a plausible locus for the biochemical mechanism of oxidative phosphorylation, which biochemists already recognized as membrane bound. But exactly how were the enzymes recognized as responsible for the different reactions bound in the membrane? As Rasmussen emphasized, one of the features of the Rockefeller approach was that it promoted a collaborative inquiry with biochemists in which techniques for studying mitochondria were complementary, not competitive: Biochemical reactions could be localized in particular parts of the cell via chemical analysis of fractions while

---

[19] In discussing the interaction between Palade and Sjöstrand at the Third International Conference on Electron Microscopy in London, England, in July 1954, Rasmussen focused on the different uses to which each put his micrographs. Sjöstrand stressed using micrographs to make quantitatively precise estimates of membrane sizes. Palade, on the other hand, emphasized the importance of drawing out the connections with findings from other techniques such as cell fractionation. Thus, Rasmussen commented, "Sjöstrand wanted to interpret his micrographs purely visually, judging fixation by the criterion of orderliness and seeking greater knowledge of molecular structure through ever-better resolution, whereas Palade, who was involved in cell fractionation himself, wanted to test micrograph interpretations against experiments on fractions" (1997, p. 139).

the cell parts and their organization were determined by electron microscopy. Contrary to Sjöstrand, Palade did not think either approach could yet yield descriptions of cell parts at the molecular level.[20] Once it was established that the processes of aerobic respiration as a whole were localized in mitochondrial fractions, the natural extension of the research was to seek to localize different enzymes in different components of the mitochondrion. The obvious way to proceed was to apply fractionation again, decomposing the whole mitochondrial system into subfractions that could carry out some but not all the operations of oxidative phosphorylation. If this were successful, researchers could hope to localize these reactions in turn in parts of the mitochondrion that appeared in different fractions. Again Green and Lehninger were the leaders in deploying this strategy, joined subsequently by Efraim Racker.

Taking advantage of the availability of beef hearts from the nearby slaughterhouses in Wisconsin, Green developed a procedure for large-scale fractionation of mitochondria. The procedures he employed routinely damaged mitochondria, but had the advantage of yielding components with different behavior. First, he obtained a *light fraction* that lacked the capacity to synthesize ATP when oxidizing succinate and a *heavy fraction* that retained that capacity. Both fractions, though, phosphorylated ATP when other citric acid cycle substrates were supplied. Green further divided the light fraction (after treating it with 15% alcohol) into subfractions, one of which carried out electron transport but not oxidative phosphorylation (he referred to these as

---

[20] Rasmussen maintained that a fundamental difference between Palade and Sjöstrand involved their respective relations to biochemistry: "In the science Sjöstrand was trying to build, the electron microscope presumed a certain authority over the territory of biochemists, who were heavily invested in their 'slick' new ultracentrifuges but were in a very weak position to establish for themselves that the cell components they were isolating had not been drastically altered by cell homogenation and the lengthy centrifugation protocols. On the other hand, the Rockefeller way posits a partnership with the fractionation biochemist, and a set of more modest goals for the electron microscopist that prevents conflict with the biochemist partner: mere description of topology of and associations among components in the unfractionated, in situ cell. The Rockefeller cell biologists had a metier whose definition did not entail conflict with the established biochemistry departments at institutions where electron microscopists were finding work in the later 1950s and 1960s" (1997, p. 148). Rasmussen developed this argument as part of a sociological explanation for the greater success of the Rockefeller approach and acceptance of the Rockefeller results. As discussed in the previous chapter, the technique of fractionation itself originated with the efforts of Bensley and especially Claude to link cell structures to biochemical operations. The Rockefeller researchers, including Palade, continued to utilize fractionation as a primary tool in their own research (see the discussion in Part 2 of this chapter of Palade and Siekevitz's collaboration on the endoplasmic reticulum). Thus, another way of viewing the Rockefeller approach is to trace its development as an internally motivated program that, as a happy side effect, minimized territorial conflicts. This differs from Rasmussen's perspective in that it subordinates sociological factors to scientific ones.

electron transport particles or ETP). Another fraction supported phosphorylation when oxidizing compounds other than succinate (he termed these *phosphorylating electron transport particles* or PETP). To understand the genesis of these particles, Green collaborated with electron microscopist Hans Ris.[21] The resulting micrographs revealed open fragments of cristae in the PETP and less functional closed fragments of cristae in the ETP particles (Green, 1957–8; Ziegler et al., 1958). This supported Palade's suggestion that the processes of oxidative phosphorylation were localized in the cristae.

Youssef Hatefi, working in Green's laboratory, developed evidence for grouping the various substances involved in the electron transport chain into four complexes:

(I) an NADH-ubiquinone reductase complex which included FMN and nonheme iron

(II) a succinate-ubiquinone reductase complex which included FAD and nonheme iron

(III) a ubiquinol-cytochrome c reductase complex which included cytochromes b and c_1, and a nonheme iron protein; and

(IV) a cytochrome c oxidase complex which included cytochrome a and copper.

Thereafter four of Green's collaborators, Hatefi, Haavik, Fowler, and Griffiths (1962), succeeded in reconstituting two systems: one capable of oxidizing NADH to carbon dioxide and water by combining complexes I, III, and IV, and another capable of oxidizing succinate to carbon dioxide and water by combining complexes II, III, and IV. Both reconstitutions revealed particulate structures, suggesting that the respiratory chain was formed into a fixed assembly electron transfer system (one in which the molecules were in advantageous spatial relations for passing electrons sequentially from molecule to molecule).

---

[21] Ris did his graduate work on mitotic division with Franz Schrader in zoology at Columbia. Subsequently, he spent a number of years working in Mirsky's laboratory at Rockefeller. Mirsky was a biochemist by background and Ris characterizes him as teaching Ris biochemistry while Ris provided cytological understanding of the structure of chromosomes as well as procedures for isolating them. Together they published a number of papers revealing, for example, the role of proteins in providing structure to chromosomes (Mirsky & Ris, 1951) and generating quantitative measures of DNA content in cell nuclei (Ris & Mirsky, 1949). Ris moved to the University of Wisconsin in 1949 and shortly thereafter began to explore the potential of electron microscopy for studying chromosomes, a project that did not fully bear fruit until the development of high-voltage electron microscopes in the late 1960s. Ris found Green's lack of sensitivity to cytological structure frustrating and soon suggested that Green work instead with Fernández-Morán at the University of Chicago (Interview, 6 November 1987, Madison).

Together with Cecil Cooper and other colleagues, Lehninger also pursued a project of "making morphologically less organized preparations from rat liver mitochondria with which the enzymatic details of oxidative phosphorylation could be more directly studied, with the ultimate goal of resolving the mechanisms of oxidative phosphorylation by enzyme separation and reconstruction approaches" (Lehninger et al., 1958, p. 450). The last clause indicates Lehninger's goal of establishing the mechanisms of oxidative phosphorylation using the traditional biochemical approaches of isolating responsible enzymes and then putting them together again into a system that performs the reaction.

The strategy Lehninger and his collaborators employed for decomposing the mechanism was to treat isolated mitochondria with digitonin to gently break down membrane lipids, which they reported caused the "virtual dissolution" of the mitochondrion, "leaving a turbid brown solution" (p. 450). They subjected this solution to centrifugation at 50,000 g for 25 minutes, removed the supernatant fluid, which contained a gelatinous material, and centrifuged it at 100,000 g for another 25 minutes. This yielded "phosphorylating membrane fragments" (p. 450), which did not catalyze most of the reactions of the citric acid cycle and generated ATP only when D-β-hydroxybutyrate or succinate (but not pyruvic acid or some other citric acid cycle intermediate) was supplied. Electron transport was evidenced by oxygen uptake, and oxidative phosphorylation by ATP synthesis (using radioactive phosphate as a tracer). Although the efficiency of the reactions was less than for intact mitochondria, they claimed that ATP synthesis did occur at all three sites along the electron transport chain. Most of the typical chemical agents that decoupled respiration from phosphorylation, such as 2,4-dinotrophenol, had the same effect on the particles, but Ca^{2+} and thyroxine did not. Appealing to electron micrographs, Lehninger and colleagues identified the particles as arising largely from the cristae.

From two facts – that the citric acid cycle enzymes could be separated in solution and that the membrane fragments isolated by Lehninger did not catalyze the reactions of that cycle – investigators could conclude that the citric acid cycle occurred in the matrix of the mitochondrion while electron transport and oxidative phosphorylation occurred in the cristae of the inner mitochondrial membrane, as Palade had suggested. The earlier determination that glycolysis occurred in the cytosol permitted localizing the three major biochemical mechanisms of cellular respiration (as detailed in Figure 3.16) in three different parts of the cell, as illustrated in Figure 6.4.

From the fact that sonic vibrations prior to centrifugation greatly reduced the sedimentation rate but not the efficacy of the particles, Lehninger and Cooper concluded that the inner membrane was comprised of repetitions

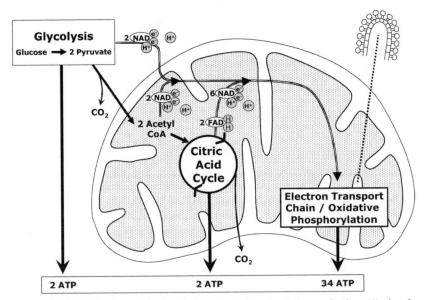

Figure 6.4. Localization of biochemical mechanisms of cellular respiration: (1) glycolysis in the cytosol; (2) citric acid cycle in the mitochondrial matrix; (3) electron transport and oxidative phosphorylation in the cristae.

of a basic respiratory assembly. Their objective was to determine the structure of one of these units, which they pursued through a series of experiments that focused selectively on the reaction between cytochrome c and oxygen (Cooper & Lehninger, 1956a; Cooper & Lehninger, 1956b; Devlin & Lehninger, 1956), on ATPase activity (Cooper & Lehninger, 1957a), and on ATP-P_i[32] and ATP–ADP exchange reactions (Cooper & Lehninger, 1957b). The exchange reactions involved the regular exchange either of a free phosphate with one in ATP or the transfer of a phosphate from an ATP molecule to an ADP molecule. All of these reactions occurred in the digitonin prepared particles and were inhibited by decoupling agents. From these studies they concluded that phosphate and ADP enter into oxidative phosphorylation in separate, sequential steps. By investigating the exchange reactions in particular, Lehninger and his colleagues claimed to determine the order of the final events in phosphorylation. The alternatives were that the responsible enzyme (they used X, Y, and Z to represent the enzymes involved in each of the phosphorylation reactions) first reacted with the phosphate, creating a high-energy intermediate X~P, etc., and then reacted with ADP to form ATP, or that they reacted first with ADP, creating X~ADP, etc., and then with the phosphate. The evidence from the exchange reaction supported the former, not the latter, possibility (see Figure 6.5).

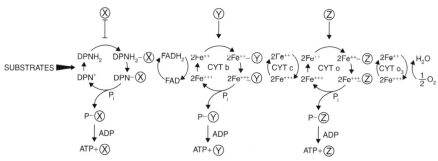

Figure 6.5. Lehninger's conception of the coupling of the electron transport chain with oxidative phosphorylation via high-energy intermediates. X, Y, and Z are three different enzymes, P_i is inorganic phosphate, $\sim$ is a high-energy bond, and other substances are carriers undergoing reversible oxidation-reduction. Reproduced from A. L. Lehninger et al. (1958), Oxidative phosphorylation, *Science, 128*, 450–6, Figure 3, p. 455.

With this account of the final step, Lehninger and his colleagues considered two possible mechanisms (their term) for the generation of X~P. By one account it was the reduced carrier (e.g., the cytochrome receiving the pair of electrons) that entered into a high-energy bond with the enzyme, whereas by the other it was the oxidized carrier (the cytochrome surrendering the electrons) that formed the high-energy bond. Appealing again to evidence from exchange reactions, specifically, that the reactions occurred at a maximal rate when the carriers were kept in a fully oxidized state, but not in a reduced one, Lehninger and his colleagues concluded that the second proposal was correct. Although expressing caution as to how decisive the evidence was, they interpreted their results as supporting the mechanism shown in Figure 6.5 according to which, at each of the three phosphorylation sites, an enzyme bound itself to the reduced carrier. When the carrier was then oxidized, the bond between the carrier and the enzyme became a high-energy one. At the first site, for example, the oxidation resulted in DPN~X. Inorganic phosphate then replaced the carrier (DPN) in the bond, yielding P~X. Finally ADP replaced the enzyme (X), yielding ATP and X. Once this type of reaction sequence had occurred at all three sites, and the energy-depleted electrons had joined with hydrogen ions and oxygen to make water, one round of oxidative phosphorylation was achieved. Energy was now stored in high-energy phosphate bonds in several molecules of ATP and was available for a variety of purposes.

At this point, both Green and Lehninger had succeeded in segregating submitochondrial particles and had proposed accounts of the operations involved in the mechanism of oxidative phosphorylation. Efraim Racker then attempted to take the endeavor a step further by isolating a soluble enzyme that was responsible for the phosphorylation of ADP.[22] He employed a technique in which he fractionated mitochondria after breaking them with glass beads in a vacuum.[23] This produced a red-brown, gelatinous residue or particulate fraction that could oxidize some citric acid cycle intermediates but not accomplish phosphorylation unless the faintly turbid yellow supernatant was re-added. Then phosphorylation increased until reaching a P:O ratio of 0.5. He set about purifying the substance in the supernatant, which he labeled *coupling factor* F_1. At this point Racker followed up on a suggestion first advanced by Henry Lardy and Conrad Elvehjem (1945) that phosphorylation might be the inverse of the breakdown of ATP to ADP, attributed to the enzyme ATPase. He found that F_1 also exhibited ATPase activity, and after showing that both F_1 coupling and ATPase activity decayed at the same rate around $0°$, identified them as the same protein (see Penefsky et al., 1960; Pullman et al., 1960; Racker, 1965, Chapter 13). Racker concluded, "F_1 catalyzes the transphosphorylation step from $X{\sim}P$ to ADP to form ATP at phosphorylation Sites 1 and 2" (1965, p. 169).

As the subscript number of the coupling factor suggests, Racker was also separating other factors – F_2, F_3, and F_4 – and investigating their role in phosphorylation or in other reactions such as ATP–P_i[32] exchange. Of particular interest was factor F_o, which Racker first identified in the context of trying to account for the sensitivity of both oxidative phosphorylation and ATPase activity to oligomycin poisoning. Addition of F_o to F_1 not only provided oligomycin sensitivity, but when treated with salt solution, generated particles. I will discuss the significance of this discovery after introducing the discovery of one more morphological structure.

One More Piece of Structure and a Proposal as to Its Function

Yet another development in electron microscopy technique, the introduction of negative staining by Humberto Fernández-Morán (1962), revealed additional structure in the mitochondrion. Negative staining uses substances

---

[22] In his initial studies Racker collaborated with Gifford Pinchot to study oxidative phosphorylation in *Escherichia coli* in hopes of finding "a system which would withstand fractionation" (Pinchot, 1953, p. 65). They used sonic vibrations to prepare extracts and separated two components, a particulate fraction that catalyzed oxidation and a soluble fraction that was required for phosphorylation (Pinchot & Racker, 1951). Pinchot went on to study the reaction in *Alcaligenes faecalis*, where he distinguished two soluble fractions, one of which was heat labile and one of which was heat stable.

[23] This is a procedure Racker had previously used with tumor cells or bacteria. He offers an interesting characterization of his work as "instrumental research: When you run out of ideas, use a new instrument" (Racker, 1965, p. 164).

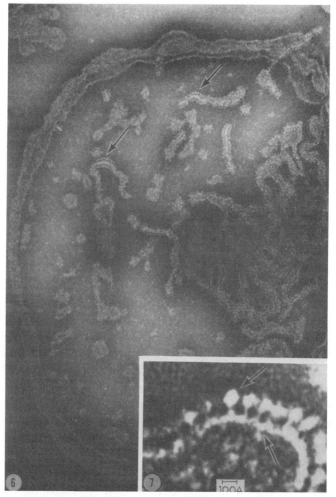

Figure 6.6. Fernández-Morán's negative-stained micrograph of cristae showing small spherical particles attached to the membrane. Reproduced from H. Fernández-Morán et al. (1964), A macromolecular repeating unit of mitochondrial structure and function: Correlated electron microscopic and biochemical studies of isolated mitochondria and submitochondrial particles of beef heart muscle, *Journal of Cell Biology*, *22*, 63–100, Figures 6 and 7, p. 73 by copyright permission of the Rockefeller University Press.

such as phosphotungstate or uranyl acetate that are electron dense but chemically inert. These substances do not react with membrane material, which then appears light against the dark background created by the electron dense material. Using this approach, Fernández-Morán discovered small particles (70–90 Å in size) located on stalks about 50 Å in length projecting from the

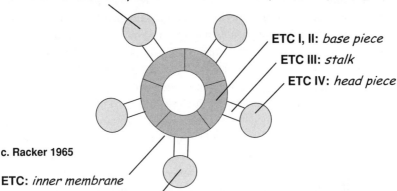

a. Green and colleagues 1963

ETC: *inner membrane sphere*

b. Green and colleagues 1964

ETC: *tripartite repeating unit*

ETC I, II: *base piece*
ETC III: *stalk*
ETC IV: *head piece*

c. Racker 1965

ETC: *inner membrane*

Oxidative phosphorylation (ATPase): *inner membrane sphere*

Figure 6.7. Three proposals as to the significance of the spherical particles on the mitochondrial inner membrane (cristae). (a) Green's 1963 proposal, in which each inner
membrane sphere contained an entire electron transport chain (ETC). (b) Green's revised
proposal of a tripartite repeating unit in which complexes I and II of the electron transport chain were contained in the base piece, complex III in the stalk, and complex IV in
the head piece. (c) Racker's account localizing ATPase in the inner membrane spheres
and the electron transport chain in the inner membrane.

cristae into the inner mitochondrial milieu. While small, these particles are
numerous, with estimates of 10,000 to 100,000 per mitochondrion. When he
applied the negative stains without prior fixation, mitochondria swelled and
burst, extruding membranous material in the form of sheets, tubules, or rib-
bons that were studded with small spherical knobs about 90 Å in diameter.
These spherical knobs can readily be seen (particularly in the inset) in Figure
6.6, which shows a micrograph Fernández-Morán published in 1964.

Green seized upon Fernández-Morán's discovery, naming the knobs *inner
membrane spheres* and proposing that they constituted the complete system
of enzymes for electron transport (Figure 6.7a). Lehninger (1964), however,
calculated that the weight of the respiratory assembly was one to two orders
of magnitude greater than that of these particles. As shown in Figure 6.7b,
Green and his collaborators then proposed a distribution of the four different complexes of enzymes involved in electron transport over the base piece
(Complexes 1 and II), stalk (Complex III), and head piece (Complex IV),
respectively (Fernández-Morán et al., 1964). Although much of their analysis
focused on the relative sizes of the stalk and head pieces and the minimum

217

sizes, based on molecular weight, of the enzyme complexes, ultimately they appeal to "biochemical considerations" to defend this localization: "Complexes I and II must interact with DPNH and succinate, respectively, both of which are localized in the interior of the crista, whereas complex IV must interact with molecular oxygen which would be more readily available in the solution outside the crista rather than in its interior" (p. 95). Green presented a popularized account of this proposal in a paper in *Scientific American* in 1964. There he also offered a speculative proposal (attributed to Robert Bock and Robert Criddle), according to which the transport of substrates between enzymes was accomplished "by means of swinging groups of atoms, mounted on the respective proteins by flexible arms, that transfer and accept the electrons" (Green, 1964). This mechanism, it should be noted, addressed the issue of electron transport, but not the accompanying phosphorylation.

Green's proposals were quickly discounted by other biochemists. I noted previously that Racker had found that F_o preparations generated particles when salt was added. Collaborating with Donald Parsons and Britton Chance, he examined these preparations with the negative staining technique and found that they were sac-shaped structures covered with "inner membrane spheres" like those found by Fernández-Morán. He then treated the preparations with trypsin, followed by urea, a treatment he had previously employed to remove ATPase activity from his preparations. Examining these preparations with negative staining, he found they had no inner membrane spheres. This established that, contrary to Green's proposal, the electron transport chain was not in the spheres. On the other hand, preparations of the F_1 factor showed spheres about 85 Å in diameter, and when such preparations were added to the trypsin-urea membrane preparation, spheres appeared on the membrane. The preparation still did not perform oxidative phosphorylation, but when F_2, F_3, and F_4 were added as well, phosphorylation was restored (Racker et al., 1965; see also Racker, 1968). He concluded that the spheres contained the ATPase and were the locus of ATP synthesis.

Racker was not fully satisfied with the demonstration, however, because the difference in number of spheres before and after adding F_1 was small enough that he had to rely on statistical analysis to establish it. His research assistant, Lawrence Horstman, tried passing mitochondrial fractions through a Sephadex column in order to remove the native spheres more effectively.[24]

---

[24] Racker (1976, p. 16) commented, "He tried these experiments without any encouragement from me because I did not think that the procedure could be carried out without damage to the particles. However, it worked, which brings us to the next lesson, Lesson 6: Progress is made by young scientists who carry out experiments old scientists said wouldn't work (F. Westheimer)."

This produced a far more definitive series of micrographs – one of submito-chondrial particles with spheres, one with the spheres removed by Sephadex and urea, and a third, resembling the first, with the spheres reconstituted with F_1 (Racker & Horstman, 1967).

Racker's research localized the electron transport chain in the inner mito-chondrial membrane and the ATPase in the spheres attached to the membrane (Figure 6.8c). This now presented a structural version of the problem bio-chemists had faced since the discovery of phosphorylation accompanying electron transport: How were the oxidation–reduction reactions of electron transport linked to ATP synthesis? The problem was now how to link activ-ities localized in the inner membrane with activities in the attached spheres. For the most part, biochemists were still seeking chemical intermediates – the hypothetical compound C that, in Slater's scheme, formed an initial high-energy bond with the substrate and then transferred that bond to ADP or, in Lehninger's version (Figure 6.8), the postulated enzymes X, Y, Z that played that role at the three phosphorylation sites along the electron transport chain.

Radical Reconceptualization of Oxidative Metabolism

In 1961 Peter Mitchell, a maverick biochemist whose background had famil-iarized him with the enzyme-catalyzed translocation of chemical groups across membranes, advanced a revolutionary reconceptualization of oxidative phosphorylation. He suggested that the crucial intermediary was not chemical in nature but rather was a proton gradient across the inner mitochondrial membrane. The enzymes of the electron transport chain were so organized in the membrane that as the respiratory substrates were oxidized, protons (H^+) were discharged on one side of the membrane (the intermembrane space of the cristae) and OH^- ions were discharged on the other side (the mito-chondrial matrix). Because the mitochondrial membrane is impermeable to H^+ and OH^- ions, a proton gradient bearing an electrical potential devel-ops. When ATPase operates normally (that is, to hydrolyze ATP to ADP and inorganic phosphate), it also pumps ions across the membrane into the inter-membrane space. But once a gradient has developed with significantly higher concentrations of protons in the intermembrane space than in the matrix, the ATPase can no longer can pump ions out of the matrix. This energy built up in the proton gradient provides the energy to drive the ATPase in reverse – synthesizing ATP from ADP and inorganic phosphate, rather than break-ing it down (Mitchell, 1961; Mitchell, 1966). Because Mitchell's hypothesis, as shown schematically in Figure 6.8, made transport across a membrane a central component of the mechanism, he called it the *chemiosmotic* hypothesis.

Mitchell's proposals were initially extremely controversial, giving rise to what are often referred to as the *ox phos wars* (Prebble, 2002). The opposition was partly empirical – for example, questioning the evidence for the claimed proton gradient – and partly theoretical. The idea of a proton gradient across a membrane was foreign to most biochemists, who were still oriented to the soluble systems model for explaining metabolic processes. Remarks of Efraim Racker reveal the combative nature of the debate: He referred to "hypothetical proton gradient and imaginary membrane potential" and compared Mitchell's claims to "pronouncements of a court jester or a prophet of doom" (Racker, 1975). Yet, at approximately the same time Racker wrote to the central figures investigating oxidative phosphorylation – Paul Boyer, Britton Chance, Lars Ernster, Tsoo E. King, Henry A. Lardy, and D. Rao Sanadi, in addition to Green, Lehninger, Mitchell, and Slater – proposing the preparation of a joint statement designed to reduce the acrimony over oxidative phosphorylation. Although negotiations over the joint review were tempestuous, several of the authors agreed on a joint introductory statement, followed by individual papers, that appeared in the *Annual Review of Biochemistry* for 1977 (Boyer et al., 1977).[25] By this time Racker had been convinced of the chemiosmotic hypothesis and his own contribution supported Mitchell's position. Although controversy continued, Mitchell was awarded the Nobel Prize in 1978, a testimony to the significance of his proposal in transforming thinking about the phosphorylation process.

With the incorporation of Mitchell's account of the linkage between electron transport and phosphorylation, the mechanism of oxidative phosphorylation was essentially resolved. Moreover, the account wove together in a fundamental way morphological structure with chemical operations. Palade's cristae not only were seen to contain the critical enzymes of electron transport in a spatially organized manner (the functional significance first attached to them), but also served to create the proton motive force that drove ATP synthesis. The stalks and spherical particles attached to the membrane housed F_0 and the ATPase and could respond to the proton gradient by synthesizing ATP. The combined contributions of studies of cell structure and biochemical function were melded into a comprehensive account of the mechanism that accounted for the phenomenon. Many details of the operation of the

---

[25] Racker (Interview, 1989, Ithaca, NY) expressed dissatisfaction with the final result since each author ended up arguing for his own position rather than engaging the others in the manner he had hoped.

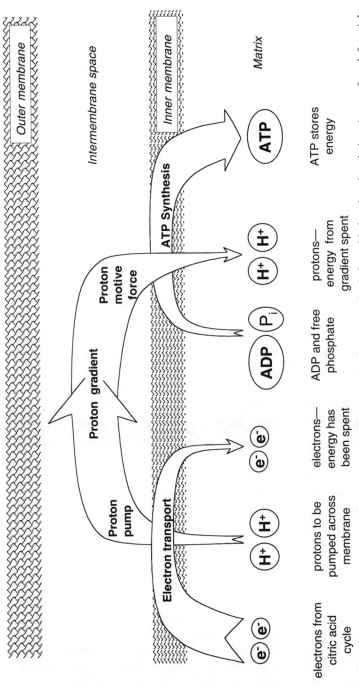

Figure 6.8. A schematic representation of Peter Mitchell's chemiosmotic hypothesis for oxidative phosphorylation. Arrows from left to right depict temporally ordered energy transfers, not spatial relations. As the width of the arrow representing each process increases (decreases), energy is gained (spent).

221

mechanism remained to be resolved, and this provided the focus of ongoing work in biochemistry. Although the locus of research on oxidative phosphorylation shifted after about 1970 to biochemistry, the research in the prior decades was an outstanding exemplar of research in the new field of cell biology. By revealing structure and organization at a level between older biochemistry and traditional cytology, de Duve's terra incognita had become an explored territory.

2. MICROSOMES, THE ENDOPLASMIC RETICULUM, AND RIBOSOMES

From Lace-like Reticulum to Endoplasmic Reticulum

As I discussed in the previous chapter, Claude first identified microsomes through cell fractionation while Porter, examining his electron micrographs, described a lace-like reticulum with granules and identified Claude's microsomes with what appeared as granules in it. Although Claude remained agnostic, many investigators proposed that microsomes were involved in protein synthesis. At the same time as research on the mitochondrion was revealing the mechanism of oxidative phosphorylation, researchers were making rapid advances in their understanding of these additional cytoplasmic structures, an endeavor that culminated in an account of how the structures figured in protein synthesis. Porter played a key role in initiating these developments. In papers published in 1952 and 1953 he, together with Frances Kallman, a postdoctoral fellow of the National Cancer Institute, captured images of this lace-like reticulum by increasing the time of fixation in osmium vapors. Porter proposed that the osmium vapor digested the "diffuse and frequently fibrous components of the ground substance" leaving "what may be thought of as a membrane skeleton of the cell" (Porter & Kallman, 1952, p. 883). With this technique, Porter and Kallman provided a further description of Porter's lace-like reticulum and gave it a new name:

> A third component uniformly present in these images is made up of vesicular or canalicular elements which sometimes constitute a complex reticulum. This material is part of the innermost cytoplasm of the cell, the endoplasm. It is referred to as the endoplasmic reticulum from its location and form. It appears to be a finely divided vacuolar system. It varies enormously in different cells in the size of its division and, though its function is not known, this variation reflects in part the physiological state of the cell at the time of fixation. Phase contrast microscopy provides evidence of its presence in the living cell. (p. 883)

Porter and Kallman then turned their attention to the particle that Porter and Thompson had observed in tumor cells and which by then had been reported by two other electron microscope laboratories (Cannan & Berger, 1951; Oberling et al., 1950). Porter and Thompson had found these particles to be limited to tumor cells and now Porter and Kallman suggested a reason: "We were . . . comparing rapidly proliferating tumor cells with 'resting' normal cells" (p. 887). They reported finding such particles in actively growing cells derived from young rat heart embryonic tissue. Because they appeared when cells, normal or tumor, were rapidly growing, they proposed "to associate these granules with growth processes in the cell, *i.e.*, in the production of new protoplasm, and they have been tentatively referred to as growth granules" (p. 887). Relying on results from absorption studies with ultraviolet microscopy (Ludford, Smiles, & Welch, 1948), which indicated a nucleotide composition of the particles, they concluded that the particles "may have a high content of ribo-nucleotides, which might be expected if they are accepted as multiplying components of the cytoplasm" (pp. 888–9). They went on to associate the particles with the microsomal fraction from fractionation studies. In what they admitted to be speculation, they continued,

> it is attractive to think of them as centers of synthesis of all cytoplasmic components. There is some preliminary evidence from the micrographs that mitochondria may begin their development in this form, but elements of the endoplasmic reticulum, the lipid granules, and inclusions, the distinctive features of differentiated cells, may be similarly derived. If such is the case, we are led to postulate that there are several subspecies among this class of cytoplasmic particles and that the complement of these in any cell would determine the type of differentiation to some extent. (p. 890)

Through the 1940s Porter's description of a lace-like reticulum and his suggestions of its function failed to attract much response from other investigators. This was largely because the structure appeared only in micrographs of whole, cultured cells, which he nearly alone was making. This changed with the improvement in techniques for making micrographs of ordinary thin-sliced cells. Albert Dalton (1951a; Dalton et al., 1950) published electron micrographs which showed a number of filamentous units in the cytoplasm, which tended to be grouped in particular areas and which were reduced in number when the animal fasted.[26] Soon thereafter, Wilhelm Bernhard and

---

[26] "Differentiation of filament-like stands are present in the cytoplasm of the proximal tubule cells but they have been found only in the basal parts of the cells and are somewhat thinner (approximately 0.05μ) than cell membranes. They are also identifiable by the fact that they terminate in the cytoplasm without returning to one of the tubule surfaces. . . . These structures

Charles Oberling (Bernhard, Gautier, & Oberling, 1951; Bernhard et al., 1952) also reported finding ergastoplasmic filamentous or lamellar structures (see Haguenau, 1958, for a review).

Noting Dalton's report of filaments in thin sections, Palade and Porter, when they started making micrographs of thin sections in 1950–1, investigated the relation between these and Porter's endoplasmic reticulum. They reported that with buffered osmium the "'filaments' were in fact fine tubules, or strands of vesicles identical in size and form to those constituting the endoplasmic reticulum of cultured cells" (*Annual Report*, 1950–1, p. 147). In his 1953 paper with Blum describing their new microtome, Porter presented a micrograph of a thin section in which he identified

> numerous 'elongated elements' and many 'granules'. These vary from 50 mμ to 150 mμ in diameter and much more in length. When they are examined carefully they are found to have a relatively 'transparent' center bounded by a single dense line apparently representing a membrane. These elements of the cytoplasm are *easily recognized* as the equivalent of the endoplasmic reticulum indicated previously in cultured cells (Porter & Thompson, 1948) and identified with the basophilic component of the cytoplasm (Palade & Porter, 1952; Porter, 1953). (Porter & Blum, 1953, p. 699, emphasis added)

Although Porter and Blum claimed the identification is easy, one thing that was clearly lost in the micrographs made of thin slices was the three-dimensional perspective offered with whole tissue-cultured cells. Without commenting on how this had already led to different interpretations of the endoplasmic reticulum, Porter and Blum stated, "In any further study of this endoplasmic reticulum it is important to determine its three-dimensional form in cells fixed *in situ* in their respective tissues. Obviously for this and similar problems serial sections of the cells are essential" (p. 700). They then presented a series of four serial sections and commented,

> it is observed that the system referred to as the endoplasmic reticulum is a complex of interconnected strands and sinusoids, the latter tending to be flattened in one dimension but otherwise very polymorphic. At certain sites in this parotid

are, as a rule, oriented in a position perpendicular to the basement membrane. An attempt has been made to determine whether they are filaments or actual lamellae" (Dalton, 1951a, p. 1167). From the fact that they stay in focus while the focus of the electron microscope is changed, Dalton concluded they are lamellar in form. In a review in 1953, Dalton argued, "In our preparations the components of this material were usually found as long, sometimes branching, parallel or concentric strands and never in cross section as solid or hollow spheres of the same diameter as the strands. This naturally suggested that they are lamellae rather than filaments or tubules" (p. 411).

cell these flattened portions of the system are organized in parallel arrays but in other regions of the cytoplasm the system is represented by finer canalicular or vesicular elements. If the serial micrographs are examined closely instances may be found where discrete elements appearing as cross sections of canaliculi come together progressively into single elements representing either longitudinal sections through canaliculi or, more likely, marginal sections through sinusoids. Also in the same series, sequences are apparent defining arborizations of unit structures. In still other regions of the cytoplasm the only elements evident appear as tiny vesicles or cross sections of canaliculi. Since certain of these can be traced from section to section they are evidently segments of canaliculi. (p. 700)

Porter's claim that the structures apparent in whole tissue-cultured cells were the same as those that could be reconstructed across a series of thin sections of cells *in situ* soon turned contentious. Critics objected that what appeared in tissue-cultured cells was an artifact of the process of growing cells in such an abnormal environment. To address this objection, Palade and Porter (1954) adopted the strategy of preparing thin sections of tissue-cultured cells to see how these would look and compared a series of them to the usual whole mounts of such cells. By demonstrating correspondences, they hoped to legitimate the use of micrographs of whole tissue-cultured cells.

Before presenting their results, they advanced a theoretical claim as to how the sections of tissue-cultured cells should look: "Under such circumstances the endoplasmic reticulum . . . cannot be expected to appear in sections as a network. Occasionally, it could be included in the thickness of a given section, but in the vast majority of sections only profiles of 'vesicles and strands' will be encountered and these will appear as independent structures because their original connections have been severed by the microtome" (p. 664). Palade and Porter presented micrographs of a sectioned chicken monocyte (white blood cell) as well as a whole-mount and sectioned macrophage grown from monocytes in tissue culture to establish the correspondences in appearance between whole mounts and thin slices. They went on to compare whole mounts of cultured cells and thin sections of various mammalian cells fixed *in situ* to make their case (see Figure 6.9).

In a second paper in the series, of which he was the sole author, Palade described the appearance of the endoplasmic reticulum in cells in several tissue types from rats and chickens – epithelial, nervous, mesenchymal, and muscular. One difference Palade observed between cells from living organisms and those from tissue culture is that in cells from living organisms the reticulum runs from the nuclear membrane to the cell membrane. Thus, it occurs in both endoplasm and exoplasm. Noting this and the fact that "the

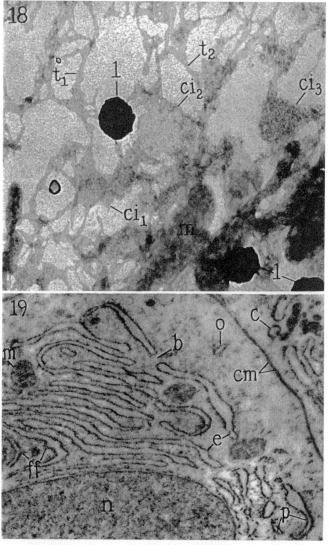

Figure 6.9. A pair of micrographs in which Palade and Porter compared the appearance of the endoplasmic reticulum in whole tissue-cultured cells (top) and in thin sections (bottom), both from glandular epithelia of the parotid of a newborn rat. Reproduced from Palade, G. E., & Porter, K. R. (1954). Studies on endoplasmic reticulum: I. Its identification in cells *in situ*. *Journal of Experimental Medicine*, 1954, *100*, 641–56, Plate 62 by copyright permission of the Rockefeller University Press.

word 'reticulum' was stretched a little to accommodate preferentially oriented dispositions of the system," Palade asserted, "the name 'endoplasmic
reticulum' has a number of admitted shortcomings. We retain it because we
do not have a better one" (1956b, p. 92).[27]

For Palade and Porter the micrographs of whole cells provided an anchoring
point to which they appealed in interpreting thin sections, and Palade noted
the problems others had in interpreting thin sections without this reference
point:

> The usual thickness of such sections, *i.e.* 20 to 40 mμ, being much smaller than
> the mesh size of the reticulum, and even smaller than the diameter of the vesi-
> cles and tubules that form its trabeculae, it follows that in sections only slices
> of these trabeculae or "profiles" can be found. Fragments of meshes, or more
> or less complete meshes are only occasionally encountered and the continuity
> of the system throughout the cytoplasm is never apparent; it has been lost by
> microtomy and can be regained only by the difficult and tedious operation of
> tridimensional reconstruction. In this respect, one may comment that the use of
> spread cells as initial specimens for the study of the endoplasmic reticulum was
> a fortunate coincidence. The main feature of the system, *i.e.* its disposition in
> a continuous reticulum that permeates the entire cytoplasm, would have been
> extremely difficult if not impossible to grasp from an exclusive study of sec-
> tions. It is then easy to understand why, at the beginning at least, the shift from
> spread to sectioned specimens caused so much confusion about the endoplasmic
> reticulum. In the few years that have elapsed, a number of conflicting descrip-
> tions and interpretations have been advanced and a corresponding number of
> names proposed for the structures belonging to the system under consideration.
> A vacuum in terminology, be it only apparent, seems effectively to lead many
> a biologist into philological temptation. (1956b, pp. 86–7)

In some cases, however, Palade and Porter found clues in the thin sections
that allowed them to reinterpret micrographs of whole cells:

> The examination of serial sections and of sections of various incidences
> indicates that the elements in question are relatively large, flattened vesicles
> of irregular outline, but of shallow and relatively constant depth for which the

[27] The name did survive even though many critics tried to displace it. In particular, French authors
such as Françoise Haguenau (1958) sought to maintain Garnier's term *ergastoplasm*. Novikoff
(1956b, p. 971) commented, "Agreement on terminology, sometimes made more difficult by
considerations of national pride and human personality, appears to be close at hand, as we rapidly
learn more about the membrane systems of cells. Porter's work (1955–6) suggests that all cells
possess a similar basic component – a vacuolar system of great complexity – and that in different
cells this system shows varying degrees of continuity and specialization. Thus, according to this
view, 'ergastoplasm' is essentially a specialized type of 'endoplasmic reticulum' characterized
by the presence of basophilic granules on its surface."

name of *cisternae* has been recently proposed (Palade & Porter, 1954). Because such elements were not originally described in cultured cells examined *in toto* (Porter, 1953), spread specimens were reexamined (Palade & Porter, 1954) and in many cases the endoplasmic reticulum was found to consist mainly of large, flat cisternae of irregular outline with only a few tubular and vesicular elements present. (Palade, 1956b, p. 89)

Just as he did in the case of mitochondria, Palade observed in his improved micrographs a new structure in the endoplasmic reticulum – he found that frequently the outer surface of the membranes of the endoplasmic reticulum were coated with fine granular material that at higher resolutions appeared as discrete structures ranging from 10 to 30 mμ in size.[28] Porter (1954), after arguing that the endoplasmic reticulum was the source of the basal staining and hence the basophilic component of the cytoplasm, concluded that the basophilia was due to these particles on the grounds that the basal staining was most prominent in tissue in which granules were attached to the endoplasmic reticulum. He further related the granules to a fraction with particles smaller than the microsomes isolated by Barnum and Huseby (1948) and to a fraction of particles of the same size as those on the endoplasmic reticulum that Petermann, Mizen, and Hamilton (1953) demonstrated to contain large amounts of RNA. Porter concluded by identifying similar particles in the nucleolus.

Dissenting Voices

As in the case of the mitochondrion, Sjöstrand criticized Porter and Palade's studies of the endoplasmic reticulum. Emphasizing the fact that in thin slices what appeared were pairs of membranes, Sjöstrand and Rhodin claimed that the ground cytoplasm was divided "into open compartments through well-defined intracellular double membranes" (1953, p. 428). The following year, in a paper with Hanzon, Sjöstrand expanded on his characterization of the intracellular cytoplasmic membranes, now identifying "'small opaque particles' attached to one side." (These are quite apparent in the micrographs they published – see Figure 6.10.) He further described,

> The smooth surfaces of the intracellular cytoplasmic membranes face each other giving the impression of the membranes being arranged in pairs. They always

---

[28] Porter's initial response to these particles on parts of the endoplasmic reticulum was that they were likely artifacts due to damage caused by the electron beam. With a denigrating intent, he initially dubbed them *Palade particles*. (Interview with George Pappas, 23 October 1995, University of Illinois Chicago.)

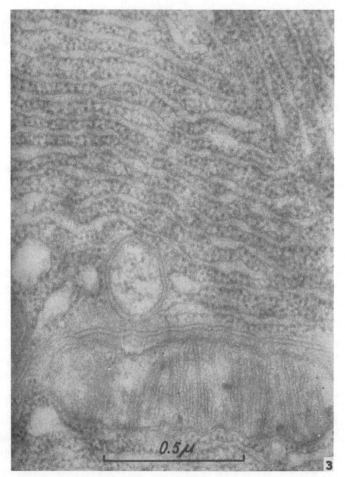

Figure 6.10. Sjöstrand and Hanzon's micrograph of paired intracellular membranes with small opaque particles attached on one side of each membrane. At the bottom a mitochondrion is shown. Reproduced from F. S. Sjöstrand and V. Hanzon (1954), Membrane structure of cytoplasm and mitochondria in exocrine cells of mouse pancreas as revealed by high resolution electron microscopy, *Experimental Cell Research, 7*, 393–414, Figure 3, p. 397, with permission of Elsevier.

face the mitochondria, the zymogen granules and the cell membrane with their rough side but the cytoplasmic membranes adjacent to the nuclear membrane always have the smooth surface directed towards the nuclear membrane. Therefore at least one of the membranes appears to be a single membrane. (Sjöstrand & Hanzon, 1954, pp. 403–4)

Sjöstrand further noted that:

> When the cytoplasmic membranes are cut obliquely or are oriented parallel to the plane of the section the basic membrane does not show up distinctly or is not visible at all due to its low electron scattering capacity. Then, only the opaque particles are observed and the cytoplasm appears as consisting only of this component. (p. 405)

From the fact that the particles do not appear in freeze-dried preparations but appear very opaque in osmium tetroxide preparations, he proposed that they react very strongly with osmium. Late in the paper Sjöstrand seemed to jump to a conclusion about the function of these particles. From the fact that they "represent the dominating structure of the cytoplasm of the exocrine pancreas cells," he inferred that "they are structures of importance for the enzyme synthesis in these cells" (412). This correct inference is surprising both in light of his failure to offer justifications for it and the fact that in the case of the mitochondrion Sjöstrand resisted any speculation about function.

Sjöstrand also linked these paired membranes to structures identified by other electron microscopists – the "interdigitating cell membranes" of Pease and Baker (1950) and the intracellular filaments or lamellae of Dalton et al. Sjöstrand commented that "the micrographs of these authors are, however, not of the quality to allow a detailed description and a correct interpretation of these structures" (p. 448). In particular, he rejected the interpretation that these structures are filaments. In his paper with Rhodin, he claimed:

> It is quite obvious that we are dealing with membranes and not filaments from the fact that they may be followed without interruption through the whole basal cell zone, the chance to hit a filament so exactly along its entire length being negligible. In addition, there have never been any indications of cross-cut filaments. (p. 448)

In the paper with Hanzon he advanced a further argument – the number of membranes was always odd, due to the membrane closest to the nucleus not having a partner. This would not be the case if the structures were filaments.

At a symposium at the Eighth Congress of Cell Biology in 1954, Sjöstrand related his views to those of Porter and Palade:

> It is of course very threatening [tempting?] to generalize regarding these different structures and to consider for instance the opaque particles as a new component of the cytoplasm of general occurrance (sic) (Palade, 1953) with a common chemical and functional significance. For the moment we may not generalize further than accepting that in the cytoplasm there exist granules

of different sizes and topographical relationships and with one property in common, that they react intensely with osmium tetroxide. But as this means a rather unspecific reaction these granules might chemically be of rather different types.

As to the membranes observed in the cytoplasm these membranes certainly are morphologically very different. It might be possible to propose a definition that would characterize some of these membrane structures and would collect them as morphologically similar. To such morphologically well defined membranes a name could be given. The naming would, however, not increase our understanding but would represent a piece of systematic work. The term "endoplasmic reticulum" (Porter, 1953) is used in a too vague way, it almost indicates anything in the ground substance of the cytoplasm. Without guarantee for homology such diffuse classification certainly does not help very much in systematizing the structural components of protoplasm. (Sjöstrand, 1955b, pp. 226-7)

Securing the Connection to Protein Synthesis

While Porter and Palade were trying to determine the structural character of the endoplasmic reticulum, biochemists in a different line of research were following up on Claude's identification of microsomes in his fractionation studies. Shortly after he discovered microsomes, Claude had noted their high RNA content and, as discussed in Chapter 5, Brachet and Caspersson linked RNA to protein synthesis. Several biochemists attempted a direct assault on the problem of protein synthesis. They saw the formation of peptide bonds between amino acids to form polypeptide chains (Figure 6.11) as the fundamental step in protein synthesis. Their strategy was to trace the uptake of radioactively labeled amino acids into protein and identify the fraction in which it appeared. Harry Borsook of the California Institute of Technology (Borsook et al., 1950) and Tore J. M. Hultin of the Wenner-Gren Institute in Stockholm (Hultin, 1950) were early pioneers, independently showing in 1950 that when the cells of a labeled tissue were broken and fractionated, the highest concentration of labeled amino acids showed up in the microsome fraction.

Paul Zamecnik, at the Huntington Laboratory of Massachusetts General Hospital, played a particularly important role in these biochemical studies. When he initiated his work, two ideas on protein synthesis dominated the landscape. Max Bergmann at the Rockefeller Institute, with whom Zamecnik worked briefly before taking up his position at the Huntington Laboratory, proposed that cathepsin enzymes, which catalyzed proteolytic reactions, might synthesize peptide bonds. Fritz Lipmann, his colleague at Huntington, proposed that a phosphorylated intermediate might play a central role in causing

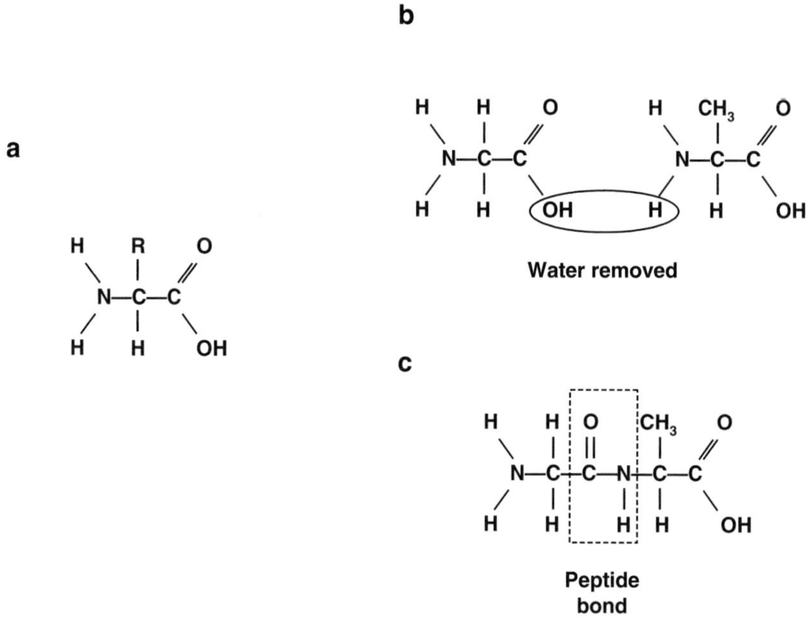

b

a

Water removed

c

Peptide
bond

Figure 6.11. Basic steps in creation of peptide bonds in proteins. (a) Generic structure of an amino acid, where R represents the portion in which different amino acids vary. (b) Glycine (left) and alanine (right) bonding together with the removal of a water molecule. (c) The resulting peptide bond formed between glycine and alanine.

amino acids to bond to one another.[29] (Lipmann, as discussed earlier in this chapter, characterized such phosphate bonds as energy-rich bonds and theorized about how they were formed in oxidative phosphorylation.)

Robert Loftfield, who joined Zamecnik's group in 1948, had developed a procedure for labeling the amino acids alanine and glycine with C^{14}. In their first studies using these labeled amino acids, Zamecnik and his collaborators (Zamecnik et al., 1948) demonstrated uptake of alanine and glycine into tissue slices from normal and malignant rats. Working with tissue slice preparations presented serious limitations, and a number of research laboratories set out to develop a cell-free system in which to study the process.[30]

[29] Zamecnik commented, "As a student of Bergmann, I felt a loyalty to the catheptic enzymes, but as a neighbor of Lipmann, I developed a feeling that his concept of a phosphorylated intermediate might be correct, and the conviction that in any case C^{14}-labeled amino acids were a tool which might resolve this dilemma" (1958–9, p. 258).

[30] In Zamecnik's group, Elizabeth Keller initially led this effort. She injected rats with low doses of labeled amino acids and later sacrificed the animals, formed a homogenate from their liver cells, and centrifuged it. In rats sacrificed within twenty minutes she found most of the labeled amino acids in the microsomal fraction, but not after longer delays (Keller, 1951).

Philip Siekevitz, who joined the Zamecnik group in 1949 as an NIH Fellow, adapted the techniques developed by Hogeboom et al. at Rockefeller to create a fraction that contained both mitochondria and microsomes in which he could study incorporation of labeled amino acids. When he added citric acid cycle intermediates, incorporation increased. If respiration were suppressed, he found adding ATP could foster incorporation (Siekevitz & Zamecnik, 1951; Siekevitz, 1952). These were important clues to the energetics of protein synthesis and offered support for Lipmann's conception.

Yet further enhancements in the fractionation procedure, especially use of the Potter-Elvehjem homogenizer, enabled Nancy Bucher (1953) in Zamecnik's laboratory to demonstrate incorporation of labeled acetate into cholesterol and subsequently into proteins. Using this preparation Zamecnik and Elizabeth Keller showed that it was the microsomal and supernatant fractions that were required for protein synthesis, and that they had to be supplemented by ATP and guanosinetriphosphate (Keller & Zamecnik, 1956). Then, in collaboration with Mahlon Hoagland, Zamecnik and Keller revealed that the ATP interacted with the amino acid, forming an amino acyl $\sim$ AMP compound, in the soluble protein fraction (Hoagland, Keller, & Zamecnik, 1956; Hoagland, 1955). They construed this as *activating* the amino acid, thus enabling it to form a peptide bond with another amino acid. With this research, the Zamecnik group (a) secured the claim that Claude's microsome fraction was the locus of protein synthesis, and (b) demonstrated the dependency of these activities on energy made available by the mitochondrial system.

Integrating Morphology and Biochemistry

In 1952 Siekevitz left Zamecnik's group to go to Wisconsin to work on oxidative phosphorylation with Van Potter. Three years later, George Palade recruited him to the Rockefeller Institute. After Hogeboom and Schneider had left, the Rockefeller group had not had a researcher primarily trained in biochemistry and had focused their efforts on electron microscopy. Those efforts had paid off handsomely, but it was now necessary to figure out the operations performed by the differentiated components, especially the laminar membranes of the endoplasmic reticulum and the small particles that dotted their surfaces in many places. With Siekevitz providing biochemical expertise, Palade immediately set out to conduct an "integrated morphological and biochemical study" of microsomes, initially from liver and subsequently from pancreatic acinar cells. They examined a portion of the specimen at each step in the fractionation process under the electron microscope. In the homogenate they identified "'hollow' profiles comparable in size, shape, and number to the profiles of the endoplasmic reticulum in intact cells" (Palade & Siekevitz, 1955, p. 178). Noting that some of the microsomes contained small dense particles attached to the outside of their limiting membrane, they argued for identifying microsomes with the rough endoplasmic reticulum. They further commented that preparing the homogenate appeared to cause "an extensive fragmentation of the networks into independent vesicles, tubules, and cisternae, which subsequently can be centrifuged down, together with other cell components, into the homogenate pellets" (p. 179). There was no evidence of a tearing of the membrane, leading them to infer that the fragments "'heal' easily" or, more likely, are formed by a "'spontaneous,' generalized pinching-off process" (p. 192). Palade and Siekevitz concluded that Claude was correct in treating the microsomes as preformed components of the cytoplasm, but wrong to think of them originally as independent particles – rather, they are parts of a "continuous, cell wide system; i.e., the endoplasmic reticulum" (p. 190). This identification provided a bridge between the biochemical investigations of microsomes and the electron microscopy studies of the endoplasmic reticulum and its particles.

Once the homogenate had been centrifuged into microsomal pellets, Palade and Siekevitz found "membrane-bound profiles of approximately the same size and shape as the profiles found in homogenate pellets and considered to be derived, by extensive fragmentation, from the endoplasmic reticula" (p. 179). They observed that particles were still attached to many of the membranes, although they were "slightly less numerous" than in the homogenates, a factor that led them to try to fractionate further the microsomal supernatant in order to isolate the particles. They failed with liver preparations but were more successful in pancreas preparations, where the resulting particles exhibited high RNA and protein content but very low phospholipid content. They had greater success treating the microsomal fraction chemically – deoxycholate treatment eliminated the membrane, leaving the particles, while versene treatment and ribonuclease treatment eliminated the particles, leaving the membrane. From this, they conclude that the RNA is found in the particles and that the protein, phospholipids, and other components associated with microsomes are found in the membrane.

A further consideration in identifying RNA with the particles was that in many cell types the particles often appeared independently of the endoplasmic membrane. Especially in developing cells in intestinal epithelium, whose cytoplasm stains broadly with basic dyes, Palade reported finding "numerous free particles evenly and randomly distributed throughout the cytoplasmic matrix." In contrast, "the endoplasmic reticulum is represented by only a few vesicles and tubules, many of them free of attached particles, and relatively large expanses of the cytoplasm contain no elements of the network" (1958a, p. 289).[31] If the particles could exist independently of the endoplasmic membrane, what was the significance of their association with the membrane in many adult cells? Palade speculated,

> it may be assumed that the reticulum provides appropriate surface for the arrangement of the small granules, but thus far we do not know whether the patterns described have any significant role in protein synthesis. Available information suggests, however, other possible reasons. In certain cells, for instance, the cavities of the endoplasmic reticulum appear to be in continuity, at least intermittently, with the extracellular medium through invaginations of the cell membrane. In such cases the supposedly fluid phase that occupies the cavities of the network may contain a variable amount of the extracellular fluid. It is possible that many raw materials, coming from outside into the cytoplasm, reach the particles through the labyrinth of the endoplasmic reticulum. In other cells the cavities of the endoplasmic reticulum appear to be used for the storage of a cell product. In plasma cells, thyroid epithelia, and fibroblasts they are frequently found distended and filled with an apparently amorphous mass of appreciable density. There are, in other words, numerous appearances suggesting that the cavities of the reticulum serve as feeding channels and storage space for the activity of the particles. Their association may have the same significance as the location of our plants along convenient ways of communication. (p. 301)

From the perspective of developing a model of a mechanism, Palade has here offered a decomposition into distinct operations: supplying materials, synthesizing proteins, and transporting the products. Palade (1958b) further developed this perspective, drawing insights from the nineteenth-century research of Rudolf Peter Heinrich Heidenhain (1875), who had found that in exocrine cells of the pancreas, granules disappear after food intake and are replaced with new granules a few hours later. Heidenhain concluded that the granules were comprised of the precursors of the digestive enzymes of pancreatic juice and that they provided temporary storage for these proteolytic enzymes. He named them zymogen granules. Following up on Heidenhain's research, Siekevitz and Palade compared the appearance of the endoplasmic reticulum in cells from guinea pigs starved for forty-eight hours with cells from guinea pigs fed an hour before. In the recently fed animals, "the cavities

---

[31] A note indicates that the text of this paper was prepared for a talk in February 1955, and was largely unaltered until publication. Palade commented on the rather unusual patterns of the particles on the endoplasmic reticulum membrane: "Sometimes, while looking at these intriguing patterns, I believe that I feel very much like the French explorers who, during Napoleon's expedition to Egypt, found themselves face to face with the hieroglyphs. Like some of them, I am recording the patterns, and I am waiting hopefully for a biochemical Champollion to decipher their meaning."

of the system are distended, the preferred orientation is lost, and relatively large, dense granules are found within the cavities of the distended cisternae" (p. 71). These granules were similar to, but smaller than, the zymogen granules. Fractionating cells from both starved and fed animals, they found that in starved animals there was no appreciable proteolytic enzyme activity in microsomes, but that in recently fed animals it approximated that of the zymogen particles. By further fractionating the microsomes after treatment with doxycholate, they showed a higher concentration of the proteolytic and ribonuclease activity in both the particles and in the whole microsome for recently fed animals than for starved animals. Palade concluded,

> To my knowledge, this is the first instance in which a product of the endoplasmic reticulum has been demonstrated in the form of well defined granules within the cavities of the system, and has been identified biochemically. (p. 73)

Palade was cautious about concluding that these enzymes were new products of protein synthesis, but did cite evidence using labeled leucine-1-C^{14} showing its earliest incorporation occurred in the particles still attached to the membrane. A little later the label was found in the intracisternal granules and only later in the zymogen granules. This pattern, Palade contended, is "compatible" with the hypothesis that the microsomal particles synthesized the new protein.

Naming the Ribosome

In February 1958, the new Biophysical Society held its first symposium at MIT, with a focus on microsomal particles and protein synthesis. One of the most important results of the meeting was the adoption of a new name, *ribosome*, for the particles. The editor, Richard Roberts, related:

> During the course of the symposium a semantic difficulty became apparent. To some of the participants, microsomes meant the ribonucleoprotein particles of the microsome fraction contaminated by other protein and lipid material; to others, the microsomes consisted of proteins and lipids contaminated by particles. The phrase "microsomal particles" does not seem adequate, and "ribonucleoprotein particles of the microsome fraction" is too awkward. During the meeting the word "ribosome" was suggested; this seems a very satisfactory name, and it has a pleasant sound. The present confusion would be eliminated if "ribosome" were adopted to designate ribonucleoprotein particles of the size range 20 to 100 Å. (1958, p. viii)[32]

---

[32] Rheinberger (1997, p. 190, n. 12) claims that the suggestion stemmed ultimately from Howard Dintzis.

This symposium also marks a threshold in the research on the endoplasmic reticulum and ribosomes. Roberts noted further in his introduction that the case for protein synthesis by the ribosomes was still inconclusive. Several of the reasons he listed had to do with technical problems in demonstrating the connection, but one focused on the absence of a conception of the responsible mechanism: "No mechanism has been suggested which shows how the structure of the particle is compatible with its function as the template for synthesis of long chains" (p. vii). The mechanism that Palade had described treated the ribosome as a unit with an operation which interacted with the operations of the endoplasmic reticulum. It had not explained the operation of the ribosome itself.

Discovering the mechanism by which the ribosome synthesized proteins required moving to a yet lower level of organization at which research could focus on the chemical structures that comprised the ribosome or interacted with it. As was the case with the mitochondrion, research at this level was chiefly the province of biochemists and practitioners of two other new disciplines, biophysics and molecular biology, not researchers affiliating with cell biology. Since the research contributing to a basic sketch of the mechanism of protein synthesis in the ribosome is illustrative of one strategy for developing an account of a mechanism, I will briefly analyze it before returning to research in cell biology focusing on the transport of the newly created proteins.[33]

Going to a Lower Level: Decomposing the RNA Machinery

The biochemical research so far focused only on the process of making peptide bonds, but synthesizing proteins required linking amino acids in appropriate orders. Zamecnik (1958) proposed that protein synthesis involved "some biological equivalent of a printing press" that would specify the order and that the "press or template . . . is very likely RNA" (p. 120). He then advanced a scheme, initially proposed by Victor Konigsberger and Theo Overbeek, according to which amino acids are first activated by binding with an enzyme and a molecule of ATP, resulting in the transfer of the phosphate bond to an amino acid. He proposed that the amino acids bonded sequentially on the RNA template, and by what he characterized as a "zipper reaction" bonds were established between adjacent amino acids. In the final step in his

---

[33] For a discussion of the discovery of the mechanism of protein synthesis that emphasizes the interaction of molecular biology with biochemistry, see Darden and Craver (2002).

schema for the mechanism, the amino acid chain then separates, binds with other chains, and folds into the protein molecule.

Zamecnik and his collaborators discovered that in preparing microsome fractions, RNA occurred both in the fraction containing ribosomes bound to the endoplasmic reticulum and in the supernatant. They started referring to the RNA in the supernatant as "soluble RNA" or "S-RNA." Moreover, they found that labeled leucine was taken up by the S-RNA (Zamecnik et al., 1957; for an analysis of the discovery of soluble RNA, see Rheinberger, 1997). Smith, Cordes, and Schweet (1959) introduced the name "transfer RNA" (tRNA) for S-RNA, and proposed that it played a role in transferring the activated amino acid to the microsomal RNA, a suggestion Zamecnik readily adopted (Hoagland, Zamecnik, & Stephenson, 1959). The notion of transfer also suggests a role for this RNA in sequencing of amino acids, a suggestion that Zamecnik began to formulate:

> We have most lately been concerned with the possibility that at least a portion of the soluble RNA molecule to which the amino acid is attached is transferred along with the amino acid to the ribonucleoprotein particle, aligning itself in some base-pairing arrangement with the microsomal RNA prior to formation of a peptide chain. This concept agrees with the proposal of Crick that the soluble RNA molecule may serve as an adaptor in a base-pairing arrangement which determines amino acid sequence. (1958–9, p. 274)

As Zamecnik conceived the mechanism at this point, ribosomal RNA remained in place to direct multiple iterations of synthesis. The tRNA brought amino acids to the template as specified; the amino acids were added to the chain; and the tRNA departed. This made sense in eukaryote cells whose synthetic activities were limited to a few specific proteins. Research on bacteria generated a very different picture, for bacterial cells are capable of generating a wide range of proteins. Especially after the Pardee, Jacob, and Monod (1959) experiment showing the inducibility of protein synthesis, attention refocused on how temporary structures could be made from DNA, then move to the cytoplasm to direct protein synthesis (Brenner, Jacob, & Meselson, 1961). Investigations by Nirenberg and Matthaei (1961) that were directed toward developing a cell-free system for performing protein synthesis revealed that synthetic RNA created with uracil resulted in the synthesis of amino acid chains comprised of phenylalanine. Although this research is most celebrated for providing the first clue to the genetic code, it also provided compelling evidence for a third form of RNA, which came to be known as *messenger RNA* (*mRNA*), which was credited with carrying information about the protein to be synthesized from the nucleus to the ribosome.

The role of RNA in the ribosomes thus became less significant as attention turned to mRNA as the template and to tRNA as the transport for bringing amino acids to the template. This is ironic, because it was RNA that had been the distinguishing feature of first microsomes and then ribosomes. Research on the ribosomes themselves emphasized instead the proteins that comprised them. Those investigating the proteins in ribosomes deployed the same strategy as those investigating the mitochondrion: That is, they attempted to decompose the ribosome into different component parts. The particles separated in a centrifuge are often reported in terms of the time required for sedimentation, measured in terms of Svedberg units (one $S = 10^{-13}$ seconds); longer times correspond to lighter weights. Separating microsomes from yeast cells, Fu-Chuan Chao and Howard Schachmann (1956) reported 80S microsomes, which in turn dissociated into 60S and 40S units unless a trace of magnesium was present. (Svedberg units are not additive because the rate of sedimentation is affected by both the mass and shape of the particle.) Mary Petermann and her collaborators (Petermann et al., 1958) found that 78S liver ribosomes decomposed into 62S and 46S units. Finally, working with *Escherichia coli*, Alfred Tissiéres and James Watson (Tissiéres & Watson, 1958; Tissiéres et al., 1959) identified 70S ribosomes that separated into 50S and 30S units. From these studies, it appeared that ribosomes generally were comprised of two subunits of slightly different sizes, referred to as large and small subunits. Electron micrographs by Palade and his collaborators subsequently provided independent evidence for the two subunits of the ribosome (Sabatini, Tashiro, & Palade, 1966).

Research on the mechanism of protein synthesis involved not just the decomposition of the system into separate types of RNA and decomposition of the ribosome itself into subunits comprised of different forms of ribosomal RNA and protein, but also research on how these parts were organized. One important clue as to the organization of the parts stemmed from theoretical speculation about the relative size of messenger RNA molecules and ribosomes. Alexander Rich, a professor of biophysics at MIT, noted that the messenger RNA would be 1,500 Å or more in length, whereas ribosomes were only about 230 Å in diameter. Although messenger RNA chains might be wrapped around the ribosome, Rich concluded this was unlikely because then it would be difficult to maintain appropriate contact between messenger RNA and the ribosome. As Rich reported, "It occurred to us that proteins might actually be made on groups of ribosomes, linked together somehow by messenger RNA" (Rich, 1963, p. 45; see also Warner, Knopf, & Rich, 1963). As he described it, "the protein 'factories' of the cell are not single ribosomes working in isolation, but collections of ribosomes working together in orderly fashion as if there were machines on an assembly line" (p. 44). Rich designated these groups *polyribosomes* or *polysomes*. To his characterization of the polysomes as constituting assembly lines, he quickly noted a difference between the ribosomal assembly line and human ones: "the polyribosome is not the usual kind of assembly line. In such an assembly line, the product moves down the line and component parts are added to it. In the polyribosome assembly line the ribosomes move down the line and each one makes a complete product."[34]

Evidence for polysomes came in two forms. The first were fractionation studies with rabbit reticulocytes, which are cells lacking a nucleus and specialized for manufacture of hemoglobin. Rather than using a medium of constant density, Rich centrifuged the contents in a solution spatially graded from 15% to 30% sucrose. Examination of the ultraviolet absorption characteristic of RNA in the centrifugation product revealed peaks in two fractions – one that corresponded to single ribosomes and the other to heavier materials, presumably polysomes. The amino acids in the preparation were labeled with C^{14}, making it possible to identify materials in which protein synthesis was occurring, and this showed a single peak corresponding to the polysome fraction. This suggested that the polysome was the locus of protein synthesis, a conclusion that was further supported by the fact that applying ribonuclease to the medium before centrifugation resulted in no fraction corresponding to polysomes and the radioactivity being transferred to the single ribosome fraction.

Rich collaborated with electron microscopists to develop a second form of evidence for polysomes. Using metal shadowing, the resulting electron micrographs clearly revealed clusters of ribosomes. In collaboration with Henry Slayter at MIT he used positive staining with uranyl acetate to develop micrographs that also revealed a thread 10 to 15 Å in diameter running between the ribosomes, which corresponds to the estimated thickness of a single strand of RNA. Calculating the diameter of the five polysomes attached to the thread and the gap between them resulted in a length of 1,500 Å, the expected length

---

[34] Rich seemed quite concerned with the appropriateness of the analogy. He returned to it again later in the paper and commented, "It is evident that protein synthesis is not really an assembly line process as it is normally understood. It would be more appropriate to compare protein synthesis with the operation of a tape-controlled machine tool. The tool will turn out an object of any shape within its range of capabilities, in response to information coded on the input tape. In factories where such tools are used each tool is provided with its own tape, but if it served any purpose a single tape could easily be fed through a battery of identical tools. The living cell evidently makes one tape serve for many tools because this is an efficient way to do the job" (pp. 50–1).

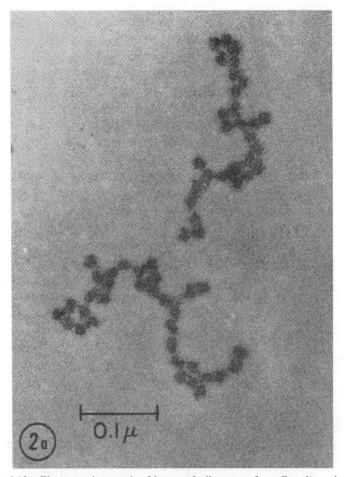

Figure 6.12. Electron micrograph of large polyribosomes from *E. coli* synthesizing β-galactosidase. Reproduced from H. Slayter et al. (1968), An electron microscopic study of large bacterial polyribosomes, *Journal of Cell Biology, 37*, 583–90, Figure 2a, p. 586, by copyright permission of the Rockefeller University Press.

of a strand of messenger RNA. Figure 6.12 shows a longer polysome from *E. coli* engaged in the synthesis of β-galactosidase (Slayter et al., 1968).

Rich proposed that the ribosomes moved along the messenger RNA, adding the appropriate amino acid to the polypeptide chain it was constructing according to the instruction at that locus on the chain. He hypothesized that a "ratchet-like mechanism" (1963, p. 49) would move the ribosomes along the chain. Thus, each ribosome attached to the messenger RNA made a copy of the

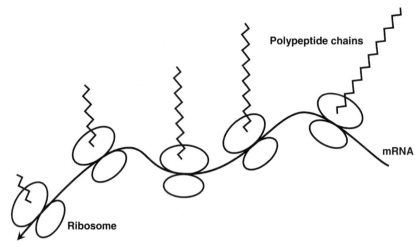

Figure 6.13. Schematic representation of a polyribosome. Five ribsosomes are attached to the mRNA. The leftmost ribosome has just been added to the mRNA string and is beginning to form a polypeptide chain, whereas the rightmost ribosome has already formed a fairly long chain.

same protein. As shown schematically in Figure 6.13, those far along the chain would have nearly completed polypeptide chains, and would drop off when they finished. New ribosomes would then join the chains and ribosomes early on the chain would have just the beginnings of the polypeptide chain completed.[35]

From the simple picture of the mid 1950s of one RNA-rich component of the cytoplasm serving as the locus of protein synthesis, a decade of research involving biochemists, cell biologists, biophysicists, geneticists, and molecular biologists had revealed a complicated structure of multiple component parts that performed different operations in the overall activity of protein synthesis. Although much remained to be filled in, a sketch of the mechanism at this lower level was established. There was no longer a mere empirical finding that ribosomes were involved in protein synthesis; rather, there was an account, in broad detail, of the mechanism responsible for protein synthesis.

---

[35] Rich also conducted a number of experiments to test his model of the polysome assembly process. First he utilized a homogenizer before centrifugation, which produced a number of peaks in terms of both ultraviolet absorption and radioactivity. He proposed that each successive peak corresponded to an additional ribosome in a cluster, a proposal he confirmed in collaboration with electron microscopist Cecil Hall by subjecting material from each fraction to electron microscopy and finding evidence of the predicted clusters.

Transporting Newly Sequenced Polypeptides

Although cell biologists' tools of electron microscopy and cell fractionation played important roles in unraveling the mechanism through which different types of RNA contributed to protein synthesis, cell biologists were also interested in the fate of newly formed proteins. Utilizing C^{14} labeling in pancreatic microsomes from pigeons, Colvin Redman, Siekevitz, and Palade (1966) found evidence that labeled amylase, the secretory protein being synthesized, would appear in the cisternal cavities of the microsome. Subsequent research by Redman and David Sabatini (1966) demonstrated that treating the ribosome with puromycin resulted in the appearance of labeled unfinished proteins in the cisternal cavities. They viewed this as showing that "from the onset of protein synthesis the growing peptide chain is directed towards the cisternal space into which it diffuses upon its release from the attached ribosome" (p. 608). Earlier research by Sabatini (Sabatini et al., 1966) had demonstrated that it was the large ribosomal subunit that was directly attached to the membrane of the endoplasmic reticulum, leading to the conclusion that the new protein was transported from the large subunit through the membrane to the cisternal space. Palade, in his 1974 Nobel Lecture, noted that the vectorial transport of newly formed proteins into the cisternal space provided the only known explanation for the complex structure of the endoplasmic reticulum:

> This conclusion provides a satisfactory explanation for the basic structural features of the endoplasmic reticulum: a cavitary cell organ of complicated geometry which endows it with a large surface. All these features make sense if we assume that one of the main functions of the system is the trapping of proteins produced for export. With the exception of Ca^{2+} accumulation in the sarcoplasmic reticulum, i.e., the equivalent cell organ of muscle fibers, no other recognized function of the endoplasmic reticulum (e.g., phosphatide and triacylglycerol synthesis, mixed function oxygenation, fatty acid desaturation) requires compellingly and directly a cavitary organ, at least according to our current knowledge. (Palade, 1992, p. 183)

As we will see, the next stage in the movement of the newly formed proteins was to the Golgi apparatus, and this discovery led Palade to reverse his earlier denial of the reality of the Golgi and to conduct landmark studies on its function.

3. TWO ADDITIONAL ORGANELLES

Research in the 1950s and early 1960s on the mitochondrion and the endoplasmic reticulum resulted in the first mechanistic models of cellular functions.

In each case, combining morphological and biochemical inquiry resulted in differentiation of component parts and proposals as to the specific operations performed by those parts. Another development over that period, though, was the discovery that yet other cell organelles performed different functions that contributed to the overall life of the cell. Particularly important in this respect was research on the Golgi apparatus and the lysosome.

The Golgi Apparatus

As I discussed in Chapter 4, in 1949 Claude and Palade challenged the existence of the Golgi apparatus, contending that it was an artifact of staining with osmium or other heavy metals. Claude and Palade's challenge to the reality of the Golgi apparatus was one of the last. A prominent exception was a charge by John Baker (1957; 1963) that many different substances were being conflated under the one label. As I argued in Chapter 3, the Golgi apparatus was a natural target for charges of artifact since, up through Claude and Palade's challenge, observations of it were generally limited to cells fixed with osmium compounds and the evidence for its functional role was far from overwhelming. But this situation changed so definitely that, when Palade together with Marilyn Farquhar (Farquhar & Palade, 1981) wrote a review of research on the Golgi apparatus they commented that "now no one questions that the Golgi apparatus is a distinct cell organelle, or is unaware of its participation in a wide variety of cellular activities. Indeed, the Golgi apparatus, or Golgi complex as it is often called, not only occupies the cell center, but it also has moved toward center stage, because it has been shown to be involved in so many cell activities" (p. 77s). Interestingly, given his earlier opposition, Palade contributed much to the vindication of the Golgi apparatus by providing crucial information about its function.[36]

---

[36] Despite the vehemence of the challenge he and Claude had issued, in the review of the history of the Golgi apparatus with Farquhar, Palade does not mention his own role as one of the last to challenge the reality of the Golgi apparatus. Farquhar and Palade (1981) write, "The period before the mid 1950s was characterized by controversy concerning the reality of the Golgi apparatus, with the scientific community divided into nonbelievers and believers. The acceptance of the status of the Golgi as a bona fide cell structure depended on whether one believed that the metallic impregnation methods (involving use of silver or OsO_4), which Golgi and others used to demonstrate the apparatus, were staining a common structure with variable form and distribution in different cell types, or alternatively, that these methods resulted in artifactual deposition of heavy metals on different cell structures in different cell types" (p. 77s). Again in 1998, on the occasion of the 100th anniversary of Golgi's discovery, Farquhar and Palade make no reference to Palade's role as challenger of the Golgi: "The debate raged because the Golgi was not visible in living cells and its visualization depended on Golgi's capricious heavy-metal staining method, called the black reaction (la reazione nera), which was difficult to reproduce reliably and stained many other structures, including whole neurons" (p. 2).

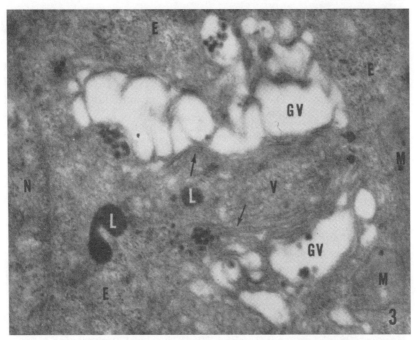

Figure 6.14. Electron micrograph of Golgi complex in a principal cell of the duodenum of a mouse. The arrows mark the Golgi membrane (Golgi lamellae or flattened cisternal sacs). Also indicated are GV: Golgi vacuoles; V: Golgi vesicles; L: Lipid droplets; M: mitochondria; N: nucleus; and E: ergastoplasm (endoplasmic reticulum). Reproduced from A. J. Dalton and M. D. Felix (1956), A comparative study of the Golgi complex, *Journal of Biophysical and Biochemical Cytology, 2* (No. 4, Part 2), 79–84, Figure 3, plate 27, by copyright permission of the Rockefeller University Press.

Before Palade returned to studying the Golgi apparatus, electron micros-copy, especially studies by Albert Dalton, played a major role in providing additional information about its structure. In an initial study, Dalton (1951b) found no evidence of the formation of myelin forms when hepatic and intesti-nal epithelial cells were fixed with Champy's fluid. Nonetheless, a membra-nous network was visible in electron micrographs in the parts of cells where the Golgi apparatus had typically been detected, thereby undercutting Palade and Claude's proposal as to how the Golgi arose as an artifact.[37] In several subsequent studies with Marie Felix, Dalton uncovered the detailed structure

---

[37] Dalton's subsequent research with Felix further challenged Palade and Claude's proposal. By examining the process of fixation, they determined that the Golgi material responded very differently than lipid droplets, thereby dispelling Palade and Claude's contention that the Golgi apparatus was an artifact produced from such lipid droplets.

of the Golgi apparatus (Dalton & Felix, 1954; Dalton & Felix, 1955; Dalton & Felix, 1956). As illustrated in Figure 6.14, they differentiated three components – a system of lamellae or flattened cisternal sacs, large vacuoles, and clusters of small vesicles. The flattened sacs (saccules) typically occur in stacks of three to seven in plant and animal tissues, and in larger numbers in unicellular organisms. The saccules are the major contributor to the traditional image of the Golgi apparatus in light microscopy. On one side, the *cis* side, the stacks abut the endoplasmic reticulum. Clusters of small vesicles are visible at the interface between the endoplasmic reticulum and the *cis* side of the stack. On the other side, the *trans* side, of the stack, larger vacuoles are found that are often referred to as condensing vacuoles. To highlight the fact that there are several different components comprising the Golgi region, Dalton introduced the term *Golgi complex*. In addition to providing much better images of the Golgi apparatus than had been available through light microscopy, Dalton and Felix also sought to address its functional significance. By comparing the images of the Golgi complex in mouse intestinal cells after fasting and forty minutes after eating, Dalton and Felix provided evidence that the Golgi apparatus is involved in storage of lipids after absorption.

Other early electron micrographs of the Golgi apparatus (Sjöstrand & Hanzon, 1954; Farquhar & Rinehart, 1954) revealed a close relation between secretory granules and the Golgi apparatus. Subsequent micrographs seemed to show secretory granules within the Golgi structures:

> Secretory granules have been seen within components of the Golgi bodies of rat pituitary acidophils and mouse pancreatic acinar cells. The fact that secretory granules are much more frequently encountered within Golgi components under conditions of increased secretory activity suggests that granule formation may occur within the Golgi apparatus in these two types of cells. (Farquhar & Wellings, 1957, p. 321)

Applying biochemical analysis required separating the Golgi apparatus from other cell components through cell fractionation. In addition to his electron microscopy of the Golgi apparatus, Dalton (Dalton & Felix, 1954) established that the Golgi apparatus could be extracted from epididymis cells, and Schneider and Kuff (1954) separated it within epididymal homogenates by gradient centrifugation. Schneider and Kuff produced evidence that the Golgi fraction was comprised of high concentrations of phospholipid as well as acid and alkaline phosphatase and RNA. However, fractionation of the Golgi apparatus was more challenging than fractionation of other organelles. Its smooth lipoprotein membranes, for instance, tended to break up and form smooth microsomes (Whaley, 1975, p. 31). A decade later Morré and his collaborators developed improved procedures for subfractionation of Golgi fractions, and Fleischer, Fleischer, and Ozawa (1969) localized galactosyl-transferase in the Golgi fraction – the first enzyme to be primarily associated with the Golgi apparatus.

After contending that the Golgi apparatus was an artifact of osmium staining, and prohibiting discussion of it in the Rockefeller laboratory for a number of years, George Palade was led back to investigating the Golgi apparatus in the 1960s as an outgrowth of his biochemical research on protein synthesis in ribosomes discussed previously. Palade and Siekevitz, using ^{14}C-leucine as a tracer, established a migration in the pancreatic exocrine cell of α-chymotrypsinogen from the ribosome into the lumen of the endoplasmic reticulum and ultimately into zymogen granules which were excreted from the cell. Detailing the path and activities occurring during this migration now became a focus of research. In an initial study in Palade's laboratory, Lucien Caro (1961; see also Caro & Palade, 1964) used ^{3}H-leucine, which appeared within three to five minutes of injection in the endoplasmic reticulum, and then after twenty to forty minutes in the condensing vacuoles on the *trans* side of the Golgi stack.[38] After an hour the label appeared in the zymogen granules. This was critical evidence in establishing the transport of secretory proteins from the ribosomes through the membranous system of the Golgi apparatus to the zymogen granules that would then be secreted.

These initial findings were further elaborated in a series of studies with graduate student James Jamieson employing tissue slices from guinea pigs beginning in 1966. They arrived at a more detailed characterization of the migration of membrane-bound vesicles from the endoplasmic reticulum through the components of the Golgi apparatus to discharge from the cell (Jamieson & Palade, 1966; Jamieson & Palade, 1967a; Jamieson & Palade, 1967b). Key to their work was the use of three-minute exposure to leucine-^{14}C, a radioactive amino acid ("labeling pulse"), followed by removal of the unincorporated label ("chase"), allowing for better time resolution of the radioactive material.

Jamieson and Palade demonstrated that proteins, after leaving the endoplasmic reticulum, are encapsulated in small peripheral vesicles on the *cis* side of the Golgi stack and appear (after about thirty minutes) in condensing vacuoles on the *trans* side of the stack. In the interval they presumably traveled through the Golgi stacks, although Jamieson and Palade do not focus on that

---

[38] At about the same time, Warshawsky, Leblond, and Droz (1961) used labeled leucine in pancreas cells to trace the sites of uptake and the path of migration. They, however, lacked the ability to examine the results with the electron microscope.

period. After the vacuoles reach the *trans* side of the stack, they migrate to the cell membrane, where the membrane of the vacuole merges with the cell membrane. There is then an opening of the joined membranes so that the contents of the vacuoles are released into the extracellular matrix without breaching the diffusion barrier provided by the cell membrane. Palade (1958b) originally termed this process "membrane fusion," but it was relabeled "exocytosis" by de Duve (1959).

Beyond tracing the course of protein transport, Jamieson and Palade set out to determine how tightly the processes of protein synthesis and transport are coupled and to identify the energy source for the process. Using cyclohexamide to block protein synthesis, Jamieson and Palade (1968a) were able to uncouple the synthesis of protein from its transport to the Golgi complex, showing that transport did not depend on new proteins entering the process. In a subsequent paper (Jamieson & Palade, 1968b), they addressed the energy requirements for the process by demonstrating that the glycolytic inhibitors (fluoride, iodoacetate) failed to block transport, but that respiratory inhibitors (N2, cyanide, antimycin A) and inhibitors of oxidative phosphorylation (dinitrophenol, oligomycin) did. Jamieson and Palade then speculated about what operation required energy:

> At present, it is clear that the energy is used to connect the RER cisternal space with that of the condensing vacuoles, and that the small peripheral vesicles of the Golgi complex participate in the connection. The details of this operation, however, remain obscure: the cell may establish intermittent connections between these two compartments or effect transport between them using the small peripheral vesicles as shuttle carriers. Both alternatives imply repeated membrane fission-fusion and this is most probably the energy-requiring event. (p. 599)

Jamieson and Palade clearly favored the hypothesis that small peripheral vesicles serve as shuttle carriers and that the nascent proteins remain membrane bound as they transverse the Golgi structure. Prior to their research, P. P. Grassé (1957), relying on early electron micrographs, had proposed a maturational or cisternal progression model according to which cisternae were continually being created on the *cis* side of the stack and matured as they moved to the *trans* side, where they disintegrated and released the newly synthesized products for further transport. Neutra and Leblond continued to espouse this view (Neutra & Leblond, 1969, p. 105), but in subsequent years evidence built for Palade's proposal and it became the dominate view.

Palade's early research did not focus on the Golgi stack, and to the degree that he investigated the function of the Golgi components, he focused on concentrating proteins in secretory granules. What did the stack, the most prominent component of the Golgi apparatus, contribute? Marian Neutra and Charles Leblond (1969, p. 103) posed the question:

> Why do these proteins pass through the Golgi apparatus? Do they undergo some essential processing operation there? One can see with the electron microscope that proteins come out of the apparatus neatly packaged in globules whose membranes have been donated by the Golgi saccules. It seems hardly likely, however, that this elaborate system exists simply for the purpose of putting the proteins in bags; nature has a way of avoiding complex solutions for simple problems. We therefore decided that a closer look had to be taken at the protein products themselves to determine if their sojourn in the Golgi apparatus was responsible for some important change in their chemical form.

An important clue was that most secretory products are not proteins alone, but rather proteins linked to carbohydrates. Leblond and his collaborators at McGill University studied the goblet cell of the colon, which secretes such a compound – mucigen. These goblet cells are long, narrow structures squeezed between other cells in the intestinal lining. Each contains several Golgi stacks of eight to ten saccules. The bottom or *cis* saccules are flattened in appearance, whereas the top or *trans* saccules are bulging with material. Above them are mucus globules that are excreted in due course. In autoradiographic studies with Marian Peterson, Leblond showed that glucose tracer first appeared in the cisternae of the Golgi apparatus five to fifteen minutes after injection and moved progressively to the more distal cisternae, with the distal cisterna being converted into mucigen granules (Peterson & Leblond, 1964a; Peterson & Leblond, 1964b). Subsequently, Neutra and Leblond (1966a; 1966b) proposed that the glucose or galactose precursors enter the goblet cell from a capillary and move directly to the Golgi apparatus, where glucose is combined with proteins synthesized in the endoplasmic reticulum to form glycoproteins. As Neutra and Leblond (1969) related, by the end of the 1960s evidence had accumulated that the Golgi apparatus performed a variety of roles in the construction of large carbohydrate molecules such as adding sulphate bonds to create polysaccharide secretion products.

After decades during which it was suspected to be an artifact, in the 1960s the Golgi apparatus came to be generally recognized as a major component of the machinery of the cell. In part its rehabilitation stemmed from the ability of the electron microscope to provide more detailed images of its structure, but of even greater significance was the development of an account of its functional significance in the generation of products that figured in cell secretions processes.

The Lysosome

The lysosome is the one cell organelle that came to prominence in the early years of cell biology that had no direct roots in classical cytology.[39] Rather, Christian de Duve discovered it in the course of research he began in 1949, when he assumed directorship of what he characterized as "the derelict laboratory associated with the Chair of Physiology at the University of Louvain," directed at isolating glucose-6-phosphatase in liver through differential centrifugation. He became interested in glucose-6-phosphatase the previous year, which he spent with Carl and Gerty Cori at Washington University.[40] The Coris had been investigating glucose metabolism and had discovered in liver a hexose phosphatase. Working with liver extracts, de Duve identified the hexose phosphatase as a specific glucose-6-phosphatase and differentiated it from acid phosphatase. When he precipitated the glucose-6-phosophatase in an acid solution, he found he could not redissolve it when he raised the pH level. De Duve had learned from Claude[41] that his large fraction would agglutinate at acid pHs. This suggested to de Duve that agglutination, not precipitation, was occurring in his glucose-6-phosphatase preparations – the enzyme was attached to a structure. He then turned to cell fractionation (helped in part by Claude, who had moved back to Belgium after leaving the Rockefeller Institute) as a gentler way of separating the enzyme than the Waring blender that he had been using, and localized 95% of its activity in the microsomal fraction. What caught de Duve's attention, though, was the fact that the homogenate he prepared before fractionation exhibited only 10% of the acid

---

[39] In part this is due to the fact it was first identified through its function, not its cytological structure, which Novikoff (1970, p. 121) identified as unusual: "Historically, study of organelles begins usually with the accumulation of morphological observations and then passes to the isolation of the organelle in relatively pure fraction and biochemical study. For lysosomes, however, this pattern was reversed."

[40] Before going to the Coris' laboratory, de Duve had spent eighteen months in Hugo Theorell's laboratory in Sweden where he mastered biochemical techniques. He was interested in visiting the Coris because his earlier research indicated that they had incorrectly ascribed to insulin actions that were not due to insulin itself but to glucagon, a contaminant in their preparations. Carl Cori initially rejected de Duve's request to spend six months in their laboratory but shortly thereafter Earl Sutherland, a postdoctoral fellow in the Coris' laboratory (who later won a Nobel Prize for the discovery of cyclic AMP), obtained evidence that glucagon was the responsible agent. Cori invited de Duve to come and collaborate with Sutherland, which de Duve was able to do with support from the Rockefeller Foundation. The identification of impurity in insulin also paid off handsomely for de Duve. He had determined that insulin prepared by Eli Lilly bore the glucagon contaminant and for alerting them to this fact, the company provided de Duve's laboratory at Louvain with a $5,000/year research budget (Interview with Christian de Duve, 5 December 1995, Rockefeller University, New York).

[41] On his return from St. Louis, de Duve had stopped to visit Claude at the Rockefeller Institute; he reports reading several of Claude's papers on the flight back from New York.

phosphatase activity found in homogenates prepared with the Waring blender. (The activity in the fractions isolated by centrifugation was higher but still lower than in Waring blender preparations.) Convinced that this represented an error in the assay, he stored the fractions in a refrigerator. When he repeated the assay five days later it was unexpectedly an order of magnitude greater in all fractions and in the range expected from Waring blender preparations in the mitochondrial fraction.[42] He concluded that the activity of the enzyme must have been masked in the fresh preparation and only activated over time.

Soon de Duve pursued the latency in the enzyme activation and proposed that upon initial fractionation the acid phosphatase was contained within a separate "baglike" particle that limited its access to the substrate. The rough treatment in the Waring blender or the gradual aging of the homogenate prepared for cell fractionation released the enzyme from this container. Only once released was it possible to assay its activity. This readily explained why acid phosphatase did not destroy the various phosphate compounds found elsewhere in the living cell. The fact that, with the four-fraction technique, aged mitochondrial fractions yielded the highest levels of acid phosphatase activity suggested that the mitochondrion itself was the sac housing the digestive enzymes (Berthet & de Duve, 1951). De Duve discovered that it was not the mitochondrion when the high-speed attachment to de Duve's centrifuge broke and François Appelmans, a medical student working with him, had to prepare mitochondria using an ordinary preparatory centrifuge with longer centrifugation times. The resulting mitochondria showed no acid phosphatase activity. This led de Duve to continue fractionation, segregating a light fraction containing acid phosphatase and a heavy fraction containing cytochrome oxidase (Berthet et al., 1951; de Duve & Berthet, 1954).

At the Second International Congress of Biochemistry in Paris in 1952, de Duve presented his claim that acid phosphatase belonged to a special cytoplasmic particle. Afterward, P. G. Walker, a British biochemist, related to him that he had found similar results with β-glucuronidase (Walker, 1952); de Duve then tested his light fraction and discovered that it contained β-glucuronidase as well. Subsequently he investigated several other enzymes – acid ribonuclease, acid deoxyribonuclease, cathepsin, urate

---

[42] In a historical rendition of the events, de Duve (1969, p. 7) wrote, "we could have rested satisfied with this result, dismissing the first series of assays as being due to one of those troublesome gremlins that so often infest laboratories, especially late at night . . . Two factors saved us. . . . The assays had been repeated with the old as well as with fresh reagents, giving identical results. The gremlin, if he was the culprit, must have been a very subtle one. Furthermore, we had noted that the greatest discrepancy between the two series of results occurred in the mitochondrial fractions, the smallest one in the supernatant fraction."

oxidase, NADH cytochrome c reductase, NADPH cytochrome c reductase, and fumarase. Although the cytochrome c reductases and fumarase did not, acid ribonuclease, acid deoxyribonuclease, cathepsin all fractionated with acid phosphatase and β-glucuronidase. Each of the enzymes that fractionated together functions, in conjunction with water, to degrade a macromolecule into its subunits through a reaction that follows the formula

$$AB + H_2O \rightarrow AH + BOH.$$

The hydrolytic nature of these enzymes led de Duve to propose the name *lysosome* (from the Greek work *lusis,* meaning to untie) for this component (de Duve et al., 1955).[43] He also proposed that there was a good reason why this group of enzymes might co-occur in a separate organelle: otherwise they would interfere with synthetic processes and disrupt cell structure. Urate oxidase had a similar distribution as the lysosome enzymes but showed no latency and was not brought into solution by the same treatments as sufficed for the hydrolases. In subsequent research, he identified it as a constituent of yet another organelle, the peroxisome.[44]

In conjunction with the Third International Congress of Biochemistry in Louvain in 1955 Alex Novikoff visited de Duve's laboratory for six weeks. Novikoff took samples of de Duve's lysosome preparation to Claude's laboratory in Brussels and then to Bernhard's laboratory in Paris to examine them with the electron microscope. The micrographs revealed particles[45] that had occasionally been seen in electron micrographs a year earlier by Charles Rouillier, who had named them "pericanalicular dense bodies" because they were structures impenetrable to electrons found along bile canaliculi. Novikoff described these structures as having a mean length of 0.37μ and as

---

[43] The choice of name was explained by de Duve (1969, p. 14): "*Lysosome* sounded too much like *lysozyme*; *lysosome* could be confused with *lyo-enzyme*, which already had a meaning; *hydrosome* brought to mind the image of some marine contraption. We finally settled for *lyso-some*, well aware of the danger of our choice." By the early 1960s a total of twelve enyzmes were associated with the lysosome, each capable of splitting important biological compounds in a slightly acid environment: acid phosphatase, cathepsin A and B, acid desoxyribonucle-ase, acid ribonuclease, β-glucuronidase, arhylsulfatase A and B, phosphoprotein phosphatase, β-galactosidase, β-N-acetylglucosamidase, and α-mannosidase (Novikoff, 1961). By 1980 the number had grown to thirty-six.

[44] In these investigations, de Duve found that urate oxidase segregated with three additional enzymes, two of which were involved in the synthesis of hydrogen peroxide (d-amino acid oxidase and α-hydroxyl acid oxidase) and one in its breakdown (catalase). He linked all four enzymes to the peroxisome, which he identified with what had previously been referred to as *microbodies.*

[45] His experience of seeing the micrographs was described by de Duve (1969, p. 16) as "like Le Verrier after the planet Neptune was discovered."

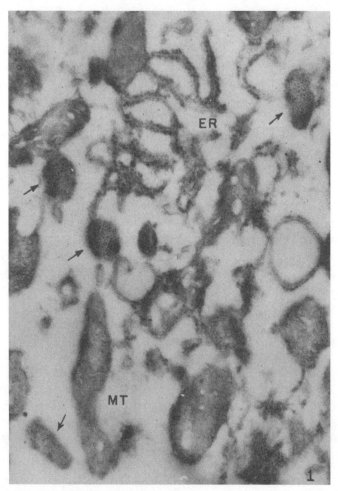

Figure 6.15. Electron micrograph of lysosome fraction from rat liver. Arrows indicate the dense bodies taken to be the lysosomes. ER designates microsomal membranes, presumably from the endoplasmic reticulum; the granules within them are much smaller than the dense bodies (lysosomes). MT designates a mitochondrion, which the authors note is not well-preserved. Reproduced from A. B. Novikoff, H. Beaufay, and C. de Duve (1956), Electron microscopy of lysosome-rich fractions from rat liver, *Journal of Biophysical and Biochemical Cytology,* 2 (No. 4, Part 2), 179–84, Figure 1, plate 60, by copyright permission of the Rockefeller University Press.

containing tiny, electron-dense granules and sometimes possessing internal cavities and external membranes (Figure 6.15). They tentatively identified the particles as the lysosomes (Novikoff, Beaufay, & de Duve, 1956). Novikoff (1961) went on to develop light and electron microscope stains for acid phosphatase and used the presence of a membrane and the positive indication of acid phosphatase as evidence for lysosomes, noting that they often differed substantially in size in different cell types.[46] Treating intracellular digestion as the defining function of lysosomes, Novikoff proposed differentiating lysosomes according to the types of exogenous or endogenous material the lysosome would digest.

Working at the State University of New York College of Medicine in Brooklyn during the same period as de Duve, Werner Straus identified droplets in the proximal tubule of the kidney that stored and broke down reabsorbed proteins. He determined that they possessed an unusually high concentration of acid phosphatase (Straus, 1954). After learning of the other enzymes de Duve associated with lysosomes, Straus (1956) determined that they were present in his droplets as well. Straus's research drew attention to the connection between lysosomes and the digestion of material brought into the cell as well as breakdown products of the cell itself. This idea of the function of the lysosome was further developed by de Duve (1958):

> Our working hypothesis will therefore be that lysosomes are involved in processes of acid hydrolysis. These may comprise: digestion of foreign material, engulfed by pinocytosis, athrocytosis or phagocytosis; physiologic autolysis, as presumably occurs to some extent in all tissues, and particularly as part of the more specialized processes of involution, metamorphosis, holocrine secretion, etc; pathological autolysis or necrosis. (p. 146)

Initially it was difficult to identify lysosomes in other cells due to the variability in their shape and size, causing de Duve to be reluctant to generalize the lysosome concept. Once their role in cellular digestion was understood, this variability made sense: "This polymorphism of the lysosomes is now perfectly understandable: their digestive activity causes them to be filled with a variety of substances and objects in an advanced state of disintegration,

---

[46] Regarding this, de Duve (1969, p. 16) commented, "I must confess that Novikoff, who pioneered this field with untiring energy, received little encouragement on my part. I objected strongly to what I considered a misappropriation of the word *lysosome*, which in my own copyrighted version implied the simultaneous presence of several acid hydrolases, and which he was now using to designate any structure giving a positive reaction for a single such enzyme." Novikoff, however, expressly agreed with de Duve that *intracellular digestion* was the defining function of lysosomes (Novikoff & Holtzman, 1970).

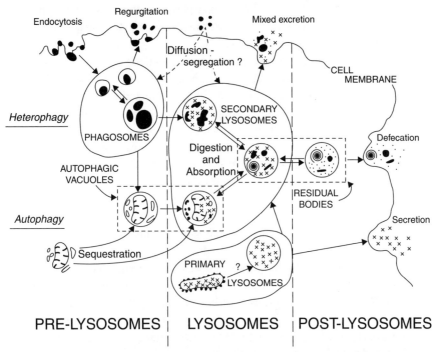

Figure 6.16. Schematic representation by de Duve and Wattriaux (1966) of the lysosomal system for eliminating cellular waste products. There are variations for material entering the cell from without (heterophagy) and from within (autophagy), both having three phases. On the left side, denoted *pre-lysosomes*, the incorporation of either endocytic vesicles into phagosomes or old cell parts into autophagic vesicles is illustrated. In the middle portion, these vesicles are shown as merging with primary lysosomes, generating secondary lysosomes. The crosses symbolize hydrolases. Finally, on the right side the packaging of the undigested material into residual bodies is illustrated. Reprinted, with permission, from the *Annual Review of Physiology*, *28*, Figure 6, p. 468, ©1966 by Annual Reviews, www.annualreviews.org.

and it is their contents that determine their shape, size, density and so on" (de Duve, 1963, p. 76).

Typically, the development of an understanding of how a mechanism works requires decomposing a system to determine the various component parts and their operations. In the case of the lysosome, the discovery that it was a sac containing hydrolytic enzymes made it clear how the key component part of the mechanism worked. What was required was to understand how it interacted with other components to perform the function of digesting and recycling both old cell components and material that had been brought into the cell. Developing the account of the mechanism thus required relating this component part and its operation to other parts and their operations. In the early 1960s de Duve and others succeeded in piecing together such an account of the lysosomal system (Bainton, 1981).

As shown in Figure 6.16, de Duve presented parallel versions of the mechanism for the digestion of material entering the cell from without (heterophagy) and from within (autophagy). In heterophagy, the material to be digested by the lysosomal enzymes was first entrapped into what he called a *phagosome*. He proposed these vesicles then fused with the lysosome sac, which he called the *primary lysosome*, creating a digestive vacuole or *secondary lysosome*. Some of the digestion products diffused back into the cytoplasm of the cell while materials resistant to attack built up in the vacuole, creating what he called a *residual body*, which either was expelled or continued to build up in the cell. In autophagy, de Duve called the digestive vacuole an *autophagic vacuole*; these were operated on in the same manner as phagosomes. He developed a technique for staining for the activity of acid phosphatase, using lead to generate an insoluble compound that has a high electron scattering potential. This yielded a dark image in the micrographs that enabled visualization not only of the lysosome itself but also of the digestive vacuole, autophagic vacuole, and residual body. Inside the autophagic vacuoles, it was possible to recognize remnants of mitochondria and the endoplasmic reticulum (de Duve, 1963).

A last piece of the story of the lysosome is an account of its formation in the cell. Because hydrolytic activity is the defining mark of the lysosome, the discovery of acid phosphatase also in some cisternae of the *trans* region of the Golgi apparatus and in adjacent smooth endoplasmic reticulum led Novikoff and his colleagues to designate the area GERL (a Golgi-related region of smooth endoplasmic reticulum from which lysosomes appear to develop) (Holtzman, Novikoff, & Villaverdi, 1967). Novikoff advanced the idea that hydrolases bypass the Golgi stack and are transported directly to the most distal area of the Golgi apparatus for incorporation into primary lysosomes.

Research on the lysosome by de Duve encapsulates the productive coordination of the results of structural and functional decomposition in discovering cell mechanisms. Modifying the techniques of cell fractionation, he identified a new fraction whose contents indicated their function. Moreover, their operation turned out to require isolation if they were not to destroy the cell itself. Comparing electron micrographs of the fraction with micrographs of cells revealed the locus of the organelle in the cell. At this stage the lysosome had a structural identity and a function, but understanding its operation required postulating other components, whose existence could also be identified in micrographs. The outcome of the coordinated work on structure and function was an ingenious account of the lysosome mechanism.

4. CONCLUSION

By the 1960s the efforts begun in the 1940s bore fruit in the articulation of a number of mechanisms operative in the cytoplasm of the cell. Understanding these mechanisms required the collaborative effort of morphologists and more functionally oriented investigators. Two of these efforts followed up on the initial forays at Rockefeller and elsewhere in the 1940s in identifying the mitochondrion as the power plant of the cell and identifying the microsomes and what came to be known as the endoplasmic reticulum as new components of the cytoplasm. In the case of the mitochondrion, these efforts led not only to an understanding of why the enzyme systems responsible for cellular respiration always involved a membrane component but also of how the membrane played a crucial role in the actual functioning of the mechanism. In the case of the endoplasmic reticulum, researchers not only identified the major steps in which the ribosomes and other RNA constituents operate to synthesize proteins, but also determined the role of the membranes in structuring environments for the newly synthesized proteins destined for export. Investigations in the 1950s and 1960s also finally established the reality and function of the Golgi apparatus, which had been the focus of bitter battles during the first half of the twentieth century. By discovering the process of migration from the rough endoplasmic reticulum through the smooth endoplasmic reticulum into the Golgi region and then into secretory vesicles, investigators came to recognize the Golgi apparatus as playing crucial roles in the preparation of proteins for export. And finally the discovery of the lysosome as a new organelle rich in hydrolytic enzymes provided the basis for developing the basic account of how cells digested either foreign substances brought into the cell or worn out cell components. By the 1960s cell biologists recognized that the cytoplasm was rich in mechanisms that play critical roles in the maintenance of cell life and understood the major parts and operations in each mechanism.

7

Giving Cell Biology an Institutional Identity

In previous chapters I have focused on how new research techniques, especially electron microscopy and cell fractionation, made it possible for researchers to investigate mechanisms within cells. I have analyzed the development of the first products of these investigations – mechanisms for oxidative metabolism in the mitochondria, for protein synthesis in the endoplasmic reticulum, for protein transport in the Golgi apparatus, and for breakdown and disposal of cellular material in the lysosome. These endeavors were the focus of a new field of science that by the 1960s called itself *cell biology*. Many researchers chose this term intentionally to mark a distinction between classical cytology, concerned primarily with morphological structure, and the new, initially interdisciplinary enterprise that took on the challenge of integrating structural and functional information about the cell. By the end of the 1960s, this new scientific field had successful occupied the terra incognita between cytology and biochemistry I identified in Chapter 3.

To become more than a temporary enterprise, cell biology needed its own institutional identity. Journals and professional societies are among the defining institutions of a scientific discipline. These provide important channels for scientists not only to disseminate their work, but also to receive credit for it. Publishing in a journal or appearing on a scientific program provides stature to scientists and evidence that their work is recognized by peers. Such institutions, though, play more than a certifying role. They also serve to direct inquiry by demarcating problems and methods for solving them that are accepted and valued by a particular community of scientists.

When a field begins by drawing upon practitioners of a number of existing disciplines, the journals and professional societies play an additional role – they serve to provide a distinctive identity for the new endeavor, enabling it to take its place among existing fields of science. Although many founding members of the field of cell biology continued to identify themselves with the institutions of the field in which they were trained, as the new institutions took root, younger researchers tended to identify themselves as cell biologists.

Scientific institutions do not create themselves. Establishing and sustaining them requires effort, and those scientists who play leading roles in such institutions often devote a significant amount of their professional energies to them. Creating and maintaining institutions also involves deciding on the character of the institutions, and this in turn will shape the area of scientific research. In particular, these decisions will determine what kinds of research will be considered as part of that area of scientific research. In cell biology, as we have already seen in Chapter 6, two different styles of research were developed by Palade, Porter, and others at the Rockefeller Institute in New York and by Sjöstrand at the Karolinska Institute in Stockholm. Researchers within the broader spheres of influence of each institute sought to shape the contours of the emerging field not only by performing, presenting, and publishing groundbreaking research but also by helping to establish new journals and societies.

In Europe, the International Society for Cell Biology was established in 1947 under the umbrella of the International Union of Biological Sciences. (Its predecessor, the International Society for Experimental Cytology, had been founded in 1933.) In the same year the Institute for Cell Research was established at the Karolinska Institute with Torbjörn Caspersson as its director. In 1950 he, together with John Runnström, created the journal *Experimental Cell Research* and two years later Geoffrey Bourne and James Danielli created the *International Review of Cytology* (both sponsored by the International Society for Cell Biology).[1] In 1957 Fritiof Sjöstrand established the *Journal of Ultrastructure Research*.

In the United States, two institutions promoting cell biology as a new inter-disciplinary field arose in the 1950s and 1960s. The *Journal of Biophysical and Biochemical Cytology* was first conceptualized in 1953 and already in 1955 had published its first issue; it switched to its current name, the *Journal of Cell Biology*, in 1962. The American Society for Cell Biology began in 1960, sponsored its first meeting in 1961, and formed an affiliation with the

---

[1] Between 1962 and 1970 the International Society for Cell Biology also sponsored a series of symposia. In 1972 the society dissolved itself in favor of the International Federation for Cell Biology, comprised of the American Society for Cell Biology, the European Cell Biology Organization (a federation of national and regional societies in Europe), and the Japanese Society for Cell Biology. The Federation held its first meeting in St. Louis in 1972, hosted by the American Society for Cell Biology.

Journal of Cell Biology in 1964. In this final chapter, I will explore in some depth the decision making that shaped these institutions and, through them, the field of cell biology in the United States and elsewhere.[2]

1. CREATION OF THE *JOURNAL OF BIOPHYSICAL AND BIOCHEMICAL CYTOLOGY*

The advent of electron microscopy was clearly one impetus for the creation of the *Journal of Biophysical and Biochemical Cytology*. The high resolution of the micrographs produced by the early 1950s required the capacity, which only a few journals possessed, to produce very high quality plates. In particular the *Anatomical Record*, one of the major journals interested in morphological issues, lacked the ability to produce sufficiently good plates.[3] Pioneers in employing electron microscopy for cell structures were therefore very interested in such a journal. Those journals that could produce the quality plates were often not interested in publishing numerous electron micrographs. This was borne out both in the event that most directly precipitated the creation of the new journal and in the subsequent correspondence. Porter, whose research I have discussed extensively in previous chapters, wrote a paper with Don Fawcett (Fawcett & Porter, 1953) identifying the 9 + 2 structure of cilia, which was rejected by the *Journal of Experimental Medicine* (published at Rockefeller).[4] Porter attributed the idea of starting a new journal to Herbert

---

[2] In part my decision to focus on these two institutions was guided by the fact that detailed records were available through the Rockefeller Archive Center and the American Society for Cell Biology's national office. I am grateful to both of these institutions and their staffs for generous and skilled assistance.

[3] In December 1953 Stanley Bennett, chair of the anatomy department at the University of Washington, wrote letters to the editors of both the *Anatomical Record* and the *American Journal of Anatomy* (both published by the Wistar Institute) complaining about the poor quality of halftone engravings appearing in those journals.

[4] Porter commented on this episode in a short account he wrote of the birth of the *Journal of Biophysical and Biochemical Cytology* on 10 October 1959: "Many of the early papers were published in the Journal of Experimental Medicine – edited largely by Peyton Rous, who had the good sense to appreciate the significance of what was coming. The other editors of the J.E.M. and most especially Rene Dubos did not feel that the character of the J.E.M. should be changed to accomodate (sic) a lot of reports on morphological studies. A real storm developed around the one on cilia (Fawcett-Porter, 1954) submitted first to the J.E.M. in the early summer of 1953. Rous was away at the time, so Dubos had the final say – and the paper was refused. I recall that Vincent Dole in giving me the manuscript said he thought it an interesting and important paper but not for the J.E.M. He went on to say he thought the few of us interested in this new area should organize a new journal. Later that summer the same paper was submitted to the J.G.P. [*Journal of General Physiology*, another Rockefeller journal] and again rejected as unsuitable in content. I later, in September, discussed the matter with Dr. Gasser who had then retired and he

S. Gasser, who had just stepped down as director of the Rockefeller Institute. Porter took the idea to Detlev Bronk, who had replaced Gasser and was given the title of President of the Rockefeller Institute. Because the Institute already published two related journals, the *Journal of General Physiology* and the *Journal of Experimental Medicine*, it was a natural candidate to publish such a journal.[5]

Bronk was extremely receptive to the idea and, after discussing it with others in the emerging field (especially H. Stanley Bennett and Francis O. Schmitt), made a proposal that the Board of Trustees of the Rockefeller Institute approved in January 1954 to establish a *Journal of Cytology*. Bronk's proposal characterized the new journal as follows:

> It is proposed that it deal with cytology from the point of view of the anatomist and histologist; the physiologist; the biochemist; and especially the histochemist, and those concerned with the development of the exciting new physical and chemical methods for investigation of the submicroscopic molecular structure of cells. This journal would deal with the results of studies made possible by the electron microscope, Xray diffraction methods, etcetera, but it certainly would not concentrate on such methods inasmuch as the course of cytology will ultimately move on with the aid of other unanticipated, undeveloped methods.[6]

From the outset, the focus of the journal was a point of contention. Porter's primary interest was electron microscopy and so a major objective for him was creating a journal able to reproduce electron micrographs with high enough quality. (Porter, for example, had several engravers prepare sample proofs so as to determine which would produce the highest quality plates.) A letter from Bennett to Bronk on 3 December 1953 emphasized a similar concern. After identifying the electron microscope as the principle instrument responsible for the recent advances in cytology, which put "the world . . . on the threshold of a classical period of advance in cytological knowledge, comparable in

was more than a little incensed because no one had asked his opinion. (He was listed as editor of both journals.) While in this mood he too said we should start a new journal" (folder 5, box 2, RU 518, Rockefeller University Archives, RAC).

5 Although Porter committed himself to the creation of his new journal, he was not so kind when Sjöstrand proposed to create another one. In a letter to Franz Schrader on 8 June 1956 Porter stated, "I heard a few days ago that F. Sjöstrand, Stockholm, feels there is adequate excuse to start another journal for papers on fine structure etc. Academic would publish it. If you have any strong feelings against such a venture I wish you would make them known to Mr. Jacoby at Academic" (folder 10, box 1, RU 518, Rockefeller University Archives, RAC). Nonetheless, Sjöstrand's new journal, *The Journal of Ultrastructure Research*, began publication in 1957.

6 Folder 10, box 1, RU 518, Rockefeller University Archives, RAC.

significance to the Vesalian advance in Anatomy, or to the classical develop-ment of microscopic anatomy in the nineteenth century," he argued,

> It is already evident that the publication outlets for this important field are inadequate. The requirements are that the publication journal be able to repro-duce high resolution electron micrographs, or other illustrative material with a wealth of fine detail, and that the editorial policies of the journal lend themselves to publication of work of this kind. At present, the *Journal of Experimental Medicine* and the *Journal of Histochemistry* in this country and *Experimen-tal Cell Research* and *Biochemica et Biophysica Acta* are preparing halftone engravings of the necessary quality, but the purposes of these journals are such that only a few cytological papers can be accommodated. The editorial prac-tices of the *Anatomical Record* and the *American Journal of Anatomy* likewise lend themselves to the accommodation of some cytological papers of the sort under discussion, but the quality of the half-tone engravings in these jour-nals is very poor, and suitable reproduction of high resolution micrographs is unsatisfactory.[7]

Schmitt (letter to Bronk of 9 February 1954), on the other hand, raised the concern that the new journal would be too focused on electron microscopy. He contended, "We now have journals which publish morphological work (J. Morph., Exp. Cell Res., J. C. C. P., J. Exp. Med., B.B.A., and many others); I take it we are not proposing to found essentially a 'Journal of Electron Microscopy.'" He characterized the new journal as "devoted to biology at the molecular (or near-molecular) level":

> Accordingly, what I should hope would be brought together in one jour-nal would be micromorphology by direct methods (electron microscopy, UV, phase contrast, interference and other types of microscopes) and by indirect methods (X-ray diffraction, polarization optics) together with physical chem-ical investigations of the shapes, sizes and properties of particles (particularly macromolecules of biological interest) *in solution* (e.g., ultracentrifuge, elec-trophoresis, light scattering, streaming birefringence, viscosity and diffusion). Of course the aim of all of this is the detailed characterization of protoplasmic constituents.[8]

Schmitt was himself an important contributor to the development of electron microscopy, although primarily of viruses. His focus, however, was more broadly biophysical.

---

[7] Folder 10, box 1, RU 518, Rockefeller University Archives, RAC.
[8] Folder 10, box 1, RU 518, Rockefeller University Archives, RAC.

In part, this debate played itself out in setting up the board of editors. Porter (letter to Bronk, 2 February 1954) initially put forward a list of eleven names: Bennett, J. Norman Davidson, Edward Dempsey, Berwind P. Kaufmann, Lehninger, Daniel Mazia, Palade, Porter, Schmitt, Arnold Seligman, and Albert Frey-Wyssling. Schmitt (letter to Bronk, 9 February 1954) objected that Porter's list overemphasized descriptive morphology; he identified Bennett, Dempsey, Palade, and Porter as all representing this area. Schmitt proposed several alternative candidates. Bronk's final choice of eight editors included six from Porter's initial list of eleven. Three were pioneers in electron microscopy who were moving beyond traditional cytology to establish cell biology as a new discipline: Bennett (University of Washington), Palade (Rockefeller), and Porter (Rockefeller). The others were Lehninger (biochemist, Johns Hopkins), Schmitt (biophysicist, MIT), and Seligman (histochemist, Sinai Hospital of Baltimore) from Porter's list; Richard Bear (crystallographer, MIT) from Schmitt's suggestions; and Franz Schrader (cytogeneticist, Columbia) as Bronk's own addition. Because the editorial offices were located at Rockefeller, Porter and Palade were the *resident editors* overseeing the ongoing operations of the journal. Palade managed the incoming manuscripts, sending each to two of the editors. If they agreed on publication or rejection, that was the final decision; if they disagreed, a third editor or a specially chosen referee decided the question.

The question of the focus of the journal, however, went beyond deciding on editors and procedures for evaluating manuscripts. When the editors met with Bronk on 12 April 1954, they debated the name of the journal. Some objected to *Journal of Cytology*, the working name for the new journal, on the grounds that *cytology* conveyed too much of a focus on morphology and not enough on the functional perspectives. Accordingly, the adjectives *biophysical* and *biochemical* were added to the title.[9] The editors also arrived at a mission statement for the new journal that reveals significant concessions to Schmitt's concerns:

> It will be the function of this new publication to provide a common medium for the presentation of morphological, biochemical and biophysical studies of the structure of cells and their components and of the functions of these components. The Journal will give special attention to investigations dealing with cellular organization at colloidal and molecular levels. Papers will be favored

---

[9] According to George Pappas (Interview, 23 October 1995, University of Illinois Chicago), Francis Schmitt and Paul Weiss were principally responsible for the new name. Hewson Swift (Interview, 24 October 1995, University of Chicago) maintained that the title was likely Porter's. This seems less likely, given Porter's later enthusiasm for changing the name.

which integrate information derived from newer approaches to cytology, such as histochemistry, cytogenetics, cytochemistry, electron microscopy and X-ray diffraction. Because of the significance of photographic evidence, the publishers of the Journal will endeavor to insure excellence of photoengravings and printing.

The first issue of the *Journal of Biophysical and Biochemical Cytology* appeared in January 1955. Including the words *biochemical* and *biophysical* in the title did not guarantee that these would figure centrally in the journal, and Schmitt's concern that the new journal not become a journal of electron microscopy continued to occupy the editors at their annual meetings during the journal's initial years. For example, in reporting to Bronk on the April 15, 1956, meeting of the editors (a little over a year into publication), Porter related, "It was agreed that morphological studies had been well represented among the published papers but that papers more properly defined as biophysical and biochemical in nature had been in short supply. Drs. Bear and Lehninger agreed to give this problem their personal attention." But the problem continued in subsequent years.[10] In their 1959 report to the other editors, Porter and Palade, the resident editors, related,

> The editors have all expressed some concern about the predominance of reports on cell and tissue fine structure, derived from electron microscopy. We have therefore made an analysis of trends in the subject matter of manuscripts contributed and are pleased to report that it shows the Journal receiving a steadily increasing number of papers of the non-E.M. variety. Early in volume 4, e.g., reports primarily dependent on electron micrographic evidence exceeded the others by 2 to 1 whereas in the last number of volume 4 they were equal. Since then this trend has continued and, of the manuscripts currently available to us, only 2/5 report solely E.M. evidence. It would seem that our contributors and readers now recognize that the Journal is interested in cell biology regardless of the experiment, device or technique used to acquire the information.[11]

[10] Schmitt, responding to a summary of the 1957 editors' meeting that he had to miss, wrote to Keith Porter on 18 May 1957: "I shall personally try to secure more biophysical and biochemical papers, for I agree that descriptive morphological papers far outweigh the more physical and chemical ones" (folder 11, box 1, RU 518, Rockefeller University Archives, RAC). Even though Porter himself identified closely with electron microscopy, he too recognized the need for a broader range of papers. In a letter welcoming Donald Fawcett as a new editor of the journal (9 October 1959) Porter commented, "Anything you can do to keep your friends from sending us long descriptive manuscripts on anything and everything they can get into the E.M. will be appreciated. More "biophysics" is what we need!" (folder 11, box 1, RU 518, Rockefeller University Archives, RAC).

[11] Folder 11, Box 1, RU 518, Rockefeller University Archives, RAC.

Despite this optimistic assessment of success in moving beyond electron microscopy, though, the topic of diversity of manuscripts continued in subsequent years.

One of the principal ways in which the journal sought to attract manuscripts in areas other than morphology was to have particular editors represent these domains. Thus, when the question of adding or replacing editors arose, one of the primary considerations was securing scientists with appropriate specialties to attract manuscripts. For example, in 1958 the editors voted to expand their number to ten. Although they advanced four names, Porter (letter to Bronk, 29 August 1958) made a special comment on two of them, Paul Doty and Daniel Mazia: "Doty was selected because he is a biophysically oriented investigator. Mazia – because he is a cell physiologist." At the urging of Bronk, the editors began to replace themselves in 1959 (drawing lots for the "privilege" of retiring). The editors recommended replacements from the same research area as the retiring member so as to continue to cover a broad range of research areas. Accordingly, when Schmitt and Porter were chosen to retire, and Bennett resigned, Bernard Davis was selected to replace Schmitt and Don Fawcett to replace Bennett. (A majority of editors wrote in Porter's name, arguing that because he was de facto the chief editor, he was needed to provide continuity.)

Another strategy the journal tried was the creation of special supplemental issues. The first of these, which appeared as a supplement to volume 2 in September 1956, consisted of papers presented at a Conference on Tissue Fine Structure in January of that year. At the first meeting of the editorial board on April 15, 1956, the members had discussed occasionally reprinting classical papers translated into English. They revisited the suggestion at several meetings and finally agreed to reprint a classic paper along with contemporary papers by major researchers on the same topic. The editors sought external funding for this effort, and received a grant for $15,000 from the United States Public Heath Service. Porter edited the first such supplement (appearing in August 1961); it focused on muscle and was organized around a translation of Emilio Veratti's 1901 paper "Investigations on the Fine Structure of Striated Muscle Fiber." Papers by C. R. Skoglund; James F. Reger; David S. Smith; Don W. Fawcett and Jean Paul Revel; Lee D. Peachey; Alexander Mauro and W. Robert Adams; U. Muscatello, Ebba Andersson-Cedergren, G. F. Azzone, and Alexandra von der Decken; and Keith Porter accompanied it. Daniel Mazia edited a second special issue on mitosis, built around Walther Flemming's 1880 "Beiträge zur Kenntnis der Zelle und Ihrer Lebenserscheinungen," which appeared in April 1965.

In one sense, the attempt to incorporate biophysics and biochemistry into the journal ended in failure. In a 1960 report as resident editors to the other editors (30 April 1960), Porter and Palade, in the context of looking toward the future, raised the question of dropping *biophysical* and *biochemical* from the name of the journal:

> First, again, is the problem of scope...As part of this it might be good to consider changing the name of the Journal, so that it more adequately describes the content, as it is now and will probably be. When the present title was decided upon, there was no *Journal of Biophysics*, and no *Journal of Molecular Biology*, and no weekly, publishing *Biochemical and Biophysical Research Communications*. As long as it makes space for them, the *J.B.B.C.* will continue to get a few manuscripts that might just as readily be published in one of these others or in *Biochemica [et] Biophysica Acta*.[12]

(Because that summer Palade wrote a letter to Porter opposing the name change[13] and cast the only vote against changing the name, it is likely that this statement represented only Porter's view.) At the 1961 meeting of the editors the name *Journal of Cell Biology* was proposed because it "better defines the purpose and scope of the Journal and is less cumbersome than the present one."

Porter formally canvassed the editors on the question of a name change in a letter of 23 June 1961 and put forward his own arguments for changing the name. He noted that the original name was a compromise, and not one that everyone endorsed: "The Journal was given its present name at the insistence of a few of the founding editors." It was chosen "to discourage a predicted flood of papers on pure morphology." While acknowledging that the choice of name may have brought a few papers properly characterized as "biophysical," Porter commented, in a pointed dig at Francis Schmitt, "Among these, it may be added, there was never one from the editor who insisted with greatest vehemence that 'biophysical' be included."[14] He also identified as a factor accounting for the paucity of submissions that were biophysical or

---

[12] Folder 3, Box 1, RU 518, Rockefeller University Archives, RAC.

[13] In a letter to Porter on 27 June 1961, Palade offered two reasons for keeping the name – the journal had already obtained a positive reputation with the name and the adjectives *biophysical* and *biochemical* characterize the trend in the field. He contended that the name *cell biology* represented "to a certain extent, a withdrawal to a more conservative position centered on cell structure and cell physiology in the traditional sense." He then proposed yet another alternative: *Cellular and Subcellular Biology*. (Folder 11, Box 1, RU 518, Rockefeller University Archives, RAC.

[14] Schmitt did publish one paper in the journal, a short note presenting electron micrographs of ultracentrifuge preparation of paramyosin (Locker & Schmitt, 1957).

biochemical that other journals were now occupying those niches, including Rockefeller's own *Biophysical Journal*. De facto, therefore, the content of the journal was concentrated in areas that he viewed as properly known as *cell biology*. Porter indicated that the name change would thus reflect the niche that the journal inhabited, partly as a result of pressure from other sources. He suggested that such external determination of the niche may not be a bad thing: "If we acknowledge this and keep the initiative we may effectively serve the increasing interest being shown in the cell as a unit and in the correlation of fine structure and function." The new name was approved later that summer and the first issue as the *Journal of Cell Biology* (volume 12, number 1) was published in January 1962. With this change the journal and, indirectly, the domain for the discipline came to fit more closely Porter's own vision.

The newly-named journal was clearly very successful. It expanded rapidly in terms of subscribers, number of submissions, and size and rate of publication. Initially the journal had appeared every other month with a volume consisting of the year's six issues. In 1959 the number of pages per issue was increased and a volume was redefined as including only three issues (thereby allowing the journal to increase the annual price without increasing the price per volume). Yet the number of submissions continued to increase so rapidly that as of April 1960 Porter reported a backlog of over 100 accepted manuscripts. Accordingly, an additional issue was published in July 1960 as a temporary stopgap, and publication expanded to monthly in 1961. The journal had reached 1700 subscriptions by January 1960 and 3000 by 1965.

Content became more diverse as the number of pages and subscribers multiplied. The journal had already been moving beyond morphological studies of cells with electron microscopy before its change of name, and less than two years later Executive Editor Raymond Griffiths[15] was able to announce in The Annual Report for 1963, "Papers on cell fine structure alone no longer predominate; a steadily increasing percentage of submitted manuscripts pertains to the investigation of other aspects of the cell and its products, and to studies utilizing various combinations of technical approaches – of electron microscopy, cytochemistry, biochemistry (cell fractionation, isolation and analysis of cell components), biophysics (autoradiography), molecular biology, as well as others – in the analysis of basic problems of the cell. Especially noticeable is the trend toward more reports focusing on the quantitative aspects of biological events."[16]

---

[15] Griffiths was hired as Executive Editor of the Journal in December 1960. He held a Ph.D. from Princeton and an M.D. from Northwestern and had previously managed and edited publications of the American Cancer Society.

[16] Folder 4, box 1, RU 518, Rockefeller University Archives, RAC.

Porter's 23 June 1961 letter arguing for the new name had brought up one final consideration: the emergence of The American Society for Cell Biology (ASCB). Porter portrayed the society as a likely usurper unless a change of name brought alliance instead: "Should we decide to make this revision in name, we must move quickly. The immediate danger is that someone else will start a journal with these specific aims and this suggested name. One new organization – The American Society for Cell Biology – is particularly menacing in this regard. If, in fact, the editors of the *J.B.B.C.* recommend the change, the new society will be encouraged more than now to adopt the Journal as its official publication; and there is much to be said for society-sponsored journals, especially in guaranteeing the publication a long life." The reference to a "menacing" threat from the American Society for Cell Biology is particularly interesting because Porter himself was playing a major, if not the major, role in shaping the society and others involved had been, to that point, reluctant to consider starting a journal. Porter achieved his goal of bringing the two institutions into alignment. Shortly after the journal's name was changed to the *Journal of Cell Biology*, the ASCB approached Bronk with the suggestion of becoming a cosponsor of the journal. I will return to the question of society sponsorship after discussing the origins of the ASCB.

2. CREATION OF THE AMERICAN SOCIETY FOR CELL BIOLOGY

The initiative to create a society for cell biology in the United States appears to have originated with Paul Weiss, who already was active in the International Society for Cell Biology. He sought action toward this goal from the U.S. National Committee for the International Union of Biological Sciences (a committee under the auspices of the U.S. National Academy of Sciences, which is officially responsible for U.S. membership in the international unions for various scientific disciplines). They complied by passing a resolution on 6 April 1959 calling for the establishment of a "national society of cell biology to act as a national representative to the International Society for Cell Biology." This resolution was transmitted to Morgan Harris, president of the Tissue Culture Association (TCA), a technique-based society devoted to fostering use of tissue culture as a research tool within biology and medicine.[17]

---

[17] The TCA was created (initially with the name Tissue Culture Commission) in 1946 at the conclusion of a three-day conference in Hershey, PA, funded by the American Cancer Society and sponsored jointly by the panels on cellular physiology, cytochemistry, and nutrition of the Committee on Growth. The commission had two aims: to prepare and make available

Such a resolution fit well with the aspirations of Harris and other members of the TCA who sought to refocus the society around a subject matter rather than a technique.[18] The TCA made a proposal to the NIH Study Section for Cell Biology to fund a committee to "improve working relations among cell biologists." The Study Section for Cell Biology, the primary vehicle in NIH for providing grants for research, was established in 1958 as a result of splitting what had been the Study Section for Morphology and Genetics into two sections, one for genetics and the other for cell biology (Copeland, 1999).[19] The Study Section unanimously approved the proposal and agreed to fund a meeting of up to ten cell biologists to consider how to promote this end. Harris was appointed chair of this committee, and he, in turn, selected Keith Porter

---

media for tissue culture at reasonable cost and to publish a bibliographic index of tissue culture methods (first published in 1952 under the title *Index of Tissue Culture*). It also ran a number of summer courses in Cooperstown, New York. Keith Porter was selected as chairman of the committee charged to establish the commission. In 1949 the name was changed to Tissue Culture Association (TCA) and George Gey was elected its first president. The TCA held annual meetings, often in conjunction with the American Association of Anatomists (see Copeland, 1999).

[18] In 1959 Morgan Harris, as President of the Tissue Culture Association, sent a Bulletin to members advancing the idea, developed by John Paul, Don Fawcett, and W. F. Scherer, of reorganizing the TCA to focus on cell biology. He first noted that since tissue culture had come to be "a standard part of the experimental armamentarium," the need to promote tissue culture as a technique had declined. He then put forward the idea of focusing on a subject matter and identified cell biology as comprising a central interest of the membership. Specifically, he proposed that "an appropriate reorganization of the present TCA might result in a timely and useful society of cell biologists – cytologists, physiologists, biochemists, morphogeneticists, etc.– including those not now in the field of tissue culture but whose interests concern the cell." At the business meeting of the TCA in Atlantic City in April, 1959, forty-seven members voted to explore such reorganization but eleven were opposed, apparently some of them quite vociferously. After the initial meeting to form the ASCB, however, the TCA elected to remain independent of it and "let nature take its course." In 1994 the society changed its named to the Society for In Vitro Biology.

[19] One of the Study Section's first initiatives was a proposal for University Laboratories for Cell Biology. The report approved by the study section on 27 September 1958 identified the first draft as having been "prepared by an *ad hoc* group including Dr. Keith Porter, Chairman, Dr. Clifford Grobstein, Dr. Heinz Herrmann, Dr. Daniel Mazia, Dr. Ernst Scharrer, Dr. Van Potter and Dr. Herbert Taylor." Their idea for these laboratories was partly modeled on the various national institutes comprising the NIH, but was explicitly to be decentralized to "college campuses, where the scientific staff would be in contact with students." This proposal was not well received by the NIH administration and the Study Section instead proposed to support long term funding of highly qualified groups of individuals in a manner less restricted than individual grants. The proposal was that the NIH would maintain such a program for at least twenty-five years. A version of this plan was implemented and some grants were given, including one to the University of California, Berkeley, for a group involving Daniel Mazia, Morgan Harris, Max Alfert, and others (Interview with Morgan Harris, 9 December 1995, University of California, Berkeley).

to serve as chair pro tempore for the meeting and identify the appropriate cell biologists. In developing his list of cell biologists, Porter emphasized the need to bring in physiologists and biochemists to ensure balance (letter of Porter to Harris, 21 October 1959).

One problem this initiative faced was that other existing societies had come to recognize the cell as a prime area of focus and began jockeying for control of the new field of cell biology. An example is the Society of General Physiologists (SGP), itself created as recently as 1946.[20] Kenneth Thiman had recently prepared a report recommending that the SGP expand to include cell biologists and geneticists. C. L. Prosser, who was asked to represent the SGP, wrote to Porter (22 December 1959):

> As you well know, the Society of General Physiologists is moving rather rapidly in the direction of becoming a society for cell biology. We feel our symposia have satisfied a real need in this area, and with the organization of a new division of comparative physiology within the American Society of Zoologists, it appears to us that the real place for the cellular approach to biological function is in the Society of General Physiologists. There is a good indication that the Society will have a journal affiliation shortly.[21]

Thus, a key question for those interested in establishing a society for cell biology was how to position it with respect to existing societies.

Porter's committee met at the Rockefeller Institute on 9 January 1960. The meeting began with remarks from representatives of the different constituencies: Morgan Harris of the TCA; E. G. Butler as chair of the Cell Biology Study Section; Paul Weiss, representing the International Society for Cell Biology; William McElroy and Teru Hayashi, representing the Society of General Physiologists; and Montrose Moses, Teru Hayashi, and A. K. Solomon, as organizers of recent Gordon Conferences on cell structure and metabolism. McElroy and Hayashi reported that by then the Society of General Physiologists had rejected the proposal to become a society for cell biology and supported forming a new society for cell biology. Thirteen of the fifteen scientists present then voted to create a new society to be named the

---

[20] According to Morgan Harris, he approached Dan Mazia when Mazia was President of the Society of General Physiologists and proposed to merge the TCA and the Society of General Physiologists into a Society for Cell Biology but was rebuffed. Interview with Morgan Harris, 9 December 1995, Berkeley, California.

[21] Indeed, in September 1960 the *Journal of General Physiology*, which had been published by the Rockefeller Institute since 1918, became the official organ of the Society of General Physiologists.

American Society for Cell Biology. Two abstained, but it is not clear who they were. The committee decided that those voting in favor would form the provisional council for the new society, and yet fourteen of the fifteen attended the next meeting on 28 May 1960. The only person who did not attend was McElroy, the representative of the SGP, but even his name appeared on the list of the provisional council, which was supplied in the October 1960 proposal to the NIH to fund the first scientific meeting of the society as well as in the official announcement of the ASCB in November 1960. Keith Porter was selected to chair the provisional council and Morgan Harris was elected secretary.

As with the *Journal of Biophysical and Biochemical Cytology*, a major question for the new society was how to ensure the desired mix of specialties in order not to find itself limited to morphology. The proposal to the NIH to fund the first scientific meeting stated, "Membership is expected to include biochemists, biophysicists, cytologists, histologists, microbiologists, physiologists, and others having a common interest in the cell" and made it clear that the program for the first meeting was to "facilitate interdisciplinary communication among cell biologists." One of the prime vehicles for accomplishing this end was the inclusion of invited symposia at the annual meetings. For the first meeting, held in Chicago in November 1961, symposia were organized on three topics: cell continuity, cell diversification, and the characters of cell interfaces. The program committee, chaired by Hewson Swift, sought to include speakers from the broad list of disciplines identified in the NIH proposal.[22] The bulk of the program consisted of contributed papers, which were accepted without review. Albert von Szent-Györgyi presented a banquet address entitled "The fusion of biological dimensions." A sense of the

---

[22] To recruit attendees for the first meeting and members for the society, the ASCB used a mailing list consisting of biologists included on the membership lists of the American Society of Anatomists, American Society of Naturalists, American Society of Zoologists, Botanical Society of America, Genetics Society of American, Gordon Research Conference on Cell Structure and Metabolism, Histochemical Society, Radiation Research Society, Society for Cell Biology (presumably the international society), Society of Protozoologists, and Tissue Culture Association. It is noteworthy that biochemical and biophysical societies are absent from this list, although additional names were later added from the membership list of the American Society of Biological Chemists. In November 1960, 2808 printed announcements were sent to the list thus compiled and by 27 January 1961, a total of 1187 replies requesting applications and additional information had been received. In the minutes of the 28 January 1961 meeting of the provisional council, it was noted that "most of the fields related to cell biology, such as biochemistry, microbiology, anatomy, pathology, histochemistry, cytology, electron microscopy, tissue culture, etc., were well represented. Botany and biophysics were two areas that were not well represented."

research interests of those attracted to the meeting is provided by the sessions into which these papers were organized in the program:

First meeting of the ASCB, 1961
Symposia

Cell Continuity

Rollin Hotchkiss, Rockefeller Institute, Continuity at the molecular level
Hans Ris, University of Wisconsin, Continuity of cytoplasmic organelles
Tracy Sonneborn, Indiana, The genetic control of cytoplasmic organization

Cell Diversification

Arthur B. Pardee, Princeton University, Diversification of bacteria in different environments
Morgan Harris, University of California, The evolution of somatic cell populations in vitro
F. C. Steward, Cornell University, Totipotency and variation in cultured cells (Some metabolic and morphogenetic manifestations)

Characteristics of Cell Interfaces

George Palade, Rockefeller Institute, The membrane systems of the cytoplasm
H. Passow, University of Hamburg, Membrane structure and ion permeability in red cells
Peter Mitchell, University of Edinburgh, The chemical asymmetry of membrane transport processes
A. D. McLaren, University of California, Effect of pH on reactions at biological interfaces

Contributed Paper Sessions

DNA and related topics (3 sessions)
Cell fine structure (2 sessions)
Cell diversification (3 sessions)
Permeability and related topics
Cell particulates (2 sessions)
Cell-parasite interaction
RNA and related topics
Enzyme location
Mitosis
Chromosome structure; protein synthesis
Cell movement
Cell culture

The society meetings were, from the outset, extremely successful. The first meeting in Chicago was attended by 844 scientists, and 744 applied for membership. The society, however, faced some of the same problems as the journal in achieving the desired interdisciplinary mix. Disciplinary mix was the topic that dominated discussions in both the council and the executive committee of the ASCB in its early years. One upshot of these discussions was the strategy of using the invited symposia to bring in more biochemists and biophysicists. After the first meeting, the ASCB also decided that accepting all submissions was not possible, so a limit was placed on the number of contributed paper slots for the 1963 meeting. It was also determined that, in addition to quality, decisions as to which papers to accept would be based in part on disciplinary approach: "It was pointed out that some control over the program content might be necessary in order to attract, for example, biochemists, who are needed by the Society, and to prevent the Society from becoming identified with any one technique or discipline (for example, fine structure)" (Minutes of ASCB Council Meeting, 4 November 1962).

The society followed much the same strategies as the journal in attempting to deal with this issue. One strategy was to include biochemists and more functionally oriented scientists on the council in hopes that they could help attract members. Thus, David Green, a biochemist and one of the codirectors of the Institute for Enzyme Research at the University of Wisconsin, was included in the second meeting of the provisional council and was elected as one of the members of the council at the first meeting of the society in 1961. (By this time, however, Green's credibility among biochemists was in some doubt. After his proposal of a cyclophorase system for oxidative phosphorylation, discussed in the previous chapter, his work was regarded as increasingly speculative.) At the same meeting Alex Novikoff, a biochemist and histochemist at Albert Einstein College of Medicine, became president-elect and accordingly served as the society's second president in 1962–3. In appointing the nominating committee in January 1963, Novikoff followed a recommendation of the council and requested that the nominating committee nominate two biochemists to run for president-elect. The committee complied by nominating Van Potter and Harry Eagle, and Potter won the election to become the fourth president of the society.[23]

For the 1964 meeting of the society David Green served as program chair, and the resulting program had a strong biochemical presence. The invited symposia became a three-part symposium on biological membranes. The first part focused on three topics: energized ion translocation in membrane systems, contractile proteins of membrane systems, and structural proteins

---

[23] Hewson Swift, whose research embraced both electron microscopy and cytochemistry and focused on processes in the cell nucleus, was elected vice president when Novikoff became president, and immediately succeeded Novikoff as the third president.

of membrane systems. The second part focused on elementary particles in membrane systems, and the third on synthesis of mitochondria. The overall topic, and especially the second symposium, was closely related to Green's own interest in particles that could be isolated from the mitochondrial membrane and figured in mechanisms of electron transport and oxidative phosphorylation. The invited portion of the program included not only a number of researchers from Green's home institution of the University of Wisconsin (especially the Enzyme Institute) but also Britton Chance and Elfraim Racker, two biochemists who often differed with Green on interpretation of the particles he had isolated from the mitochondrial membrane. The following year another biochemist, Lee Peachey, served as program chair and he followed the same strategy as Green, organizing a three-part symposium. Its topic was subcellular movement, with one part focusing on physical and physical-chemical aspects (electron and proton mobility, movement of functional groups, and molecular movement, including conformational changes and movement of molecules), another part on translocation of ions and small molecules (ionic movement in mitochondria and other membrane systems), and a third on recent progress on biological movement (chromosome movement, spindle fibers, flagella, cilia, etc.).

These efforts did provide a significant presence of biochemistry on the programs of the ASCB, and a few biochemists became affiliated with the ASCB over the long term (e.g., Novikoff, Rollin Hotchkiss, Philip Siekevitz) The broader goal, however, was not achieved. Biochemists, like biophysicists, already had their own institutions and journals, which provided their primary institutional base. As cell biology became more institutionalized and developed its own departments in universities as well as journals and societies, cell biologists incorporated those tools of biochemistry that were important to them in the practice of cell biology. Biochemists had already become the primary users of cell fractionation techniques, adopted from early cell biology. Accordingly, the concern with reaching out to biochemists declined over time.

One of the main issues that confronted the new society was the question of sponsoring a journal. It arose at the first meeting of the permanent ASCB council on 9 December 1961 (the provisional council had deferred all matters that might imply long-term commitments for the society). Philip Siekevitz advanced a proposal for affiliation with the *Journal of Cellular and Comparative Physiology*, published by the Wistar Institute of Anatomy and Biology. That journal, which had begun publication in 1932, was widely viewed as in need of reinvigorization. Hewson Swift and Alex Novikoff reported on discussions with James Danielli, editor of *International Review of Cytology*, about how the society might help improve the quality of that journal. Later in the meeting the suggestion of affiliating with the *Journal of Biophysical and Biochemical Cytology* was raised. To encourage interest in this third option, Porter reported on the nature of the linkage between the *Journal of General Physiology* (also published by Rockefeller University Press) and the Society of General Physiologists. At the meeting a motion was passed to appoint a committee to study publication options, but apparently no action was taken immediately.

The next spring Runnström invited the ASCB to appoint one or more representatives to the editorial board of the international journal he edited, *Experimental Cell Research*. At this stage Fawcett asked Swift to chair a Committee on Publication (the other members were Novikoff, Porter, Prescott, Siekevitz, and Solomon). The committee moved quickly and by the annual meeting in November 1962 made its recommendation that the society associate with the newly renamed *Journal of Cell Biology*. Bronk accepted the proposal the next month and in the spring of 1963 the ASCB membership approved it. A procedure was established so that, starting in 1964, the ASCB began to nominate half of the editors for the journal (the society nominated two candidates for each society slot on the editorial board, and the current editors selected among them). As part of its commitment to involve biochemists more centrally, the society emphasized biochemists among its nominees.

The activities of a professional society such as ASCB extend beyond sponsoring annual conferences; another important function is to promote knowledge of cell biology in the broader community. With the third meeting, held in New York City in November 1963, the society began to sponsor a pre-meeting for high school teachers. The first pre-meeting was held at the Rockefeller Institute and was co-sponsored by the High School Biology Teacher's Association.[24] Organized by Philip Siekevitz, the meeting featured presentations by George Palade, Van Potter, Wolfgang Joklik, and David Prescott. Starting with the seventh meeting in 1967, the society sponsored a symposium for high school students that involved both a number of talks and a guided tour of the commercial exhibits designed to educate students as to the kinds of instruments used in research (not to mention making the public more generally aware of their cost).[25]

---

[24] This was the name stated in the minutes but presumably the cosponsor was the National Association of Biology Teachers.

[25] The sessions for high school students appear to have been very successful. Joseph Gall urged society members not to attend the session because a very large turnout was expected and there might not be enough space. A second symposium for high school students was held in 1968 with talks by R. Perry, H. Lyman, and D. Beattie.

The idea of a brochure outlining opportunities for careers in cell biology was first put forward in the context of this symposium. Birgit Satir took on the responsibility for writing "Cell Biology: A Guide to Opportunities." The guide provided a brief introduction to the history of and current opportunities for new discoveries in cell biology, as well as information about training and job opportunities (including likely salaries) in cell biology. It was published in 1969 and, in a letter to the membership on 25 June 1969, ASCB Secretary George Pappas reported that they were filling orders rapidly and "undoubtedly there will have to be another printing."

The young society clearly struck a chord with a significant number of researchers, including many just beginning their careers. Winston Anderson, for example, commented, "Consider the national environment during this period – the ASCB was only 2 years old, and the founders, including Fawcett, Palade, Porter, and Swift, ruled supreme. We were all caught up in this national enthusiasm and as cell biologists considered ourselves 'cytonauts' – exploring the cell – discovering and defining, rediscovering, and redefining the structure and functions of organelles" (2000, p. 795).

Although it is beyond the time focus of this book, a survey of the membership of the ASCB in 1976 provides a good sense of how the organization grew during its first fifteen years. The survey was sent to all 2959 members of the ASCB in 1975; 1584 useable responses were returned. The membership was overwhelmingly white and male, although 23% were women. Over half had received their Ph.D. in the last ten years, and 69% in the last fifteen years. Universities and medical schools employed 75% of the members – 65% of those mainly on soft money.

3. CONCLUSION

The two brief historical sketches in this chapter are intended to illustrate the sort of decision making activities that go into establishing scientific institutions.[26] The founders of both the *Journal of Biophysical and Biochemical Cytology* and the American Society for Cell Biology were explicitly attempting to shape the discipline of cell biology. In both cases, the founders worried about being predominantly focused on morphology or tied to electron microscopy as a specific technique. Accordingly, they adopted a variety of

---

[26] I do not take this case to be unique. For example, in describing the founding of the American Physiological Society, Toby Appel (1987) showed how the founders set out to restrict membership to scientists publishing research in physiology, but also drew broad boundaries for the discipline of physiology.

strategies to increase participation by more functionally oriented investigators (for example, they invited such investigators to serve as editors, officers, or symposium speakers).

As it turned out, however, it was not possible to structure the journal or the society precisely as the organizers had envisioned. Including Schmitt, Bear, Davis, etc., on the board of editors and having them encourage submission of papers in biophysics failed to attract significant numbers of biophysically oriented papers to the *Journal of Biophysical and Biochemical Cytology*. Likewise, including Green, Novikoff, Siekevitz, etc., as officers and program chairs of the ASCB failed to attract significant numbers of biochemists to join or attend meetings regularly. The existence of other institutions for biochemistry and biophysics constrained the efforts of cell biologists to position themselves in the ways they sought. Yet, in another respect, those shaping the new institutions were successful. They did avoid the narrowness of becoming a journal or society solely concerned with morphological structure, and they did succeed in establishing a new scientific domain in which the structure and function of cell constituents were well-integrated. The point I want to emphasize is that this did not just happen; it was the product of conscious attention and strategic efforts by those scientists who participated in creating these institutions.

Afterword

With the development of a mature institutional identity in the 1960s, cell biology joined the ranks of the biological disciplines. Although its roots were interdisciplinary, I have described how it developed into a distinct and enduring discipline. A critical element in this achievement was that it deployed new research techniques, especially cell fractionation and electron microscopy, which enabled its practitioners to explore mechanisms that were inaccessible to existing disciplines such as cytology and biochemistry. Using these tools the pioneers in cell biology, sometimes in collaboration with biochemists and molecular biologists, developed mechanistic explanations of numerous cell functions at multiple levels of organization. The discovery of these mechanisms, as described in Chapters 5 and 6, exemplifies the project of explaining phenomena mechanistically that I presented more abstractly in Chapter 2.

Cell biology, like any discipline, continues to develop and adjust its niche relative to other disciplines. It is most distinctive in (1) the attention it gives to variations in structure and function across cells from different organs and organisms; and (2) its status as an interdisciplinary nexus in which findings from physical, chemical, developmental, and other types of investigation are integrated toward an overall goal of understanding the cell. The emphasis given to different contributing disciplines has changed over time, however. In the 1950s and 1960s, collaborations with biochemists were of crucial importance. More recently, cell biology has drawn closer to molecular biology. In November 1989 the American Society for Cell Biology established a new journal, initially named *Cell Regulation*, whose mission statement provides a sense of how the scope of cell biology had grown and shifted toward molecular biology:

> CELL REGULATION . . . publishes papers describing outstanding research contributions in molecular cell biology. The scope of the journal covers a broad

range of topics including all aspects of cell regulation and the reception, trans-
duction, and integration of information. CELL REGULATION deals with the
molecular biology and physiology of receptors, transducing systems, informa-
tional molecules, and their relevance to the growth, development, and physiology of prokaryotic and eukaryotic cells.

In 1992 the journal changed its name to *Molecular Biology of the Cell* and
broadened its focus to include "original research concerning the molecular
aspects of cell structure and function."

Although more focused on processes at the molecular level, today's cell
biologists continue the program of explaining cell phenomena by discovering
the responsible mechanisms. The pioneers in the period 1940–70 established
this agenda and created the initial means of achieving it. Their story not only
provides historical perspective on contemporary cell biology but also provides an instructive exemplar of how new disciplines emerge in the domains
between existing disciplines. Even more, it provides the foundation for more
generally rethinking fundamental issues in philosophy of science. Cell biology in the period I have examined did not advance any new scientific laws,
traditionally thought to be the prime measure of success in science. Yet it
extended scientific knowledge by advancing explanations of previously inex-
plicable phenomena. These explanations are of a type that is finally being
recognized as ubiquitous in the biological (and many other) sciences: the
articulation of mechanisms that account for phenomena. Working with other
philosophers of science to turn our own field in a direction that can give adequate attention to this crucial type of explanation, I have returned time and
time again to cell biology as a wellspring of insight into the true nature of
biological science. The revolution that created cell biology half a century ago
is now providing fuel for another revolution – one that promises to reenergize
philosophy of science by relating it to the actual enterprise of biology.

References

Afzelius, Björn A. (1962). Chemical fixatives for electron microscopy. In R. J. C. Harris (Ed.), *The Interpretation of Ultrastructure* (Vol. 1, pp. 1–19). New York: Academic Press.

Afzelius, Björn A. (1966). *Anatomy of the Cell* (B. Satir, Trans.). Chicago: University of Chicago Press.

Allchin, Douglas (1996). Cellular and theoretical chimeras: Piecing together how cells process energy. *Studies in the History and Philosophy of Science, 27*, 31–41.

Allchin, Douglas (1997). A twentieth-century phlogiston: Constructing error and differentiating domains. *Perspectives on Science, 5*, 81–127.

Allchin, Douglas (2002). To err and win a Nobel Prize: Paul Boyer, ATP synthase and the emergence of bioenergetics. *Journal of the History of Biology, 35*, 149–72.

Allfrey, Vincent G. (1959). The isolation of subcellular components. In J. Brachet & A. E. Mirsky (Eds.), *The Cell* (pp. 193–290). New York: Academic Press.

Altmann, Richard (1889). Ueber Nucleinsäuren. *Archiv für Anatomie Physiologie und wissenschaftliche Medicin*, 524–36.

Altmann, Richard (1890). *Die Elementarorganismen und ihre Beziehungen zu den Zellen.* Leipzig: von Veit.

Anderson, Thomas F. (1956). Electron microscopy of microorganisms. In G. Oster & A. W. Pollister (Eds.), *Physical Techniques in Biological Research, Vol. 3: Cells and Tissues* (pp. 177–240). New York: Academic Press.

Anderson, Winston A. (2000). The value of mentoring in the career of a young scientist. *Molecular Biology of the Cell, 11*, 795–7.

Appel, Toby A. (1987). Founding. In J. R. Brobeck, O. E. Reynolds, & T. A. Appel (Eds.), *History of the American Physiological Society*. Bethesda, MD: The American Physiological Society.

Bainton, Dorothy F. (1981). The discovery of lysosomes. *Journal of Cell Biology, 91*, 66s–76s.

Baker, John R. (1942). Some aspects of cytological technique. In G. H. Bourne (Ed.), *Cytology and Cell Physiology* (First ed.). Oxford: Clarendon Press.

Baker, John R. (1944). The structure and the chemical composition of the Golgi element. *Quarterly Journal of Microscopical Science, 85*, 1.

Baker, John R. (1951). *Cytological Technique*. London: Methuen & Co. Ltd.

References

Baker, John R. (1957). The Golgi controversy. *Symposia of the Society for Experimental Biology, 10*, 1–10.

Baker, John R. (1963). New developments in the Golgi controversy. *International Review of Cytology, 19*, 183–201.

Barnes, Barry (1977). *Interests and the Growth of Knowledge*. London: Routledge and Kegan Paul.

Barnum, C. P., & Huseby, R. A. (1948). Some quantitative analyses of the particulate fractions from mouse liver cell cytoplasm. *Archives of Biochemistry, 19*, 17–23.

Barsalou, Lawrence W. (1999). Perceptual symbol systems. *Behavioral and Brain Sciences, 22*, 577–660.

Battelli, Federico, & Stern, Lina Salomonovna (1911). Die Oxydation der Bersteinsäure durch Tiergewebe. *Biochemische Zeitschrift, 30*, 172–94.

Beams, H. W., & King, R. L. (1934). The effects of ultracentrifuging upon the Golgi apparatus in the uterine gland cells. *Anatomical Record, 59*, 363.

Beams, Jesse W. (1938). High speed centrifuging. *Review of Modern Physics, 10*, 245–63.

Bechtel, William (1984). The evolution of our understanding of the cell: A study in the dynamics of scientific progress. *Studies in the History and Philosophy of Science, 15*, 309–56.

Bechtel, William (1986a). The nature of scientific integration. In W. Bechtel (Ed.), *Integrating Scientific Disciplines* (pp. 3–52). Dordrecht: Martinus Nijhoff.

Bechtel, William (1986b). Biochemistry: A cross-disciplinary endeavor that discovered a distinctive domain. In W. Bechtel (Ed.), *Integrating Scientific Disciplines* (pp. 77–100). Dordrecht: Martinus Nijhoff.

Bechtel, William (1995). Deciding on the data: Epistemological problems surrounding instruments and research techniques in cell biology. In D. Hull, M. Forbes, and R. M. Burian (Eds.), *PSA 1994* (Vol. 2, pp. 167–78). East Lansing, MI: Philosophy of Science Association.

Bechtel, William (2000). From imaging to believing: Epistemic issues in generating biological data. In R. Creath & J. Maienschein (Eds.), *Biology and Epistemology* (pp. 138–63). Cambridge, England: Cambridge University Press.

Bechtel, William (2001). Decomposing and localizing vision: An exemplar for cognitive neuroscience. In W. Bechtel, P. Mandik, J. Mundale, & R. S. Stufflebeam (Eds.), *Philosophy and the Neurosciences: A Reader* (pp. 225–49). Oxford: Basil Blackwell.

Bechtel, William (2002a). Decomposing the mind-brain: A long-term pursuit. *Brain and Mind, 3*, 229–42.

Bechtel, William (2002b). Aligning multiple research techniques in cognitive neuroscience: Why is it important? *Philosophy of Science, 69*, S48–S58.

Bechtel, William, & Richardson, Robert C. (1993). *Discovering Complexity: Decomposition and Localization as Strategies in Scientific Research*. Princeton, NJ: Princeton University Press.

Belitzer, Vladimir A., & Tsibakova, Elena T. (1939). [The mechanism of phosphorylation associated with respiration]. *Biokhimiya, 4*, 516–35.

Bell, L. G. E. (1952). The application of freezing and drying techniques in cytology. *International Review of Cytology, 1*, 35–63.

References

Benda, Carl (1898). Über die Spermatogenese der Vertebraten und hoeherer Everte-
braten. II. Theil. Die Histiogenese der Spermien. *Archiv für Anatomie und Physiologie
(Physiologische Abteilung)*, 393–8.

Benda, Carl (1899). Weitere Mitteilungen über die Mitochondria. *Archiv für Anatomie
und Physiologie (Physiologische Abteilung)*, 376–83.

Bensley, Robert R. (1937). On the fat distribution of mitochondria in the guinea pig
liver. *Anatomical Record, 69*, 341–53.

Bensley, Robert R. (1943). Chemical structure of cytoplasm. In N. L. Hoerr (Ed.), *Frontiers in Cytochemistry: The Physical and Chemical Organization of the Cytoplasm*
(Vol. 10, pp. 323–34). Lancaster, PA: Jacques Cattell Press.

Bensley, Robert R. (1951). Facts versus artifacts in cytology: The Golgi apparatus.
Experimental Cell Research, 2, 1–9.

Bensley, Robert R. (1953). Introduction and greetings: Symposium on the structure
and biochemistry of mitochondria. *Journal of Histochemistry and Cytochemistry, 1*,
179–82.

Bensley, Robert R., & Gersh, Isidore (1933a). Studies on the cell structure by the
freezing-drying method. I. Introduction. *Anatomical Record, 57*, 205–15.

Bensley, Robert R., & Gersh, Isidore (1933b). Studies on the cell structure by the
freezing-drying method. II. Mitochondria. *Anatomical Record, 57*, 217.

Bensley, Robert R., & Hoerr, Normand L. (1934a). Studies on cell structure by the
freezing-drying method. V. The chemical basis of the organization of the cell.
Anatomical Record, 60, 251–66.

Bensley, Robert R., & Hoerr, Normand L. (1934b). Studies on cell structure by the
freeze-drying method. VI. The preparation and properties of mitochondria. *Anatomical Record, 60*, 449–55.

Bernard, Claude (1848). De l'origine du sucre dans l'économic animale. *Archives
générales de médecine, 18*, 303–19.

Bernard, C. (1858). *Leçons sur les propriétés physiologiques et les altérations
pathologiques des liquides de l'organisme*. Paris: Baillière.

Bernard, Claude (1865). *An Introduction to the Study of Experimental Medicine*. New
York: Dover.

Bernard, Claude (1878a). *Leçons sur les phénomènes de la vie communs aux animaux
et aux végétaux*. Paris: Baillière.

Bernard, Claude (1878b). La fermentation alcoolique. Dernières expériences de Claude
Bernard. Edited posthumously by M. Berthelot. *Revue scientifique de la France et de
l'étranger, Paris, 16*, 49–56.

Bernhard, Wilhelm, Gautier, A., & Oberling, Charles (1951). Fibrillary elements of
probable ergastoplasmic nature in cytoplasm of hepatic cells revealed by electron
microscopy. *Comptes rendus des séances de la Société de biologie et de ses filial,
145*, 566.

Bernhard, Wilhelm, Haguenau, Francoise, Gautier, A., & Oberling, Charles (1952). La
structure submicroscopique des elements basolphes cytoplasmiques dans le foie, le
pancreas, et les glandes salivaires. *Zeitschrift fur Zellforschung, 37*, 281–300.

Berthet, Jacques, Berthet, Lucie, Appelmans, Françoise, & de Duve, Christian (1951).
Tissue fractionation studies: 2. The nature of the linkage between acid phosphatase
and mitochondria in rat-liver tissue. *Biochemical Journal, 50*, 182–9.

283

References

Berthet, Jacques, & de Duve, Christian (1951). Tissue fractionation studies. I. The existence of a mitochondria-linked enzymatically inactive form of acid phosphatase in rat liver tissue. *Biochemical Journal, 50,* 174–81.

Berthollet, Claude Louis (1780). Recherches sur la nature des substances animales et sur leurs rapports avec les substances végétales. *Mémoires de l'Acadeâmie royale des sciences,* 120–5.

Bertrand, Gabriel (1895). Sur la laccase et sur le pouvoir oxydant de cette diastase. *Comptes rendus de l'Académie des sciences, 120,* 266–9.

Berzelius, Jöns Jacob (1836). Einige Ideen über bei der Bildung organischer Verbindungen in der lebenden Naturwirksame, aber bisher nicht bemerke Kraft. *Jahres-Berkcht über die Fortschritte der Chemie, 15,* 237–45.

Bichat, Xavier (1805). *Recherches Physiologiques sur la Vie et la Mort.* (3rd ed.). Paris: Machant.

Bittner, John J. (1936). Some possible effects of nursing on the mammary gland tumor incidence in mice. *Science, 84,* 162.

Bloor, David (1991). *Knowledge and Social Imagery* (2nd ed.). Chicago: University of Chicago Press.

Boas, Marie (1952). The establishment of the mechanical philosophy. *Osiris, 10,* 412–541.

Bogen, J., & Woodward, J. (1988). Saving the phenomena. *Philosophical Review, 97,* 303–52.

Bonner, John T. (1952). *Morphogenesis.* Princeton: Princeton University Press.

Borsook, Henry, Deasy, Clara L., Hagen-Smit, Arie J., Keighley, Geoffrey, & Lowy, Peter H. (1950). The uptake in vitro of C^{14}-labeled glycine, L-leucine, and L-lysine by different components of guinea pig liver homogenate. *Journal of Biological Chemistry, 184,* 529.

Bourne, Geoffrey H. (1942). Mitochondria and the Golgi apparatus. In G. Bourne (Ed.), *Cytology and Cell Physiology* (pp. 99–138). Oxford: Oxford University Press.

Bourne, Geoffrey H. (1962). *Division of Labor in Cells.* New York: Academic Press.

Bowen, Robert H. (1924). On a possible relation between the Golgi apparatus and secretory products. *American Journal of Anatomy, 33,* 197–217.

Bowen, Robert H. (1929). The cytology of glandular secretion. *Quarterly Review of Biology, 4,* 299–324 and 484–51.

Boyer, Paul D., Chance, Britton, Ernster, Lars, Mitchell, Peter, Racker, Efraim, & Slater, Edward Charles (1977). Oxidative phoshorylation and photophosphorylation. *Annual Review of Biochemistry, 46,* 955–1026.

Brachet, Jean (1942). La localisation des acides pentosenucléiques dans les tissues animaux et dans les oeufs d'Amphibiens en voie de développement. *Archive de Biologie, 43,* 207–57.

Brachet, Jean (1957). *Biochemical Cytology.* New York: Academic Press.

Brachet, Jean, & Jeener, Raymond (1944). Recherches sur les particules cytoplasmiques de dimensions macromoléculaires riches en acide pentosenucléique. Pt. I. Propriétés générales, relations avec les hydrolases, les hormones, les protéines de structure. *Enymologia, 13,* 196–212.

Brenner, Sydney, Jacob, Francois, & Meselson, Matthew (1961). An unstable intermediate carrying information from genes to ribosomes for protein synthesis. *Nature, 190,* 576–81.

284

References

Bretschneider, L. H. (1952). The electron-microscopic investigation of tissue sections. *International Review of Cytology, 1*, 305–22.

Brodmann, Korbinian (1909/1994). *Vergleichende Lokalisationslehre der Grosshirnrinde* (L. J. Garvey, Trans.). Leipzig: J. A. Barth.

Brown, Robert (1833). On the organs and mode of fecundation in Orchideae and Ascledieae. *Transactions of the Linnean Society, 16*, 685–745.

Bucher, Nancy (1953). The formation of radioactive cholesterol and fatty acids from C^{14}-labeled acetate by rat liver homogenates. *Journal of the American Chemical Society, 75*, 498.

Buchner, Eduard (1897). Alkoholische Gärung ohne Hefezellen (Vorläufige Mittheilung). *Berichte der deutschen chemischen Gesellschaft, 30*, 117–24.

Buchner, Eduard, & Meisenheimer, J. (1904). Die chemische Vorgänge bei der alkoholischen Gärung. *Berichte der deutschen chemischen Gesellschaft, 37*, 417–28.

Burian, Richard M. (1996). Underappreciated pathways toward molecular genetics as illustrated by Jean Brachet's cytochemical embryology. In S. Sarkar (Ed.), *The Philosophy and History of Molecular Biology: New Perspectives* (pp. 67–85). Dordrecht: Kluwer.

Cagniard-Latour, Charles (1838). Memoire sur la fermentation vineuse. *Annales de chimie et de physique, 68*, 206–23. Cajal, Santiago Ramón. See Ramón y Cajal, Santiago.

Campbell, Peter N., & Epstein, Michael A. (1966). *The Structure and Function of Animal Cell Components*. Oxford: Pergamon Press.

Cannan, C. M., & Berger, R. (1951). Quantitative comparison of submicroscopic cytoplasmic particles observed in normal and malignant cells with the electron microscope. *Cancer Research, 2*, 242.

Cannon, Walter B. (1929). Organization of physiological homeostasis. *Physiological Reviews, 9*, 399–431.

Caro, Lucien (1961). Electron microscopic radioautography of thin sections: The Golgi zone as a site of protein concentration in pancreatic acinar cells. *Journal of Cell Biology, 10*, 37–45.

Caro, Lucien, & Palade, George E. (1964). Protein synthesis, storage, and discharge in the pancreatic exocrine cell. An autoradiographic study. *Journal of Cell Biology, 20*, 473–95.

Caspersson, Torbjörn O. (1936). Über den chemischen Aufbau der strukturen des Zellkernes. *Skandinavisches archiv für physiologie, 73*(Suppl. nr. 8), 1–151.

Caspersson, Torbjörn O. (1950). *Cell Growth and Cell Function*. New York: W. W. Norton & Co.

Caspersson, Torbjörn O., & Schultz, Jack (1938). Nucleic acid metabolism of the chromosomes in relation to gene reproduction. *Nature, 142*, 294.

Caspersson, Torbjörn O., & Schultz, Jack (1940). Ribonucleic acids in both nucleus and cytoplasm, and the function of the nucleolus. *Proceedings of the National Academy of Sciences, USA, 26*, 507–15.

Causey, Robert L. (1977). *Unity of Science*. Dordrecht: D. Reidel Publishing Company.

Champy, Christian (1911). *Archives d'anatomie microscopique, 13*, 55.

Chance, Britton, & Williams, G. R. (1956). The respiratory chain and oxidative phosphorylation. *Advances in Enzymology, 17*, 65–134.

Chantrenne, Hubert (1947). Hétérogénéité des granules cytoplasmiques du foie de souris. *Biochimica et Biophysica Acta, 1*, 437–48.

Chao, Fu-Chuan, & Schachmann, Howard K. (1956). The isolation and characterization of a macromolecular ribonucleoprotein from yeast. *Archives of Biochemistry and Biophysics, 61*, 220–30.

Chubin, Daryl E. (1982). *Sociology of Sciences: An Annotated Bibliography on Invisible Colleges, 1972–1981*. New York: Garland.

Churchland, Patricia S., & Sejnowski, Terrence J. (1992). *The Computational Brain*. Cambridge, MA: MIT Press.

Claude, Albert (1935). Properties of the causative agent of a chicken tumor. XI. Chemical composition of purified chicken tumor extracts containing the active principle. *Journal of Experimental Medicine, 61*, 41–57.

Claude, Albert (1937). Preparation of an active agent from inactive tumor extracts. *Science, 85*, 294–5.

Claude, Albert (1938a). A fraction from normal chick embryo similar to the tumor producing fraction of chicken tumor I. *Proceedings of the Society for Experimental Biology and Medicine, 39*, 398–403.

Claude, Albert (1938b). Concentration and purification of Chicken Tumor I agent. *Science, 87*, 467–8.

Claude, Albert (1939). Chemical composition of the tumor-producing fraction of chicken tumor 1. *Science, 90*, 213–5.

Claude, Albert (1940). Particulate components of normal and tumor cells. *Science, 91*, 77–8.

Claude, Albert (1941). Particulate components of cytoplasm. *Cold Springs Harbor Symposia on Quantitative Biology, 9*, 263–71.

Claude, Albert (1943a). Distribution of nucleic acids in the cell and the morphological constitution of cytoplasm. In N. L. Hoerr (Ed.), *Frontiers in Cytochemistry: The Physical and Chemical Organization of the Cytoplasm* (Vol. 10, pp. 111–29). Lancaster, PA: Jacques Cattell Press.

Claude, Albert (1943b). The constitution of protoplasm. *Science, 97*, 451–6.

Claude, Albert (1944). The constitution of mitochondria and microsomes and the distribution of nucleic acid in the cytoplasm of a leukemic cell. *Journal of Experimental Medicine, 80*, 19–29.

Claude, Albert (1946). Fractionation of mammalian liver cells by differential centrifugation. II. Experimental procedures and results. *Journal of Experimental Medicine, 84*, 61–89.

Claude, Albert (1948). Studies on cells: morphology, chemical constitution, and distribution of biochemical functions. *Harvey Lectures, 43*, 121–64.

Claude, Albert (1950). Studies on cell morphology and functions: Methods and results. *Annals of the New York Academy of Sciences, 50*, 854–60.

Claude, Albert, & Fullam, Ernest F. (1945). An electron microscope study of isolated mitochondria. *Journal of Experimental Medicine, 81*, 51–61.

Claude, Albert, & Fullam, Ernest F. (1946). The preparation of sections of guinea pig liver for electron microscopy. *Journal of Experimental Medicine, 89*, 499–503.

Claude, Albert, Porter, Keith R., & Pickels, E. G. (1947). Electron microscope study of chicken tumor cells. *Cancer Research, 7*, 421–30.

286

References

Cleland, K. W., & Slater, Edward Charles (1953). Respiratory Granules of heart muscles. *Biochemical Journal, 53*, 547–56.

Collins, Harry (1981). What is TRASP? The radical programme as a methodological imperative. *Philosophy of the Social Sciences, 11*, 215–24.

Cooper, Cecil, & Lehninger, Albert L. (1956a). Oxidative phosphorylation by an enzyme complex from extracts of mitochndria. I. The span ß-hydroxybutyrate to oxygen. *Journal of Biological Chemistry, 219*, 489–506.

Cooper, Cecil, & Lehninger, Albert L. (1956b). Oxidative phosphorylation by an enzyme complex from extracts of mitochondria. III. The span cytochrome c to oxygen. *Journal of Biological Chemistry, 219*, 519–29.

Cooper, Cecil, & Lehninger, Albert L. (1957a). Oxidative phosphorylation by an enzyme complex from extracts of mitochondria. IV. Adenosinetriphosphatase activity. *Journal of Biological Chemistry, 224*, 547–60.

Cooper, Cecil, & Lehninger, Albert L. (1957b). Oxidative phosphorylation by an enzyme complex from extracts of mitochondria. V. The adenosine triphosphate-phosphate exchange reaction. *Journal of Biological Chemistry, 224*, 561–78.

Copeland, D. Eugene (1999). Origins of cell biology in the United States. *FASEB Journal, 13*, S181-S4.

Corner, George W. (1964). *A History of the Rockefeller Institute, 1901–1953, Origins and Growth*. New York: The Rockefeller Institute Press.

Correns, Carl (1900). G. Mendel's Regel über das Verhalten der Nachkommenschaft der Rassenbastarde. *Berichte der deutschen botanischen Gesellschaft, 18*, 158–68.

Cosslett, Vernon Ellis (1955). Electron microscopy. In G. Oster & A. W. Pollister (Eds.), *Physical Techniques in Biological Research, Vol. 1 Optical Techniques* (pp. 461–531). New York: Academic Press.

Cowdry, Edmund Vincent (1918). The mitochondrial constituents of protoplasm. *Contributions to Embryology. Carnegie Institution of Washington, Washington, DC, 8*, 39–160.

Cowdry, Edmund Vincent (1924). Cytological constituents – mitochondria, Golgi apparatus, and chromidial substance. In E. V. Cowdry (Ed.), *General Cytology: A Textbook of Cellular Structure and Function for Students of Biology and Medicine* (pp. 313–82). Chicago: University of Chicago Press.

Cowdry, Edmund Vincent (1943). In appreciation of Dr. R. R. Bensley. In N. L. Hoerr (Ed.), *Frontiers in Cytochemistry: The Physical and Chemical Organization of the Cytoplasm* (Vol. 10, pp. 7–8). Lancaster, PA: Jacques Cattell Press.

Crane, Diana (1972). *Invisible Colleges*. Chicago: University of Chicago Press.

Crane, Frederick L., Hatefi, Youssef, Lester, R. L., & Widmer, C. (1957). Isolation of a quinone from beef heart mitochondria. *Biochemica et Biophysica Acta, 25*, 220–1.

Cranefield, Paul (1957). The organic physics of 1847 and the biophysics of today. *Journal of the History of Medicine, 12*, 407–23.

Craver, Carl (forthcoming). *Explaining the brain: What a science of the mind-brain could be*.

Creath, Richard (1988). The pragmatics of observation, *PSA 1988* (Vol. 1, pp. 149–53).

Crick, Francis H. C. (1988). *What Mad Pursuit: A Personal View of Scientific Discovery*. New York: Basic Books.

Dalton, Albert J. (1951a). Structural details of some of the epithelial cell types in the kidney of the mouse as revealed by the electron microscope. *Journal of the National Cancer Institute, 11*, 1163–85.

Dalton, Albert J. (1951b). Observations of the Golgi substance with the electron microscope. *Nature, 168*, 244.

Dalton, Albert J. (1953). Electron microscopy of tissue sections. *International Review of Cytology, 2*, 403–17.

Dalton, Albert J. (1955). A chrome-osmium tetroxide fixative for electron microscopy. *Anatomical Record, 121*, 281A.

Dalton, Albert J., & Felix, Marie D. (1954). Cytological and cytochemical characteristics of the Golgi substance of epithelial cells of the epididymis – in situ, in homogenates and after isolation. *American Journal of Anatomy, 94*, 171–208.

Dalton, Albert J., & Felix, Marie D. (1955). A study of the Golgi substance and ergastoplasm in a series of mammalian cell types. In *Fine Structure of Cells. Symposium Held at the VIIIth Congress of Cell Biology, Leiden, 1954* (pp. 274–93). New York: Interscience Publishers.

Dalton, Albert J., & Felix, Marie D. (1956). A comparative study of the Golgi complex. *Journal of Biophysical and Biochemical Cytology, 2*(No. 4, Part 2), 79–84.

Dalton, Albert J., Kahler, H., Streibich, M. J., & Lloyd, B. (1950). Finer structure of hepatic, intestinal and retinal cells of the mouse as revealed by the electron microscope. *Journal of the National Cancer Institute, 11*, 439–61.

Danielli, James F. (1953). *Cytochemistry: A Critical Approach*. New York: John Wiley and Sons.

Danielli, James F., & Davson, Hugh (1935). A contribution to the theory of permeability of thin films. *Journal of Cellular and Comparative Physiology, 5*, 495–508.

Darden, Lindley (1990). Diagnosing and fixing faults in theories. In J. Shrager & P. Langley (Eds.), *Computational Models of Scientific Discovery and Theory Formation* (pp. 319–53). San Mateo, CA: Morgan Kaufmann.

Darden, Lindley (1991). *Theory Change in Science: Strategies from Mendelian Genetics.* New York: Oxford University Press.

Darden, Lindley (1992). Strategies for anomaly resolution. In R. Giere (Ed.), *Cognitive Models of Science* (pp. 251–73). Minneapolis, MN: University of Minnesota Press.

Darden, Lindley (2005). Relations among fields: Mendelian, cytological and molecular mechanisms. *Studies in the History and Philosophy of Biological and Biomedical Science, 36*, 349–71.

Darden, Lindley, & Craver, Carl (2002). Strategies in the interfield discovery of the mechanism of protein synthesis. *Studies in the History and Philosophy of the Biological and Biomedical Sciences, 33*, 1–28.

Darden, Lindley, & Maull, Nancy (1977). Interfield theories. *Philosophy of Science, 43*, 44–64.

de Duve, Christian (1958). Lysosomes, a new group of cytoplasmic particles. In T. Hayashi (Ed.), *Subcellular Particles* (pp. 128–59). New York: The Road Press Company.

de Duve, Christian (Ed.) (1959). *Lysosomes: A New Group of Cytoplasmic Particles.* New York: Ronald Press Company.

de Duve, Christian (1963). The lysosome. *Scientific American, 208*(5), 64–72.

References

de Duve, Christian (1963–4). The separation and characterization of subcellular particles. *Harvey Lectures, 59*, 49–87.

de Duve, Christian (1969). The lysosome in retrospect. In J. T. Dingle & H. B. Fell (Eds.), *Lysosomes in Biology and Pathology* (pp. 3–40). Amsterdam: North Holland.

de Duve, Christian (1971). Tissue fractionation: Past and present. *Journal of Cell Biology, 50*, 20d–55d.

de Duve, Christian (1984). *A Guided Tour of the Living Cell*. New York: Scientific American Library.

de Duve, Christian, & Berthet, Jacques (1954). The use of differential centrifugation in the study of tissue enzymes. *International Review of Cytology, 3*, 225–75.

de Duve, Christian, Pressman, Burton D., Gianetto, Robert, Wattiaux, Robert, & Appelmans, Françoise (1955). Tissue fractionation studies. 6. Intracellular distribution patterns of enzymes in rat liver tissue. *Biochemical Journal, 60*, 604–17.

de Robertis, Eduardo D. P., Nowinski, Wiktor W., & Saez, Francisco A. (1949). *General Cytology*. Philadelphia: W. B. Saunders Company.

de Robertis, Eduardo D. P., Nowinski, Wiktor W., & Saez, Francisco A. (1954). *General Cytology*. (Second ed.). Philadelphia: W. B. Saunders Company.

Devlin, Thomas M., & Lehninger, Albert L. (1956). Oxidative phosphorylation by an enzyme complex from extracts of mitochondria II. The span hydroxbutyrate to cytochrome *c*. *Journal of Biological Chemistry, 219*, 507–18.

Dowe, Phil (1995). Causality and conserved quantities: A reply to Salmon. *Philosophy of Science, 62*, 321–33.

Dröscher, Ariane (1998). Camillo Golgi and the discovery of the Golgi apparatus. *Histochemistry and Cell Biology, 109*, 425–30.

Drummond, D. G. (1950). The practice of electron microscopy. *Journal of the Royal Microscopical Society, 70*, 1–158.

du Bois-Reymond, Emil Heinrich (1859). Über die Angeblick saure Reaktion des Muskelfleisches. *Monatsbericht der Königlich-Preussischen Akademie der Wissenschaften zu Berlin*, 288–324.

Dujardin, Felix (1835). Recherches sur les organismes inférieurs. *Annales des Science Naturelles: Zoologie, 2nd series, 4*, 343–77.

Dumortier, Barthélemy Charles (1832). Recherches sur la structure comparée et le développement des animaux et des végéetaux. *Nova Acta Physico Medica Academiae Caesareae Leopoldino Carolinae, 16*, 217.

Dutrochet, René Henri Joachim (1826). *L'agent immédiat du mouvement vital dévoilé dans la nature et dans son mode d'action chez les végétaux et les animaux*. Paris: JB Ballière.

Dutrochet, René Henri Joachim (1828). *Nouvelles recherches sur l'endosmose et l'exosmose*. Paris: JB Ballière.

Eggleton, Philip, & Eggleton, Marion Grace Palmer (1927). The physiological significance of phosphagen. *Biochemical Journal, 63*, 155–61.

Einbeck, Hans (1914). Über das Vorkommen der Fumarsäure im freschen Fleische. *Zeitschrift für physiologische Chemie, 90*, 303–7.

Elman, Jeffrey L., Bates, Elizabeth A., Johnson, Mark H., Karmiloff-Smith, Annette, Parisi, Dominico, & Plunkett, Kim (1996). *Rethinking Innateness: A Connectionist Perspective on Development*. Cambridge, MA: MIT Press.

References

Embden, Gustav, Deuticke, H. J., & Kraft, G. (1933). Über die intermediaren Vorgänge bei der Glykolyse in der Muskulatur. *Klinische Wochenschrift, 12*, 213–5.

Embden, Gustav, Kalberlah, F., & Engel, H. (1912). Über Milchsäurebildung im Muskelpreßsaft. *Biochemische Zeitschrift, 45*, 45–62.

Embden, Gustav, & Laquer, Fritz Oscar (1914). Über der Chemie des Lactacidogens. I. Mitteilung. Isolierungsversuche. *Zeitschrift für Physiologische Chemie, 93*, 94–123.

Embden, Gustav, & Laquer, Fritz Oscar (1921). Über die Chemie des Lactacidogens. III. *Zeitschrift für Physiologische Chemie, 113*, 1–9.

Englehardt, Vladimir Aleksandrovich (1932). Die Beziehunger zwischen Atmung und Pyrophosphatumsatz in Vogelerythrocyten. *Biochemische Zeitschrift, 251*, 343–68.

Farah, Martha (1988). Is visual imagery really visual? Overlooked evidence from neuropsychology. *Psychological Review, 95*, 307–17.

Farquhar, Marilyn Gist, & Palade, George E. (1981). The Golgi apparatus (complex) – (1954–1981) – from artifact to center stage. *Journal of Cell Biology, 91*, 77s–103s.

Farquhar, Marilyn Gist, & Palade, George E. (1998). The Golgi apparatus: 100 years of progress and controversy. *Trends in Cell Biology, 8*, 2–10.

Farquhar, Marilyn Gist, & Rinehart, J. F. (1954). Cytologic alterations in the anterior pituitary gland following thyroidectomy: An electron microscope study *Endocrinology, 55*, 857–76.

Farquhar, Marilyn Gist, & Wellings, Robert S. (1957). Electron microscopic evidence suggesting secretory granule formation within the Golgi apparatus. *Journal of Biophysical and Biochemical Cytology, 3*(No. 2), 319–22.

Fawcett, Don W., & Porter, Keith R. (1953). A study of the fine structure of ciliated epithelia. *Journal of Morphology, 94*, 221–64.

Fernández-Morán, Humberto (1952). The submicroscopic organization of vertebrate nerve fibres: An electron microscope study of myelinated and unmyelinated nerve fibres. *Experimental Cell Research, 3*, 282–359.

Fernández-Morán, Humberto (1962). Cell-membrane ultrastructure. Low-temperature electron microscopy and X-ray diffraction studies of lipoprotein components in lamellar systems. *Circulation, 26*, 1039–65.

Fernández-Morán, Humberto, Oda, T., Blair, P. V., & Green, David E. (1964). A macromolecular repeating unit of mitochondrial structure and function: Correlated electron microscopic and biochemical studies of isolated mitochondria and submitochondrial particles of beef heart muscle. *The Journal of Cell Biology, 22*, 63–100.

Feulgen, Robert Joachim, & Rossenbeck, Heinrich (1924). Mikrokopisch-chemischer Nachweis einer Nukleinsäure vom Typus Thymusnukleinsäure und die darauf beruhene elektive Fäbung von Zellkernen in mikrokopischen Präparaten. *Zeitschrift für physiologische Chemie, 135*, 203–48.

Fischer, Alfred (1899). *Fixierung, Färbung und Bau des Protoplasmas. Kritische Untersuchungen über Technik und Theorie in der neueren Zellforschung.* Jena: Gustav Fischer.

Fischer, Emil (1894). Einfluss der Konfiguration auf die Wirkung der Enzyme. *Berichte der deutschen chemischen Gesellschaft, 27*, 2985–93.

Fiske, Cyrus Hartwell, & Subbarow, Yellapragada (1929). Phosphorus compounds of muscle and liver. *Science, 70*, 381–2.

Fleischer, Becca, Fleischer, Sidney, & Ozawa, Hidehiro (1969). Isolation and characterization of Golgi membranes from bovine liver. *Journal of Cell Biology, 43*, 59–79.

References

Flemming, Walther (1878). Zur Kenntnis der Zelle und ihrer Theilungserscheinungen. *Schriften des naturwissenschaftlicher Verein für Schleswig-Holstein, 3,* 23–7.

Flemming, Walther (1879). Ueber das Verhalten des Kerns bei der Zellteilung und über die Bedeutung mehrkerniger Zellen. *Archiv für pathologische Anatomie und Physiologie und für klinische Medizin, 77,* 1–28.

Flemming, Walther (1887). Neue Beiträge zur Kenntniss der Zelle. *Archiv für mikroskopische Anatomie und Entwirklungsgeschichte, 29,* 389–463.

Fletcher, Walter Morley, & Hopkins, Frederick Gowland (1907). Lactic acid in amphibian muscle. *Journal of Physiology, 35,* 247–309.

Florkin, Marcel (1972). *A History of Biochemistry. Comprehensive Biochemistry.* (Vol. 30). Amsterdam: Elsevier.

Fodor, Jerry A. (1980). Fixation of belief and concept acquisition. In M. Piatelli-Palmarini (Ed.), *Language and Learning: The Debate between Jean Piaget and Noam Chomsky.* Cambridge, MA: Harvard University Press.

Fol, Hermann (1873). Le premier développement de l'oeuf chez les Géronidés. *Archives des sciences physiques et naturelles, 2nd series, 48,* 335–40.

Fourcroy, Antoine François de (1789). *Elémens d'histoire naturelle et de chimie* (Third ed.) (Vol. Three). Paris: Cuchet.

Frederic, J. (1956). Study of cytoplasms by highly enlarged microscopy with an anoptral device; photography of living cells and after osmic fixation of in vitro cultured cells. *Experimental Cell Research, 11,* 18–35.

Frédéricq, Léon (1884). *Theodore Schwann: sa vie et ses travaux.* Liege.

Friedkin, Morris, & Lehninger, Albert L. (1949). Oxidation-coupled incorporation of inorganic radiophosphate into phospholipide and nucleic acid in a cell-free system. *Journal of Biological Chemistry, 177,* 775–88.

Friedmann, Herbert (1997). From Friedrich Wöhler's urine to Eduard Buchner's alcohol. In A. Cornish-Bowden (Ed.), *New Beer in an Old Bottle: Eduard Buchner and the Growth of Biochemical Knowledge* (pp. 67–122). Valencia: Universitat de València.

Fruton, Joseph S. (1972). *Molecules and Life: Historical Essays on the Interplay of Chemistry and Biology.* New York: Wiley Interscience.

Fullam, Ernest F., & Gessler, Albert E. (1946). A high speed microtome for the electron microscope. *Review of Scientific Instruments, 17,* 23–5.

Galilei, Galileo (1638(1914). *Dialogues Concerning Two New Sciences.* New York: MacMillan.

Gall, Joseph G. (1996). *Views of the Cell.* Bethesda, MD: The American Society for Cell Biology.

Gánti, T. (2003). *The principles of life.* New York: Oxford.

Garfield, Eugene (1979). *Citation Indexing.* New York: Wiley.

Garnier, Charles (1897). Les 'filaments basaux' des cellules glandulaires. *Bibliographie anatomique, 5,* 278–89.

Garnier, Charles (1900). Contribution à l'étude de la structure et du fonctionnement des cellules glandulaires séreuses. Du role de l'ergastroplasme dans la sécrétion. *Journal de l'Anatomie et de la Physiologie normal es et pathologique de l'homme et des animaux, 36,* 22–98.

Gaudillière, Jean-Paul (1996). Molecular biologists, biochemists, and messenger RNA: The birth of a scientific network. *Journal of the History of Biology, 29,* 417–45.

References

Gay-Lussac, Joseph Louis (1810). Extrait d'un mémoire sur la Fermentation. *Annales de chimie, 76*, 245–59.

Gersh, Isidore (1932). The Altmann technique for fixation by drying with freezing. *Anatomical Record, 53*, 309–37.

Gersh, Isidore (1959). Fixation and staining. In J. Brachet & A. E. Mirsky (Eds.), *The Cell: Biochemistry, Physiology, and Morphology*. (Vol. 1, pp. 21–66). New York: Academic Press.

Gicklhorn, Josef (1932). Intracelluläre Myelinfiguren und ähnliche Bildungen bei der reversiblen Entmischung des Protoplasmas. *Protoplasma, 15*, 90–108.

Giere, Ronald G. (1999). *Science without Laws*. Chicago: University of Chicago Press.

Glennan, Stuart (1996). Mechanisms and the nature of causation. *Erkenntnis, 44*, 50–71.

Glennan, Stuart (2002). Rethinking mechanistic explanation. *Philosophy of Science, 69*, S342–S53.

Glick, David (1953). A critical survey of current approaches in quantitative histo- and cytochemistry. *International Review of Cytology, 2*, 447–74.

Golgi, Camilo (1898). Intorno alla struttura delle cellule nervose. *Bollettino della Società Medico-Chirurgica di Pavia, 13*, 3–16.

Gorter, Evert, & Grendel, F. (1925). On bimolecular layers of lipoids on the chromocytes of the blood. *Journal of Experimental Medicine, 41*, 439–43.

Graham, Thomas (1861). Liquid diffusion applied to analysis. *Philosophical Transactions of the Royal Society, London, 151*, 183–224.

Grassé, P. P. (1957). Ultrastructure, polarité et reproduction de l'appareil de Golgi. *Comptes rendus de l'Académie des sciences, 245*, 1278–81.

Green, David E. (1936a). α-Glycerophosphate dehydrogenase. *Biochemical Journal, 30*, 629–44.

Green, David E. (1936b). The malic dehydrogenase of animal tissue. *Biochemical Journal, 30*, 2095–110.

Green, David E. (1951a). The cyclophorase system of enzymes. *Biological Reviews, 26*, 410–55.

Green, David E. (1951b). The cyclophorase system. In J. T. Edsall (Ed.), *Enzymes and Enzyme Systems* (pp. 17–46). Cambridge, MA: Harvard University Press.

Green, David E. (1957–8). Studies in organized enzyme systems. *Harvey Lectures, 53*, 177–227.

Green, David E. (1964). The mitochondrion. *Scientific American, 210*(1), 63–74.

Green, David E., & Brosteaux, Jeanne (1936). The lactic dehydrogenase of animal tissue. *Biochemical Journal, 30*, 1489–508.

Green, David E., Dewan, John G., & Leloir, Luis F. (1937). The ß-hydroxybutyric dehydrogenase of animal tissues. *Biochemical Journal, 31*, 934–49.

Green, David E., & Dixon, Malcolm (1934). Studies on xanthine oxidase. XI. Xanthine oxidase and lactoflavine. *Biochemical Journal, 28*, 237–43.

Green, David E., Loomis, W. F., & Auerbach, V. H. (1948). Studies on the cyclophorase system. I. *Journal of Biological Chemistry, 172*, 389–402.

Gregory, Richard L. (1961). The brain as an engineering problem. In W. H. Thorpe & O. L. Zangwill (Eds.), *Current Problems in Animal Behavior* (pp. 307–30). Cambridge: Cambridge University Press.

References

Gregory, Richard L. (1968). Models and the localization of function in the central nervous system. In C. R. Evans & A. D. J. Robertson (Eds.), *Key Papers: Cybernetics* (pp. 91–102). London: Butterworths.

Hacking, Ian (1983). *Representing and Intervening*. Cambridge: Cambridge U.P.

Haguenau, Françoise (1958). The ergastoplasm: Its history, ultrastructure, and biochemistry. *International Review of Cytology, 7*, 425.

Hanson, Norwood Russell (1958). *Patterns of Discovery*. Cambridge: Cambridge.

Harden, Arthur (1903). Über alkoholische Gärung mit Hefe-Presstoff (Buchners zymase) bein Gegenwart von Blutserum. *Berichte der deutschen chemischen Gesellschaft, 36*, 715–6.

Harden, Arthur, & Young, William J. (1908). The alcoholic fermentation of yeast-juice, Part III – The function of phosphates in the fermentation of glucose. *Proceedings of the Royal Society, London, B80*, 299–311.

Hardy, W. B. (1899). Structure of cell protoplasm. *Journal of Physiology, 24*, 158–210.

Harman, J. W. (1950a). Studies on mitochondria: I. The association of cyclophorase with mitochondria. *Experimental Cell Research, 1*, 382–93.

Harman, J. W. (1950b). Studies of mitochondria: II. The structure of mitochondria in relation to enzymatic activity. *Experimental Cell Research, 1*, 394–402.

Harris, Henry (1999). *The Birth of the Cell*. New Haven: Yale University Press.

Hatefi, Youssef, Haavik, A. G., Fowler, L. R., & Griffiths, D. E. (1962). Studies on the electron transfer system. 42. Reconstitution of the electron transfer system. *Journal of Biological Chemistry, 237*, 2661–9.

Hegarty, Mary (1992). Mental animation: Inferring motion from static displays of mechanical systems. *Journal of Experimental Psychology: Learning, Memory, and Cognition, 18*, 1084–102.

Heidenhain, Rudolf (1875). Beiträge zur Kenntniss des Pancreas. *Archiv für die gesammte Physiologie des Menschen und der Thiere, 10*, 557–632.

Hempel, Carl G., & Oppenheim, Paul (1948). Studies in the logic of explanation. *Philosophy of Science, 15*, 137–75.

Hewson, William (1773). On the figure and composition of the red particles of the blood, commonly called the red globules. *Philosophical Transactions of the Royal Society of London, 63*, 306–24.

Hill, Archibald Vivian (1910). The heat produced by contracture and muscular tone. *Journal of Physiology, 40*, 389–403.

Hill, Archibald Vivian (1913). The energy degraded in the recovery processes of stimulated muscles. *Journal of Physiology, 46*, 28–80.

Hill, Arthur Croft (1898). Reversible zymohydrolysis. *Journal of the Chemical Society, 73*, 634–58.

Hoagland, Mahlon B. (1955). An enzymatic mechanism for amino acid activation in animal tissues. *Biochimica et Biophysica Acta, 16*, 288–9.

Hoagland, Mahlon B., Keller, Elizabeth B., & Zamecnik, Paul C. (1956). Enzymatic carboxyl activation of amino acids. *Journal of Biological Chemistry, 218*, 345–58.

Hoagland, Mahlon B., Zamecnik, Paul C., & Stephenson, Mary L. (1959). A hypothesis concerning the roles of particulate and soluble ribonucleic acids in protein synthesis. In R. E. Zirkle (Ed.), *A Symposium on Molecular Biology* (pp. 105–14). Chicago: University of Chicago Press.

Hoerr, Normand L. (1943). Methods of isolation of morphological constituents of the liver cell. In N. L. Hoerr (Ed.), *Frontiers in Cytochemistry: The Physical and Chemical Organization of the Cytoplasm* (Vol. 10, pp. 185–231). Lancaster, PA: Jaques Cattell Press.

Hofmeister, Franz (1901). *Der chemische Organization der Zelle.* Braunschweig: Vieweg.

Hofmeister, Wilhelm (1849). *Die Entstehung des Embryos der Phanerogamen.* Leipzig: Friedrich Hofmeister.

Hogeboom, George H., & Adams, Mark H. (1942). Mammalian tyrosinase and dopa oxidaase. *Journal of Biological Chemistry, 145,* 273–9.

Hogeboom, George H., Claude, Albert, & Hotchkiss, Rollin D. (1946). The distribution of cytochrome oxidase and succinoxidase in the cytoplasm of the mammalian liver cell. *Journal of Biological Chemistry, 165,* 615–29.

Hogeboom, George H., Schneider, Walter C., & Palade, George E. (1948). Cytochemical studies of mammalian tissues. I. Isolation of intact mitochondria from rat liver; some biochemical properties of mitochondria and submicroscopic particulate matter. *Journal of Biological Chemistry, 172,* 619–35.

Holmes, Frederic Lawrence (1963). Elementary analysis and the origins of physiological chemistry. *Isis, 54,* 50–81.

Holmes, Frederic Lawrence (1986). Intermediary metabolism in the early 20th century. In W. Bechtel (Ed.), *Integrating Scientific Disciplines* (pp. 59–76). Dordrecht: Martinus Nijhoff.

Holmes, Frederic Lawrence (1992). *Between Biology and Medicine: The Formation of Intermediary Metabolism.* Berkeley, CA: Office for History of Science and Technology, University of California at Berkeley.

Holmgren, Emil (1902). Einige Worte über das "Trophospongium" verschiedener Zellarten. *Anatomischer Anzeiger, 20,* 433–40.

Holtzman, Eric, Novikoff, Alex B., & Villaverdi, H. (1967). Lysosomes and GERL in normal chromatolytic neurons of the rat ganglion nodosum. *Journal of Cell Biology, 33,* 419–35.

Hooke, Robert (1665). *Micrographia: or some physiological descriptions of minute bodies made by magnifying glasses with observations and inquiries thereupon.* London: John Martin and James Allestry.

Huennekens, Frank M., & Green, David E. (1950). Studies on the cyclophorase system. XI. The effect of various treatments on the requirement for pyridine nucleotide. *Archives of Biochemistry, 27,* 428–40.

Hughes, Arthur (1959). *A History of Cytology.* London: Abelard-Schuman.

Hultin, T. (1950). Incorporation in vivo of ^{15}N-labeled glycine into liver fractions of newly hatched chicks. *Experimental Cell Research, 1,* 376–81.

Huxley, Thomas H. (1869). On the physical basis of life. *The Fortnightly Review, 26,* 129–45.

Jamieson, James D., & Palade, George E. (1966). Role of the Golgi complex in the intracellular transport of secretory proteins. *Proceedings of the National Academy of Sciences, USA, 55,* 424–31.

Jamieson, James D., & Palade, George E. (1967a). Intracellular transport of secretory proteins in the pancreatic exocrine cell. I. Role of the peripheral elements of the Golgi complex. *Journal of Cell Biology, 34,* 577–96.

References

Jamieson, James D., & Palade, George E. (1967b). Intracellular transport of secretory proteins in the pancreatic exocrine cell. II. Transport of condensing vacuoles and zymogen granules. *Journal of Cell Biology, 34*, 597–615.

Jamieson, James D., & Palade, George E. (1968a). Intracellular transport of secretory proteins in the pancreatic exocrine cell. III. Dissociation of intracellular transport from protein synthesis. *Journal of Cell Biology, 39*, 580–8.

Jamieson, James D., & Palade, George E. (1968b). Intracellular transport of secretory proteins in the pancreatic exocrine cell. IV. Metabolic requirements. *Journal of Cell Biology, 39*, 589–603.

Jonker, Catholijn, Treur, Jan, & Wijngaards, Wouter C. A. (2002). Reductionist and anti-reductionist perspectives on dynamics. *Philosophical Psychology, 15*, 381–409.

Kalckar, Herman (1939). The nature of phosphoric esters formed in kidney extracts. *Biochemical Journal, 33*, 631–41.

Kauffman, Stuart A. (1971). Articulation of parts explanation in biology and the rational search for them. In R. C. Bluck & R. S. Cohen (Eds.), *PSA 1970* (pp. 257–72). Dordrecht: Reidel.

Keilin, David (1925). On cytochrome, a respiratory pigment, common to animals, yeast, and higher plants. *Proceedings of the Royal Society B, 98*, 312–39.

Keilin, David (1966). *The History of Cell Respiration and Cytochrome*. Cambridge: Cambridge University Press.

Keilin, David, & Hartree, Edward F. (1939). Cytochrome and cytochrome oxidase. *Proceedings of the Royal Society B, 127*, 167–91.

Keilin, David, & Hartree, Edward F. (1940). Succinic dehydrogenase-cytochrome system of cells. Intracellular respiratory system catalyzing aerobic oxidation of succinic acid. *Proceedings of the Royal Society B, 129*, 277–306.

Keilin, David, & Hartree, Edward F. (1949). Activity of the succinic dehydrogenase-cytochrome system in different tissue preparation. *Biochemical Journal, 44*, 205–18.

Keller, Elizabeth B. (1951). Turnover of proteins of cell fractions of adult rat liver *in vivo*. *Federation Proceedings, 10*, 206.

Keller, Elizabeth B., & Zamecnik, Paul C. (1956). The effect of guanosine diphosphate and triphosphate on the incorporation of labeled amino acids into proteins. *Journal of Biological Chemistry, 221*, 45–9.

Kennedy, Eugene P., & Lehninger, Albert L. (1949). Oxidation of fatty acids and tricarboxylic acid cycle intermediaries by isolated rat liver mitochondria. *Journal of Biological Chemistry, 179*, 957–72.

Kingsbury, B. F. (1913). Cytoplasmic Fixation. *Anatomical Record, 6*, 39–52.

Kirchhoff, Gottlieb Sigismund (1816). Formation du sucre dans les graines cereals converties en malt et dans la farine infusée dans l'eau bouillante. *Journal de Pharmacie et de Chimie, 2*, 250–8.

Kirkman, Hadley, & Severinghaus, Aura E. (1938a). Review of the Golgi apparatus. Part I. *Anatomical Record, 70*, 413–30.

Kirkman, Hadley, & Severinghaus, Aura E. (1938b). Review of the Golgi apparatus. Part III. *Anatomical Record, 71*, 79–103.

Kitcher, Philip (1989). Explanatory unification and the causal structure of the world. In P. Kitcher & W. C. Salmon (Eds.), *Scientific Explanation*. (Vol. XIII, pp. 410–505). Minneapolis, MN: University of Minnesota Press.

Kitcher, Philip (2001). *Science, Truth, and Democracy*. Oxford: Oxford University Press.

References

Knoop, Franz (1904). *Der Abbau aromatischer Fettsäuren im Tierkörper.* Freiburg: Kuttruff.

Koertge, Noretta (Ed.) (1998). *A House Built on Sand: Exposing Postmodernist Myths about Science.* New York: Oxford University Press.

Kohler, Robert E. (1971). The background to Eduard Buchner's discovery of cell-free fermentation. *Journal of the History of Biology, 4,* 35–61.

Kohler, Robert E. (1973). The enzyme theory and the origin of biochemistry. *Isis, 64,* 181–96.

Kohler, Robert E. (1982). *From Medical Chemistry to Biochemistry: The Making of a Biomedical Discipline.* Cambridge: Cambridge University Press.

Kosslyn, Stephen Michael (1981). The medium and the message in mental imagery: A theory. *Psychological Review, 88,* 46–66.

Kosslyn, Stephen Michael (1994). *Image and Brain: The Resolution of the Imagery Debate.* Cambridge, MA: MIT Press.

Krebs, Hans Adolf, & Johnson, William Arthur (1937). The role of citric acid in intermediate metabolism in animal tissues. *Enzymologia, 4,* 148–56.

Kuhn, Thomas S. (1962(1970). *The Structure of Scientific Revolutions.* (Second ed.). Chicago: University of Chicago Press.

Kühne, Wilhelm Friedrich (1877a). Erfahrungen und Bemerkungen über Enzyme und Fermente. *Untersuchungen aus dem physiologischen Institut Heidelberg, 1,* 291–324.

Kühne, Wilhelm Friedrich (1877b). Ueber das Verhalten verschiedener organisirter und sogenannte ungeformter Fermente. *Verhandlungen des naturhistorisch-medicinischen Vereins zu Heidelberg, new series, 1,* 190–3.

Kützing, Friedrich Traugott (1837). Microscopische Untersuchungen über die Hefe und Essigmutter, nebst mehreren andern dazu gehörigen vegetabilischen Gebilden. *Journal für praktische Chemie, 11,* 385–409.

Lakatos, Imre (1970). Falsification and the methodology of scientific research programmes. In I. Lakatos & A. Musgrave (Eds.), *Criticism and the Growth of Knowledge* (pp. 91–196). Cambridge: Cambridge University Press.

Lardy, Henry A., & Elvehjem, Conrad A. (1945). Biological oxidations and reductions. *Annual Review of Biochemistry, 14,* 1–30.

Larkin, Jill H., & Simon, Herbert A. (1987). Why a diagram is (sometimes) worth ten thousand words. *Cognitive Science, 11,* 65–99.

Latour, Bruno, & Woolgar, Steven (1979). *Laboratory Life: The Social Construction of Scientific Facts.* Beverly Hills: Sage Publications.

Latta, H., & Hartman, J. F. (1950). Use of a glass edge in thin sectioning for electron microscopy. *Proceedings of the Society for Experimental Biology and Medicine, 74,* 436–9.

Laudan, Larry (1977). *Progress and Its Problems.* Berkeley: University of California Press.

Laudan, Larry (1981). The pseudo-science of science. *Philosophy of the Social Sciences, 11,* 173–98.

Lavoisier, Antoine Laurent, & LaPlace, Pierre Simon de (1780). Mémoire sur la Chaleur. *Mémoires de l'Acadeâmie royale des sciences,* 35–408. Article IV reprinted in *Oeuvres de Lavoisier,* Vol. II, pp. 318–33. Paris, Imprimerie Impériale (1886).

References

Lavoisier, Antoine Laurent. (1781). Mémoire sur la formation de l'acide nommé air fixe ou acide crayeux, que je désignerai désormais sous le nom d'acide du charbon. *Mémoires de l'Acadeâmie royale des sciences*, 448–58.

Lavoisier, Antoine Laurent. (1789). *Traité élémentaire de chimie, présenté dans un ordre nouveau et d'après les découvertes modernes.* Paris: Cuchet.

Ledingham, C. G., & Gye, W. E. (1935). On the nature of the filterable tumour-exciting agent in avian sacromata. *Lancet, 228*, 376–7.

Lehninger, Albert L. (1951). The organized respiratory activity of isolated rat-liver mitochondria. In J. Edsall (Ed.), *Enzymes and Enzyme Systems* (pp. 1–14). Cambridge, MA: Harvard University Press.

Lehninger, Albert L. (1954). Oxidative phosphorylation. *Harvey Lectures, 49*, 176–215.

Lehninger, Albert L. (1964). *The Mitochondrion: Molecular Basis of Structure and Function.* New York: W. A. Benjamin, Inc.

Lehninger, Albert L., Wadkins, Charles L., Cooper, Cecil, Devlin, Thomas M., & Gamble, James L., Jr. (1958). Oxidative phosphorylation. *Science, 128*, 450–6.

Levene, Phoebus A., & Mori, Takajiro (1929). Ribodesose and xylodesose and their bearing on the structure of the thyminose. *Journal of Biological Chemistry, 83*, 803–16.

Lewontin, Richard C. (1970). The units of selection. *Annual Review of Ecology and Systematics, 1*, 1–18.

Liebig, Justus (1831). Ueber einen neuen Apparat zur Analyse orgaischer Körper, und über die Zusammensetzung einiger organischer Substanzen. *Annalen der Physik und Chemie, 21*, 1–43.

Liebig, Justus (1842). *Animal Chemistry: Or Organic Chemistry in Its Application to Physiology and Pathology.* Cambridge: John Owen.

Lipmann, Fritz (1939). An analysis of the pyruvic acid oxidation system. *Cold Spring Harbor Symposium, 7*, 248–59.

Lipmann, Fritz (1941). Metabolic generation and utilization of phosphate bond energy. *Advances in Enzymology, 1*, 99–160.

Lipmann, Fritz (1945). Acetylation of sulfanilamide by liver homogenates and extracts. *Journal of Biological Chemistry, 160*, 173–90.

Lipmann, Fritz (1946). Metabolic process patterns. In D. E. Green (Ed.), *Currents in Biochemical Research* (pp. 137–48). New York: Interscience.

Locker, Ronald H., & Schmitt, Francis O. (1957). Some chemical and structural properties of paramyosin. *Journal of Biophysical and Biochemical Cytology, 3*, 889–96.

Lohmann, Karl (1929). Über die Pyrophosphatfraktion im Muskel. *Naturwissenschaften, 17*, 624–5.

Longino, Helen (1990). *Science as Social Knowledge.* Princeton: Princeton University Press.

Longino, Helen (2002). *The Fate of Knowledge.* Princeton: Princeton University Press.

Ludford, R. J., Smiles, J., & Welch, F. V. (1948). The study of malignant cells by phase contrast and ultra-violet microscopy. *Journal of the Royal Microscopical Society, 68*, 1.

Lynen, Feodor, & Reichert, Ernestine (1951). Zur chemischen Struktur der 'aktivierten Essigsäure.' *Angewandte Chemie, 63*, 47–8.

Machamer, Peter, Darden, Lindley, & Craver, Carl (2000). Thinking about mechanisms. *Philosophy of Science, 67*, 1–25.

References

MacMunn, Charles A. (1884). On myohaematin, an intrinsic muscle-pigment of vetrebrates and invertebrates, on histohaematin, and on the spectrum of the supra-renal bodies. *Journal of Physiology, 5*, xxiv.

MacMunn, Charles A. (1886). Researches on myohaematin and the histohaematins. *Philosophical Transactions of the Royal Society of London, 177*, 267–98.

Manasseïn, Marie Mikhailovna (1872). Zur Frage von der alkoholischen Gährung ohne lebende Hefezellen. *Berichte der deutschen chemischen Gesellschaft, 30*, 3061–2.

Marton, Ladislaus (1934). Electron microscopy of biological objects. *Nature, 133*, 911.

Maxwell, J. C. (1868). On governors. *Proceedings of the Royal Society of London, 16*, 270–83.

McIntosh, James (1935). The sedimentation of the virus of Rous sarcoma and the bacteriophage by a high-speed centrifuge. *Journal of Pathology and Bacteriology, 41*, 215–7.

Mercer, Edgar H. (1962). *Cells: Their Structure and Function*. Garden City, NY: Anchor Books.

Mercer, Edgar H., & Birbeck, M. S. C. (1972). *Electron Microscopy: A Handbook for Biologists*. (Third ed.). Oxford: Blackwell Scientific Publications.

Merton, Robert K. (1973). *The Sociology of Science*. Chicago: University of Chicago Press.

Merzenich, Michael M., Recanzone, Gregg H., Jenkins, William M., & Grajski, K. A. (1990). Adaptive mechanisms in cortical networks underlying cortical contributions to learning and nondeclarative memory. *Cold Spring Harbor Symposia on Quantitative Biology, 55*, 873–87.

Meyerhof, Otto (1918). Über das Vorkommen des Coferments der alkoholischen Hefegärung im Muskelgewebe und sein mutmass Bedeutung im Atmungsmechanismus. *Zeitschrift für physiologische Chemie, 101*, 165–75.

Meyerhof, Otto (1920). Die Energieumwandlungen im Muskel. *Archiv für die gesammte Physiologie des Menschen und der Thiere, 182*, 232–83.

Meyerhof, Otto (1924). *Chemical Dynamics of Life Phenomena*. Philadelphia: Lippincott.

Meyerhof, Otto, Lohmann, Karl, & Meyer, Kurt Otto Hans (1931). Über anaerobe Bildung und Schwund von Brenztraubensäure in der Muskulatur. *Biochemische Zeitschrift, 260*, 417–45.

Meyerhof, Otto, Ohlmeyer, Paul, & Möhle, Walter (1938). Über die Koppelung zwischen Oxydoreuktion und Phosphatveresterung bei der anaeroben Kohlenhydratspaltung. *Biochemische Zeitschrift, 297*, 90–133.

Michaelis, Leonor (1899). Die vitale Färbung, eine Darstellungsmethode der Zellgranula. *Archiv für mikrokopische Anatomie, 55*, 558–75.

Miescher, Johann Friedrich (1871). Ueber die chemische Zusammensetzung des Eiters. *Hoppe-Seyler's medicinisch-chemische Untersuchungen, 4*, 441–60.

Milne-Edwards, Henri (1823). *Mémoire sur la structure élémentaire des pincipaux tissus organiques des animaux*. Paris: Lejeune.

Mirsky, Alfred E., & Ris, Hans (1951). The composition and structure of isolated chromosomes. *Journal of General Physiology, 34*, 475–92.

Mitchell, Peter (1961). Coupling of phosphorylation to electron and hydrogen transfer by a chemi-osmotic type of mechanism. *Nature, 191*, 144–8.

298

References

Mitchell, Peter (1966). *Chemiosmotic Coupling in Oxidative and Photosynthetic Phosphorylation*. Bodmin: Glynn Research Ltd.

Mohl, Hugo von (1837). Ueber die Vermehrung der Pflanzenzellen durch Theilung. *Flora, 20*, 1–16.

Mohl, Hugo von (1852). *Principles of the anatomy and physiology of the vegetable cell* (A. Henfrey, Trans.). London: J. Van Voorst.

Morelle, J. (1927). *La Cellule, 37*, 178.

Müller, Johannes (1835). *Vergleichende Anatomie der Myxinoiden, der Cyclostomen mit durchbohrten Gaumen*. Berlin: Königliche Academie der Wissenschaften.

Mundale, Jennifer (1998). Brain mapping. In W. Bechtel & G. Graham (Eds.), *A Companion to Cognitive Science*. Oxford: Basil Blackwell.

Murphy, James B., Helmer, Oscar M., & Sturm, Ernest (1928). Association of the causative agent of a chicken tumor with a protein fraction of the tumor filtrate. *Science, 68*, 18–9.

Nagel, Ernst (1961). *The Structure of Science*. New York: Harcourt, Brace.

Nägeli, Carl Wilhelm von, & Cramer, Carl (1855). *Pflanzenphysiologische Untersuchungen*. Zürich: Schulthess.

Nassonov, Dimitry (1923). Das Golgische Binnennetz und seine Beziehungen zu der Sekretion. Untersuchungen über einige Amphibiendrüssen. *Archiv für mikroskopische Anatomie, 97*, 136–86.

Nassonov, Dimitry (1924). Das Golgischem Binnennetz und seine Beziehungen zu der Sekretion. *Archiv für mikroskopische Anatomie und Entwirklungsgeschichte, 100*, 433–72.

Needham, Dorothy Moyle (1937). Chemical cycles in muscle contraction. In J. Needham & D. Green (Eds.), *Perspectives in Biochemistry* (pp. 200–14). Cambridge: Cambridge University Press.

Needham, Joseph (1942). *Biochemistry and Morphogenesis*. London: Cambridge University Press.

Needham, Joseph, & Needham, Dorothy Moyle (1930). On phosophorus metabolism in embryonic life. I. Invertebrate frogs. *Journal of Experimental Biology, 7*, 317–47.

Negelein, Erwin Paul, & Brömel, Heinz (1939). R-Diphophoglycerinsäure, ihre Isolierung und Eigenschaften. *Biochemische Zeitschrift, 303*, 132–44.

Neubauer, Otto, & Fromherz, Konrad (1911). Über den Abbau der Aminosäuer bei der Hefegärung. *Zeitschrift für physiologische Chemie, 70*, 326–50.

Neuberg, Carl, & Kerb, Johannes Wolfgang (1914). Über zukerfreie Hefegärungen. *Biochemische Zeitschrift, 58*, 158–70.

Neuberg, Carl, & Kobel, M. (1925). Zur Frage der künstlichen und natürlichen Phosphoryleirung des Zuckers. *Biochemische Zeitschrift, 155*, 499–506.

Neutra, Marian, & Leblond, Charles P. (1966a). Synthesis of the carbohydrate of mucus in the Golgi complex as shown by electron microscope radioautography of goblet cells from rats injected with glucose-H^3. *Journal of Cell Biology, 30*, 119–36.

Neutra, Marian, & Leblond, Charles P. (1966b). Radioautographic comparison of the uptake of galactose-H^3 and glucose-H^3 in the Golgi region of various cells secreting glycoproteins or mucopolysaccharides. *Journal of Cell Biology, 30*, 137–50.

Neutra, Marian, & Leblond, Charles P. (1969). The Golgi apparatus. *Scientific American, 222*(2), 100–7.

Newman, M. E. J. (2001). The structure of scientific collaboration networks. *Proceedings of the National Academy of Sciences, USA, 98,* 404–9.

Newman, Sanford B., Borysko, Emil, & Swerdlow, Max (1949a). Ultra-microtomy by a new method. *Journal of Research of the National Bureau of Standards, 43,* 183–99.

Newman, Sanford B., Borysko, Emil, & Swerdlow, Max (1949b). New sectioning techniques for light and electron microscopy. *Science, 110,* 66–8.

Nickles, Thomas (Ed.) (1980a). *Scientific Discovery: Case Studies.* Dordrecht: Reidel.

Nickles, Thomas (Ed.) (1980b). *Scientific Discovery: Logic and Rationality.* Dordrecht: Reidel.

Nicolis, Grégoire, & Prigogine, Ilya (1977). *Self-organization in Nonequilibrium Systems: From Dissipative Structures to Order through Fluctuations.* New York: Wiley.

Nirenberg, Marshall W., & Matthaei, Heinrich J. (1961). The dependence of cell-free protein synthesis in E. Coli upon naturally occurring or synthetic polyribonucleotides. *Proceedings of the National Academy of Sciences, USA, 47,* 1588–602.

Novikoff, Alex B. (1956a). Preservation of the fine structure of isolated liver cell particulates with polyvinylpyrrollidone-sucrose. *Journal of Biophysical and Biochemical Cytology, 2*(No. 4, Part 2), 65–6.

Novikoff, Alex B. (1956b). Electron microscopy: Cytology of cell fractions. *Science, 124*(3229), 969–72.

Novikoff, Alex B. (1959). Approaches to the in vivo function of subcellar particles. In H. Teru (Ed.), *Subcellular Particles* (pp. 1–22). New York: The Ronald Press.

Novikoff, Alex B. (1961). Lysosomes. In J. Brachet & A. E. Mirsky (Eds.), *The Cell* (Vol. II). New York: Academic Press.

Novikoff, Alex B., Beaufay, Henri, & de Duve, Christian (1956). Electron microscopy of lysosome-rich fractions from rat liver. *Journal of Biophysical and Biochemical Cytology, 2*(No. 4, Part 2), 179–84.

Novikoff, Alex B., & Holtzman, Eric (1970). *Cell and Organelles.* New York: Holt, Rinehart, and Winston, Inc.

Novikoff, Alex B., Podber, E., Ryan, J., & Noe, E. (1953). Biochemical heterogeneity of the cytoplasmic particles isolated from rat liver homogenate. *Journal of Histochemistry and Cytochemistry, 1,* 27–46.

Oberling, Charles, Bernhard, Wilhelm, Guérin, M., & Harrel, J. (1950). Images de cellules cancereuses au microscope electronique. *Bulletin du Cancer, 37,* 97.

O'Brien, H. C., & McKinley, G. M. (1943). New microtome and sectioning method for electron microscopy. *Science, 98,* 455–6.

Ochoa, Severo (1940). Nature of oxidative phosphorylation in brain tissue. *Nature, 146,* 267.

Ochoa, S. (1943). Efficiency of aerobic phosphorylation in cell-free heart extracts. *Journal of Biological Chemistry, 151,* 493–505.

Ochoa, Severo, & Rossiter (1939). Flavin-Adenine-Dinucleotide in rat tissues. *Biochemical Journal, 33,* 2008–16.

Olesko, Kathryn (1991). *Physics as a Calling: Discipline and Practice in the Königsberg Seminar for Physics.* Ithaca, NY: Cornell University Press.

Oppenheim, Paul, & Putnam, Hillary (1958). The unity of science as a working hypothesis. In H. Feigl & G. Maxwell (Eds.), *Concepts, Theories, and the Mind-body Problem* (pp. 3–36). Minneapolis: University of Minnesota Press.

References

Oppenheimer, Carl (1909). *Handbuch der Biochemie*. Jena: Gustav Fischer.

Ostwald, Wolfgang (1909). *Grundriss der Kolloidchemie*. Dresden: T. Steinkopff.

Overton, Ernest (1895). Über die osmotischen Eigenschaften der lebenden Pflanzen und Thierzelle. *Vierteljahrsschrift der Naturforschenden Gesellschaft in Zürich, 40*, 159–201.

Overton, Ernest (1896). Über die osmotischen Eigenschaften der Zelle und ihre Bedeutung für die Toxikologie und Pharmakologie. *Vierteljahrsschrift der Naturforschenden Gesellschaft in Zürich, 41*, 383–406.

Overton, Ernest (1899). Über die allgemeinen osmotischen Eigenschaften der Zelle, ihre vermutliche Ursachen und ihre Bedeutung für die Physiologie. *Vierteljahrsschrift der Naturforschenden Gesellschaft in Zürich, 44*, 88–135.

Palade, George E. (1952a). The fine structure of mitochondria. *Anatomical Record, 114*, 427–51.

Palade, George E. (1952b). A study of fixation for electron microscopy. *Journal of Experimental Medicine, 95*, 285–97.

Palade, George E. (1953). An electron microscope study of mitochondrial structure. *Journal of Histochemistry and Cytochemistry, 1*, 188–211.

Palade, George E. (1956a). Electron microscopy of mitochondria and other cytoplasmic structures. In O. H. Gaebler (Ed.), *Enzymes: Units of Biological Structure and Function* (pp. 185–215). New York: Academic Press.

Palade, George E. (1956b). The endoplasmic reticulum. *Journal of Biophysical and Biochemical Cytology, 2*(Supplement), 85–99.

Palade, George E. (1956c). The fixation of tissues for electron microscopy. In *Proceedings of the Third International Conference on Electron Microscopy* (pp. 129–42). London: Royal Microscopical Society.

Palade, George E. (1958a). A small particulate component of the cytoplasm. In S. L. Palay (Ed.), *Frontiers in Cytology* (pp. 283–304). New Haven: Yale University Press.

Palade, George E. (1958b). Functional changes in the structure of cell components. In T. Hayashi (Ed.), *Subcellular Particles* (pp. 64–83). New York: The Ronald Press Company.

Palade, George E. (1971). Albert Claude and the beginnings of biological electron microscopy. *The Journal of Cell Biology, 50*, 5D–19D.

Palade, George E. (1987). Cell fractionation. In J. E. Pauly (Ed.), *The American Association of Anatomists, 1888–1987. Essays on the History of Anatomy in America and a Report on the Membership – Past and Present*. Baltimore: Wilkin and Wilkins.

Palade, George E. (1992). Intracellular aspects of the process of protein secretion. In J. Lindsten (Ed.), *Nobel Lectures, Physiology or Medicine: 1971–1980* (pp. 177–206). Singapore: World Scientific Publishing.

Palade, George E., & Claude, Albert (1949a). The nature of the Golgi apparatus. I. Parallelism between intercellular myelin figures and Golgi apparatus in somatic cells. *Journal of Morphology, 85*, 35–69.

Palade, George E., & Claude, Albert (1949b). The nature of the Golgi apparatus. II. Identification of the Golgi apparatus with a complex of myelin figures. *Journal of Morphology, 85*, 71–111.

Palade, George E., & Porter, K. R. (1952). The endoplasmic reticulum of cells *in situ*. *Anatomical Record, 112*(2), 68.

301

Palade, George E., & Porter, Keith R. (1954). Studies on the endoplasmic reticulum: I. Its identification in cells *in situ. Journal of Experimental Medicine, 100,* 641–56.

Palade, George E., & Siekevitz, Philip (1955). Liver microsomes: An integrated morphological and biochemical study. *Journal of Biophysical and Biochemical Cytology, 2,* 171–200.

Parat, Marcel (1928). Contributions a l'étude morphologique et physiologique du cytopasme. *Archives d'anatomie microscopique et de morphologie experimentale, 24,* 73–357.

Pardee, Arthur B., Jacob, François, & Monod, Jacques (1959). The genetic control and cytoplasmic expression of 'inducibility' in the synthesis of ß-galactosidase by E. coli. *Journal of Molecular Biology, 1,* 165–78.

Parnas, Jacob Karol, Ostern, Pawel, & Mann, Thaddeus (1934). Über die Verkettung der chemischen Vorgäange im Muskel. *Biochemische Zeitschrift, 272,* 64–70.

Pascual-Leone, Alvaro, & Hamilton, Roy (2001). The metamodal organization of the brain. In C. Casanova & M. Ptito (Eds.), *Progress in Brain Research* (Vol. 134, pp. 425–45). New York: Elsevier.

Pasteur, Louis (1857). Mémoire sur la fermentation appelée lactique. *Comptes rendus de l'Académie des sciences, 45,* 913–6.

Pasteur, Louis (1858). Mémoire sur la fermentation appelée lactique. *Annales de Chimie, 3e Ser, 52,* 404–18.

Pasteur, Louis (1860). Mémoire sur la fermentation alcoolique. *Annales de Chimie, 3e Ser, 58,* 323–426.

Pasteur, Louis (1861). Sur la fermentation visqueuse et la fermentation butyrique. *Bulletin Société chimique de Paris, 11,* 30–1.

Payen, Anselme, & Persoz, Jean F. (1833). Mémoire sur la diastase, les principaux produits de ses réactions, et leurs applications aux arts industriels. *Annales de Chemie et de Physique, 53,* 73–92.

Pease, Daniel C. (1987). The development of cytological transmission electron microscopy. In J. Pauly (Ed.), *The American Association of Anatomists, 1888–1987: Essays on the History of Anatomy in America and a Report on the Membership – Past and Present.* Baltimore: Williams and Williams.

Pease, Daniel C., & Baker, Richard F. (1950). Electron microscopy of the kidney. *American Journal of Anatomy, 87,* 349–70.

Pease, Daniel C., & Porter, Keith R. (1981). Electron microscopy and ultramicrotomy. *Journal of Cell Biology, 91,* 287s–92s.

Penefsky, Harvey S., Pullman, Maynard E., Datta, Anima, & Racker, Efraim (1960). Partial resolution of the enzymes catalyzing oxidative phosphorylation. *Journal of Biological Chemistry, 235,* 3330–6.

Petermann, Mary L., Hamilton, Mary G., Balis, M. Earl, Samarth, Kumud, & Pecora, Pauline (1958). Physicochemical and metabolic studies on rat liver nucleoprotein. In R. B. Roberts (Ed.), *Microsomal Particles and Protein Synthesis* (pp. 70–5). London: Pergamon Press.

Petermann, Mary L., Mizen, N. A., & Hamilton, Mary G. (1953). The macromolecular particles of normal and regenerating rat liver. *Cancer Research, 13,* 372–5.

Peters, Rudolpf A. (1930). Surface structure in the integration of cell activity. *Faraday Society Transactions, 26,* 797–809.

References

Petersen, Steven E., Fox, Peter T., Snyder, Abraham Z., & Raichle, Marcus E. (1990). Activation of extrastriate and frontal cortical areas by visual words and word-like stimuli. *Science, 249,* 1041–4.

Peterson, Marian, & Leblond, Charles P. (1964a). Uptake by the Golgi region of glucose labelled with tritium in the 1 or 6 position, as an indicator of synthesis of complex carbohydrates. *Experimental Cell Research, 34,* 420–3.

Peterson, Marian, & Leblond, Charles P. (1964b). Synthesis of complex carbohydrates in the Golgi region, as shown by radioautography after injection of glucose. *Journal of Cell Biology, 21,* 143–8.

Pfeffer, Wilhelm Friedrich Philipp (1887). *Osmotische Untersuchungen: Studien zur Zellmechanik.* Leipzig: Wilhelm Engelmann.

Pflüger, Eduard (1872). Über die Diffusion des Sauerstoffs, den Ort und die Gesetze der Oxydationsprocesse im thierischen Organismus. *Pflüger's Archiv für die gesammte Physiologie des Menschen und der Thiere, 6,* 43–64.

Pflüger, Eduard (1875). Beiträge zur Lehre von der Respiration: I. Ueber die physiologische Verbrennung in den lebendigen Organismen. *Pflüger's Archiv für die gesammte Physiologie des Menschen und der Thiere, 10,* 251–367.

Piaget, Jean (1971). *The Science of Education and the Psychology of the Child.* London: Longmans.

Pickels, Edward G. (1942). An improved type of electrically driven high speed laboratory centrifuge. *Review of Scientific Instruments, 13,* 93–100.

Picken, L. E.R (1940). The fine structures of biological systems. *Biological Review, 15,* 133–67.

Pinchot, Gifford B. (1953). Phosphorylation coupled to electron transport in cell-free extracts of alcaligenes faecalis. *Journal of Biological Chemistry, 205,* 65–74.

Pinchot, Gifford B., & Racker, Efraim (1951). Ethyl alcohol oxidation and phosphorylation in extracts of *E. coli* (Vol. 1, pp. 366–69). In W. D. McElroy & B. Glass (Eds.), *Phosphorus Metabolism.* Baltimore: Johns Hopkins University Press.

Polanyi, Michael (1966). *The Tacit Dimension.* New York: Doubleday.

Porter, Keith R. (1941a). Diploid and androgenetic haploid hybridization between two forms of Rana pipiens. *Biological Bulletin, 80,* 238.

Porter, Keith R. (1941b). Developmental variations resulting from the androgenetic hybridization of four forms of Rana pipiens. *Science, 93,* 439.

Porter, Keith R. (1953). The fine structure of a submicroscopic, basophilic component of cytoplasm. *Journal of Experimental Medicine, 97,* 727–50.

Porter, Keith R. (1954). Electron microscopy of basophilic components of cytoplasm. *Journal of Histochemistry and Cytochemistry, 2,* 346–75.

Porter, Keith R. (1955–6). The submicroscopic morphology of protoplasm. *The Harvey Lectures, 51,* 175–228.

Porter, Keith R. (1987). Electron microscopy of cultured cells. In J. E. Pauly (Ed.), *The American Association of Anatomists, 1888–1987. Essays on the History of Anatomy in America and a Report on the Membership –Past and Present* (pp. 59–67). Baltimore: Williams and Wilkins.

Porter, Keith R., & Blum, Joseph (1953). A study in microtomy for electron microscopy. *The Anatomical Record, 117,* 685–707.

Porter, Keith R., Claude, Albert, & Fullam, Ernest F. (1945). A study of tissue culture cells by electron microscopy. *Journal of Experimental Medicine, 81,* 233–55.

References

Porter, Keith R., & Kallman, Frances L. (1952). Significance of cell particulates as seen by electron microscopy. *Annals of the New York Academy of Science, 54*, 882–91.

Porter, Keith R., & Kallman, Frances L. (1953). The properties and effects of osmium tetroxide as a tissue fixative with special reference to its use for electron microscopy. *Experimental Cell Research, 4*, 127–41.

Porter, Keith R., & Thompson, Helen P. (1947). Some morphological features of cultured rat sarcoma cells as revealed by the electron microscope. *Cancer Research, 7*, 431–8.

Porter, Keith R., & Thompson, Helen P. (1948). A particulate body associated with epithelial cells cultured from mammary carcinomas of mice of a milk-factor strain. *Journal of Experimental Medicine, 88*, 15–23.

Potter, Van Rensselaer, & Elvehjem, Conrad A. (1936). A modified method for the study of tissue oxidations. *Journal of Biological Chemistry, 114*, 495–504.

Prebble, John (2002). Peter Mitchell and the ox phos wars. *Trends in Biochemical Sciences, 27*, 209–12.

Prescott, David M. (Ed.) (1973). *Methods in Cell Biology*. New York: Academic Press.

Price, Derek J. de Solla (1961). *Science since Babylon*. New Haven: Yale University Press.

Prout, William (1827). On the ultimate composition of simple alimentary substances; with some preliminary remarks on the analysis of organised bodies in general. *Philosophical Transactions of the Royal Society of London, 117*, 355–88.

Pullman, Maynard E., Penefsky, Harvey S., Datta, Anima, & Racker, Efraim (1960). Partial resolution of the enzyme catalyzing oxidative phosphorylation. *Journal of Biological Chemistry, 235*, 3322–9.

Pylyshyn, Zenon W. (1981). The imagery debate: Analogue media versus tacit knowledge. *Psychological Review, 88*, 111–33.

Pylyshyn, Zenon W. (2003). *Seeing and Visualizing: It's Not What You Think*. Cambridge, MA: MIT Press.

Racker, Efraim (1965). *Mechanisms in Bioenergetics*. New York: Academic Press.

Racker, Efraim (1968). The membrane of the mitochondrion. *Scientific American, 218*, 32–9.

Racker, Efraim (1975). Reconstitution, mechanism of action and control of ion pumps. *Biochemical Society Transactions, 3*, 785–802.

Racker, Efraim (1976). *A New Look at Mechanisms in Bioenergetics*. New York: Academic Press.

Racker, Efraim, & Horstman, Lawrence L. (1967). Partial resolution of the enzymes catalyzing oxidative phosphorylation XIII. Structure and function of submitochondrial particles completely resolved with respect to coupling factor 1. *Journal of Biological Chemistry, 242*, 2547.

Racker, Efraim, Tyler, D. D., Estabrook, Ronald W., Conover, Thomas E., Parsons, D. F., & Chance, Britton (1965). Correlations between electron-transport activity, ATP-ase and morphology of submitochrondrial particles. In T. E. King, H. S. Mason, & M. Morrison (Eds.), *Oxidases and Related Redox Systems* (pp. 1077–101). New York: Wiley.

Ramón y Cajal, Santiago (1907). L'appareil reticulaire de Golgi-Holmgren coloré par le nitrate d'argent. *Trabajos del Laboratorio de Investigaciones Biológicas, 5*, 151–4.

References

Ramón y Cajal, Santiago (1908). Les conduits de Golgi-Holmgren du protoplasme nerveux et le réseau péricéllulaire de la membrane. *Trabajos del Laboratorio de Investigaciones Biológicas, 6*, 123–35.

Ramón y Cajal, Santiago (1914). Algunas variaciones fisiológicas y patológicas del aparato reticular de Golgi. *Trabajos del Laboratorio de Investigaciones Biológicas, 12*, 127–227.

Rasmussen, Nicolas (1995). Mitochondrial structure and the practice of cell biology in the 1950s. *Journal of the History of Biology, 28*, 381–429.

Rasmussen, Nicolas (1997). *Picture Control: The Electron Microscope and the Transformation of Biology in America.* Stanford, CA: Stanford University Press.

Raspail, François-Vincent (1825). Développement de la fécule dans les organes de la fructification des céréales. *Annales des Science Naturelles, 6*, 224–39.

Redman, Colvin, & Sabatini, David D. (1966). Vectorial discharge of peptides released by puromycin from attached ribosomes. *Proceedings of the National Academy of Sciences, USA, 56*, 608–15.

Redman, Colvin, Siekevitz, Philip, & Palade, George E. (1966). Synthesis and transfer of amylase in pigeon pancreatic microsomes. *Journal of Biological Chemistry, 241*, 1150–8.

Regaud, Claudius (1909). Attribution aux 'formations mitochondriales' de la fonction generale d'extraction et de fixation electives, exercee par les cellules vivantes sur les substances dissouten dans le milieu ambiant. *Comptes Rendus de Societe Biologique, 66*, 919–21.

Reichenbach, Hans (1966). *The Rise of Scientific Philosophy.* Berkeley: University of California Press.

Reichert, Karl Bogislaus (1847). Bericht über die Leistungen in der mikroskopischen Anatomie des Jahres 1846. *Archiv für Anatomie, Physiologie und wissenschaftliche Medicin*, 1–67.

Remak, Robert (1852). Ueber extracellulare Entstehung thierischer Zellen und über Vermehrung derselben durch Theilung. *Archiv für Anatomie, Physiologie und wissenschaftliche Medicin*, 47–57.

Remak, Robert (1855). *Untersuchungen über die Entwicklung der Wirbelthiere.* Berlin: Reimer.

Rheinberger, Hans-Jörg (1995). From microsomes to ribosomes: "Strategies" of "representation." *Journal of the History of Biology, 28*, 49–89.

Rheinberger, Hans-Jörg (1997). *Toward a History of Epistemic Things: Synthesizing Proteins in the Test Tube.* Stanford, CA: Stanford University Press.

Rhodin, Johannes. (1954). *Correlation of Ultrastructural Organization and Function in Normal and Experimentally Changed Proximal Convoluted Tublule Cells of the Mouse Kidney.* Karolinska Institut, Aktiebolaget Godvil, Stockholm.

Rich, Alexander (1963). Polyribosomes. *Scientific American, 209* (December), 44–53.

Richardson, K. C. (1934). The Golgi apparatus and other cytoplasmic structures in normal and degenerate cells in vitro. *Archiv für experimentelle Zellforschung, 16*, 100–15.

Ris, Hans, & Mirsky, Alfred E. (1949). Quantitative cytochemical determination of desoxyribonucleic acid with the Feulgen nucleal reaction. *Journal of General Physiology, 32*, 125–46.

305

Roberts, Richard B. (1958). *Microsomal Particles and Protein Synthesis*. New York: Pergamon.

Robertson, J. David (1987). The early days of electron microscopy of nerve tissues and membranes. *International Review of Cytology, 100*, 129–201.

Rosenberg, Charles (1979). Toward an ecology of knowledge: On discipline, context, and history. In A. Oleson & J. Voss (Eds.), *The Organization of Knowledge in Modern America*. Baltimore: John Hopkins.

Rosenblueth, Arturo, Wiener, Norbert, & Bigelow, Julian (1943). Behavior, purpose, and teleology. *Philosophy of Science, 10*, 18–24.

Ruiz-Mirazo, Kepa, Peretó, Juli, & Moreno, Alvaro (2004). A universal definition of life: Autonomy and open-ended evolution. *Origins of Life and Evolution of the Biosphere, 34*, 323–46.

Ruska, Helmut (1941). Die Sichtbarmachung der bakteriophagen Lyse im Übermikroskop and Über ein neues bei der bakteriophagen Lyse auftretendes Formelement. *Naturwissenschaften, 29*, 367–8.

Sabatini, David D., Tashiro, Yukata, & Palade, George E. (1966). On the attachment of ribosomes to microsomal membranes. *Journal of Molecular Biology, 19*, 503.

Salmon, Wesley C. (1984). *Scientific Explanation and the Causal Structure of the World*. Princeton: Princeton University Press.

Salmon, Wesley C. (1989). Four decades of scientific explanation. In P. Kitcher & W. C. Salmon (Eds.), *Scientific Explanation. Minnesota Studies in the Philosophy of Science, Volume XIII* (pp. 3–219). Minneapolis: University of Minnesota Press.

Salmon, Wesley C. (1994). Causality without counterfactuals. *Philosophy of Science, 61*, 297–312.

Sanders, F. K. (1951). Cytological techniques: B. Special methods. In G. H. Bourne (Ed.), *Cytology and Cell Physiology* (pp. 20–38). Oxford: Oxford.

Schleiden, Mathias Jacob (1838). Beiträge zur phytogenesis. *Archiv für Anatomie, Physiologie und wissenschaftliche Medecin*, 137–76.

Schleiden, Matthias J. (1842). *Grundzüge der wissenschaftlichen Botanik*. Leipzig: Wilhelm Engelmann.

Schlenk, Fritz (1997). Early research on fermentation – a story of missed opportunities. In A. Cornish-Bowden (Ed.), *New Beer in an Old Bottle: Eduard Buchner and the Growth of Biochemical Knowledge*. Valencia, Spain: Universitat de València.

Schmitt, Francis O. (1944–5). Ultrastructure and the problem of cellular organization. *The Harvey Lectures, 40*, 249.

Schmitt, Francis O., Hall, Cecil E., & Jakus, Marie A. (1942). Electron microscope investigations of the structure of collagen. *Journal of Cellular and Comparative Physiology, 20* (1), 11–33.

Schmitt, Francis O., Hall, Cecil E., & Jakus, Marie A. (1943). The ultrastructure of protoplasmic fibrils. In N. L. Hoerr (Ed.), *Frontiers in Cytochemistry: The Physical and Chemical Organization of the Cytoplasm* (Vol. 10, pp. 261–76). Lancaster, PA: The Jaques Cattell Press.

Schneider, Walter C. (1948). Intracellular distribution of enzymes. III. The oxidation of octanoic acid by rat liver fractions. *Journal of Biological Chemistry, 176*, 259–66.

Schneider, Walter C., & Hogeboom, George H. (1951). Cytochemical studies of mammalian tissue: The isolation of cell components by differential centrifugation: A review. *Cancer Research, 11*, 1–22.

References

Schneider, Walter C., & Kuff, Edward L. (1954). On the isolation and some bio-chemical properties of the Golgi substance. *American Journal of Anatomy, 94*, 209–24.

Schneider, Walter C., & Kuff, Edward L. (1964). Centrifugal isolation of subcellular components. In G. H. Bourne (Ed.), *Cytology and Cell Physiology* (3rd ed., pp. 19–89). New York: Academic Press.

Schultze, Max (1861). Über Muskelkörperchen und das, was man eine Zelle zu nennen habe. *Müllers Archiv für Anatomie, Physiologie, und wissenschaftliche Medizin,* 1–27.

Schultze, Max (1865). *Archiv für mikroskopische Anatomie und Entwirklungsgeschichte, 1*, 124.

Schwann, Theodor (1836). Über das Wesen des Verdauungsprocesses. *Archiv für Anatomie, Physiologie und wissenschaftliche Medecin*, 90–138.

Schwann, Theodor (1837). Vorläufige Mitteilung, betreffend Versuche über die Weingärung und Faulnis. *Poggendorf's Annalen der Physik und Chemie, 41*, 184–93.

Schwann, Theodor (1839(1947). *Microscopical Researches into the Accordance in the Structure and Growth of Animals and Plants* (H. Smith, Trans.). London: Sydenham Society.

Shapere, Dudley (1974). Scientific theories and their domains. In F. Suppe (Ed.), *The Structure of Scientific Theories*. Urbana: University of Illinois Press.

Shapere, Dudley (1984). *Reason and the Search for Knowledge*. Dordrecht: Reidel.

Shapin, Steven (1982). History of science and its sociological reconstructions. *History of Science, 20*, 157–211.

Siekevitz, Philip (1952). Uptake of radioactive alanine in vitro into the proteins of rat liver fractions. *Journal of Biological Chemistry, 195*, 549–65.

Siekevitz, Philip, & Zamecnik, Paul C. (1951). *In vitro* incorporation of $1\text{-}C^{14}\text{-}DL$-alanine into proteins of rat-liver granular fractions. *Federation Proceedings, 10*, 246.

Simon, Herbert A. (1996). *The Sciences of the Artificial*. (Third ed.). Cambridge, MA: MIT Press.

Sjöstrand, Fritiof S. (1943). Electron microscopic examination of tissues. *Nature, 151*, 725–26.

Sjöstrand, Fritiof S. (1953a). A new microtome for ultra-thin sectioning for high resolution electron microscopy. *Experientia, 9*, 114–5.

Sjöstrand, Fritiof S. (1953b). Electron microscopy of mitochondria and cytoplasmic double membranes. *Nature, 171*, 30–2.

Sjöstrand, Fritiof S. (1955a). The ultrastructure of mitochondria. In *Fine Structure of Cells. Symposium Held at the VIIIth Congress of Cell Biology* (pp. 16–30). Leiden: Interscience Publishers.

Sjöstrand, Fritiof S. (1955b). The ultrastructure of the ground substance of the cytoplasm. In *Fine Structure of Cells. Symposium Held at the VIIIth Congress of Cell Biology* (pp. 222–8). Leiden: Interscience Publishers.

Sjöstrand, Fritiof S. (1956a). The ultrastructure of cells as revealed by the electron microscope. *International Review of Cytology, 5*, 455–533.

Sjöstrand, Fritiof S. (1956b). Recent advances in the biological application of the electron microscope. In *Third International Conference on Electron Microscopy* (pp. 26–37). London: Royal Microscopical Society.

Sjöstrand, Fritiof S., & Hanzon, Viggo (1954). Membrane structures of cytoplasm and mitochondria in exocrine cells of mouse pancreas as revealed by high resolution electron microscopy. *Experimental Cell Research, 7*, 393–414.

Sjöstrand, Fritiof S., & Rhodin, Johannes (1953). The ultrastructure of the proximal convoluted tubules of the mouse kidney as revealed by high resolution electron microscopy. *Experimental Cell Research, 4*, 426–56.

Slater, Edward Charles (1950). Phosphorylation coupled with the reduction of cytochrome *c* by α-ketoglutarate in heart muscle granules. *Nature, 166*, 982–3.

Slater, Edward Charles (1953). Mechanism of phosphorylation in the respiratory chain. *Nature, 172*, 975–8.

Slator, Arthur (1906). Studies in fermentation. I. The chemical dynamics of alcoholic fermentation by yeast. *Journal of the Chemical Society, London, 89*, 128–42.

Slayter, Henry, Kiho, Yukio, Hall, Cecil E., & Rich, Alexander (1968). An electron microscopic study of large bacterial polyribosomes. *Journal of Cell Biology, 37*, 583–90.

Smith, Kendric C., Cordes, Eugene, & Schweet, Richard S. (1959). Fractionation of transfer ribonucleic acid. *Biochemica et Biophysica Acta, 33*, 286–7.

Solomon, Miriam (2001). *Social Empiricism.* Cambridge, MA: MIT Press.

Stanley, Wendell M, & Anderson, Thomas F. (1941). Study of purified viruses with electron microscope. *Journal of Biological Chemistry, 139*, 325–38.

Stenning, Keith, & Lemon, Oliver (2001). Aligning logical and psychological perspectives on diagrammatic reasoning. *Artificial Intelligence Review, 15*, 29–62.

Strangeways, Thomas S. P., & Canti, R. G. (1927). The living cell in vitro as shown in dark-ground illumination and the changes induced in such cells by fixing reagents. *Quarterly Journal of Microscopical Science, 71*, 1–14.

Strasburger, Eduard (1884). *Neue Untersuchungen uber den Befruchtungsvorgang bei den Phanerogamen als Grunglagefur eine Theorie der Zeugung.* Jena: Gustav Fischer.

Straus, Werner (1954). Isolation and biochemical properties of droplets from the cells of rat kidney. *Journal of Biological Chemistry, 207*, 745–55.

Straus, Werner (1956). Concentration of acid phosphatase, ribonuclease, desoxyribonuclease, ß-glucoronidase, and cathepsin in "droplets" isolated from the kidney cells of normal rats. *Journal of Biophysical and Biochemical Cytology, 2*, 513–21.

Svedberg, Theodor, & Fhraeus, Robin (1926). A new method for the determination of the molecular weight of the proteins. *Journal of the American Chemical Society, 48*, 430–8.

Svedberg, Theodor, & Rinde, Herman (1924). The ultra-centrifuge, a new instrument for the determination of size and distribution of size of particle in amicroscopic colloids. *Journal of the American Chemical Society, 46*, 2677–93.

Swift, Hewson (1953). Quantitative aspects of nuclear nucleoproteins. *International Review of Cytology, 2*, 1–76.

Szent-Györgyi, Albert von (1924). Über den Mechanismus der Succin- und Paraphenylen-diaminoxydation. Ein Beitrag zur Theorie der Zellatmung. *Biochemische Zeitschrift, 150*, 195–210.

Szent-Györgyi, Albert Von (1937). *Studies on Biological Oxidation and Some of Its Catalysts.* Leipzig: Johann Ambrosius Barth.

Tabery, James (2004). Synthesizing activities and interactions in the concept of a mechanism. *Philosophy of Science, 71*, 1–15.

References

Thénard, Louis Jacques (1803). Mémoire sur la Fermentation vineuse. *Annales de Chimie, 46,* 294–320.

Thunberg, Torsten Ludvig (1913). Zur Kenntnis einiger autoxydabler Thioverbindungen. *Skandinavisches Archiv für Physiologie, 20,* 289–90.

Thunberg, Torsten Ludvig (1916). Über die vitale Dehydrierung der Bernsteinsäure bei Abwesenheit von Sauerstoff. *Zentralblatt für Physiologie, 31,* 91–3.

Thunberg, Torsten Ludvig (1920). Zur Kenntnis des intermediären Stoffwechsels und der dabei wirksamen Enzyme. *Skandinavisches Archiv für Physiologie, 40,* 9–91.

Tissiéres, Alfred, & Watson, James D. (1958). Ribonucleoprotein particles from Escherichia coli. *Nature, 182,* 778–80.

Tissiéres, Alfred, Watson, James D., Schlessinger, David, & Hollingworth, B. R. (1959). Ribonucleoprotein particles from Escherichia coli. *Journal of Molecular Biology, 1,* 221–33.

Toulmin, Stephen (1972). *Human Understanding: The Collective Use and Evolution of Concepts.* Princeton: Princeton University Press.

Turpin, Pierre J. F. (1838). Mémoire sur la cause et les effets de la fermentation alcoolique et acéteuse. *Annales de chimie et de physique, 7,* 369–402.

van Beneden, Edouard (1875). La maturation de l'uf, la fécondation et les premières phases du développement embryonnaire des mammifères d'après des recherches faites chez le Lapin. *Annuaire de l'Académie Royale de la Belgique, 40,* 686–736.

van Beneden, Edouard, & Neyt, Adolphe (1887). Nouvelles recherches sur la fécondation et la division mitosique chez l'Ascaride mégalocephale. *Annuaire de l'Académie Royale de la Belgique, 14,* 238.

van Essen, David C., & Gallant, Jack L. (1994). Neural mechanisms of form and motion processing in the primate visual system. *Neuron, 13,* 1–10.

Varela, Francisco J. (1979). *Principles of Biological Autonomy.* New York: Elsevier.

Virchow, Rudolf (1855). Cellular-Pathologie. *Archiv für pathologische Anatomie und Physiologie und für klinische Medizin, 8,* 3–39.

Virchow, Rudolf (1858). *Die Cellularpathologie in ihrer Begründung auf physiologische und pathologische Gewebelehre.* Berlin: August Hirschwald.

von Ardenne, Manfred (1939). Die Keilschnittmethode, ein Weg zur Herstellung von Mikrotomschnitten mit weniger als Stärke für elektronenmikroskopische Zwecke. *Zeitschrift für wissenschaftliche Mikroscopie, 56,* 8–23.

Waldeyer, Wilhelm (1888). Ueber Karyokinese und ihre Bezeihung zu den Befruchtungsvorgängen. *Archiv für mikroskopische Anatomie und Entwirklungsgeschichte, 32,* 1–122.

Walker, C. E., & Allen, M. (1927). On the nature of the "Golgi bodies" in fixed material. *Proceedings of the Royal Society B, 101,* 468.

Walker, P. G. (1952). The preparation and properties of ß-glucuronidase: 3. Fractionation and activity of homogenate in isotonic media. *Biochemical Journal, 51,* 223–32.

Warburg, Otto Heinrich (1910). Über die Oxidationen in lebenden Zellen nach Versuchen am Seeigelei. *Zeitschrift für physiologische Chemie, 59,* 305–40.

Warburg, Otto Heinrich (1911). Über Beeinflussung der Sauerstoffatmung. *Zeitschrift für physiologische Chemie, 70,* 413–32.

Warburg, Otto Heinrich (1913a). *Über die Wirkung der Struktur auf chemische Vorgänge in den Zellen.* Jena: Gustav Fischer.

References

Warburg, Otto Heinrich (1913b). Über sauerstoffatmende Körnchen aus Leberzellen und über Sauerstoffatmung in Berkefeld-Filtraten wässriger Leberextrakte. *Pflüger's Archiv für Gesammte Physiologie des Menschen und der Thiere, 154*, 599–617.

Warburg, Otto Heinrich (1914). Beitrage zur Physiologie der Zelle, insbesondere über die Oxydationsgewchwindigkeit in Zellen. *Erggebnisse der Physiologie, 14*, 253–337.

Warburg, Otto Heinrich (1923). Versuche an überlebendem Carcinomgewebe. *Biochemische Zeitschrift, 142*, 317–33.

Warburg, Otto Heinrich (1925a). Monometrische Messung des Zellstoffwechsels in Serum. *Biochemische Zeitschrift, 164*, 481.

Warburg, Otto Heinrich (1925b). Iron, the oxygen carrier of the respiration ferment. *Science, 61*, 575–82.

Warburg, Otto Heinrich (1929). Atmungsferment und Oxydasen. *Biochemische Zeitschrift, 214*, 1–3.

Warburg, Otto Heinrich (1932). Das sauerstoffübertragende Ferment der Atmung. *Zeitschrift für angewandte Chemie, 45*, 106.

Warburg, Otto Heinrich, & Christian, Walter (1932). Über ein neues Oxydationsferment und sein Absorptionsspektrum. *Biochemische Zeitschrift, 254*, 438–58.

Warburg, Otto Heinrich, & Christian, Walter (1935). Co-Fermentproblem. *Biochemische Zeitschrift, 275*, 364.

Warburg, Otto Heinrich, & Christian, Walter (1936). Pyridin, der wasserstoffübertragene Bestandteil von Gärungsfermenten. *Biochemische Zeitschrift, 287*, 291–328.

Warburg, Otto Heinrich, & Christian, Walter (1938). Bemerkung über gelbe Fermente. *Biochemische Zeitschrift, 298*, 368–77.

Warburg, Otto Heinrich, Christian, Walter, & Griese, Alfred (1935). Wasserstoffübertragendes Co-Ferment, seine Zusammensetzung und Wirkungsweise. *Biochemische Zeitschrift, 282*, 157–65.

Warner, Jonathan R., Knopf, Paul M., & Rich, Alexander (1963). A multiple ribosomal structure in protein synthesis. *Proceedings of the National Academy of Sciences, USA, 49*, 122–9.

Warshawsky, Hershey, Leblond, Charles P., & Droz, Bernard (1961). Synthesis and migration of proteins in the cells of the exocrine pancreas as revealed by specific activity determination from radioautographs. *Journal of Cell Biology, 16*, 1–23.

Wasserman, Stanley, & Faust, Katherine (1994). *Social Network Analysis: Methods and Applications.* New York: Cambridge.

Weismann, August (1885). *Die Kontinuität des Keimplasmas als Grundlage einer Theorie der Vererbung.* Jena: Gustav Fischer.

Wersäll, J. (1956). Studies on the structure and innervation of the sensory epithelium of the crista ampullares in the guinea pig. *Acta Oto-Laryngologica, 126*(Supplement), 1–85.

Whaley, William G. (1975). *The Golgi Apparatus.* (Vol. 2). New York: Springer-Verlag.

Whewell, William (1840). *The Philosophy of the Inductive Sciences, Founded upon Their History.* London: J. W. Parker.

Whitley, Richard (1980). The context of scientific investigation. In K. D. Knoor, R. Krohn, & R. Whitley (Eds.), *The Social Process of Scientific Investigation* (pp. 297–321). Dordrecht: Reidel.

Wieland, Heinrich (1913). Über den Mechanismus der Oxydationsvorgänge. *Berichte der deutschen chemischen Gesellschaft, 46*, 3327–42.

References

Wiener, Norbert (1948). *Cybernetics: Or, Control and Communication in the Animal Machine*. New York: Wiley.

Williams, Robley C., & Wyckoff, Ralph W. G. (1946). Applications of metallic-shadow-casting to microscopy. *Journal of Applied Physics, 17*, 23–33.

Wilson, Edmund B. (1896). *The Cell in Development and Inheritance*. New York: MacMillan.

Wilson, Edmund B. (1923). *The Physical Basis of Life*. New Haven, CT: Yale University Press.

Wimsatt, William C. (1976). Reductive explanation: A functional account. In R. S. Cohen, C. A. Hooker, A. C. Michalos, & J. van Evra (Eds.), *PSA 1974* (pp. 671–710). Dordrecht: Reidel.

Witter, Robert F., Watson, Michael L., & Cottone, Mary A. (1955). Morphology and ATP-ase of isolated mitochondria. *Journal of Biophysical and Biochemical Cytology, 1*(No. 2), 127–38.

Wöhler, Frederich (1828). Ueber künstliche Bildung des Harnstoffs. *Annalen der Physik und Chemie, 12*, 253–6.

Wöhler, Frederich (1839). Das enträthselte Geheimniss der geistigen Gährung. *Annalen der Pharmacie, 29*, 100–4.

Wyckoff, Ralph W. G. (1959). Optical methods in cytology. In J. Brachet & A. E. Mirsky (Eds.), *The Cell: Biochemistry, Physiology, Morphology*. New York: Academic Press.

Zamecnik, Paul C. (1958). The microsome. *Scientific American, 198*(3), 118–24.

Zamecnik, Paul C. (1958–9). Historical and current aspects of the problem of protein synthesis. *Harvey Lectures, 54*, 256–81.

Zamecnik, Paul C., Frantz, Ivan D., Loftfield, Robert B., & Stephenson, Mary L. (1948). Incorporation *in vitro* of radioactive carbon from carboxyl-labeled DL-alanine and glycine into proteins of normal and malignant rat livers. *Journal of Biological Chemistry, 175*, 299–314.

Zamecnik, Paul C., Stephenson, Maryl L., Scott, Jesse F., & Hoagland, Mahlon B. (1957). Incorporation of C^{14}-ATP into soluble RNA isolated from 105,000 x g supernatant from rat liver. *Federation Proceedings, 16*, 275.

Zetterqvist, H. (1956). *The Ultrastructural Organization of the Columnar Absorbing Cells of the Mouse Jejunum*. Stockholm: PhD Thesis, Karolinska Institute.

Ziegler, Daniel M., Linnane, Anthony W., Green, David E., Dass, C. M. S., & Ris, Hans (1958). Studies on the electron transport system: Correlation of the morphology and enzymic properties of mitochondrial and sub-mitochondrial particles. *Biochimica et Biophysica Acta, 28*, 524–39.

Index

Index

Index

Hydén, Holger, 176
hydrolytic enzymes, 95, 131, 172, 252, 255–7

inner membrane spheres, 217–18
Institute for Enzyme Research (Wisconsin),
  11, 196, 273
Interchemical Corporation, 182, 184
interdisciplinary research clusters, 13
interfield theories, 12–13
internal environment, 45–6, 92
International Federation for Cell Biology,
  259
International Review of Cytology, 259,
  274
International Society for Cell Biology, 259,
  268, 270

Jacob, Francoise, 238
Jakus, Marie, 143–4
Jamieson, James D., 247–8
Janus green, 81, 130, 165, 170, 181
Jeener, Raymond, 172–3
Jenkins, William, 53
Johnson, William, 111–12
Jonker, Catholijn, 39
*Journal of Biophysical and Biochemical
  Cytology*, 18, 259–60, 264, 271, 275–7
Journal of Cell Biology, 12, 18, 259–60,
  266–8, 275
*Journal of Cellular and Comparative
  Physiology*, 274
Journal of Experimental Medicine, 260–2
Journal of General Physiology, 260–1, 270,
  275
Journal of Ultrastructure Research, 259, 261
journals, professional, 9–10, 13, 18, 94, 258

Kahler, H., 223
Kalberlah, F., 100
Kalckar, Hermann, 115, 192
Kallman, Frances L., 150, 154, 222
Kaplan, Nathan, 119
Karolinska Institute, Stockholm, 169, 173,
  201–2, 207, 259
Kauffman, Stuart, 29
Kaufmann, Berwind P., 263
Keighley, Geoffrey, 231
Keilin, David, 112, 114, 116, 192, 197
Keller, Elizabeth B., 232–3
Kennedy, Eugene, 194–5
Kerb, Johannes Wolfgang, 98

Kiho, Yukio, 241
King, R. L., 87, 159
King, Tsoo E., 220
Kingsbury, B. F., 82
Kirchhoff, Gottlieb Sigismund, 90
Kirkman, Hadley, 86, 88
Kitcher, Philip, 9, 25
Knoll, Max, 141
Knoop, Franz, 96–7, 106
Knopf, Paul M., 239
Kobel, M., 99
Koertge, Noretta, 9
Kohler, Robert E., 8, 9, 89, 93, 95–6, 109
Kölliker, Rudolf Albert von, 80
Konigsberger, Victor, 237
Kosslyn, Stephen, 38–9
Kraft, G., 102
Krebs cycle. *See* citric acid cycle
Krebs, Hans, 111–12
Ktesibios's water clock, 47
Kuff, Edward L., 129, 246
Kuhn, Thomas S., 7, 16, 27
Kühne, Wilhelm, 93, 95
Kützing, Friedrich Traugott, 92

lace-like reticulum, 148, 186–7, 189, 222–3.
  See also endoplasmic reticulum
lactacidase, 98
lactacidogen, 100
lactic acid, 97–8, 100–1, 103–5, 108
lactic acid cycle, 47
Lakatos, Imre, 7, 62
LaPlace, Pierre Simon, 54, 90
Laquer, Fritz Oscar, 100
Lardy, Henry, 215, 220
large granule fraction, 169–70, 179–81. *See
  also* mitochondrion
Larkin, Jill H., 40
Latour, Bruno, 9, 119
Latta, Harrison, 146
Laudan, Larry, 7, 9
Lavoisier, Antoine L., 54, 89–90, 93
laws of nature, 2–3, 7, 19, 23–6, 33, 61, 280
Lazarow, Arnold, 167
Leblond, Charles P., 247–9
Ledingham, C. G., 164
Lehninger, Albert L., 192, 194–8, 201, 210,
  212–14, 217, 219–20, 263–4
Leloir, Louis F., 196
Lemon, Oliver, 40
Lester. R. L., 115

318